THE UNIVERSITY SERIES IN MATHEMATICS

Series Editor: Joseph J. Kohn
Princeton University

THE CLASSIFICATION OF FINITE SIMPLE GROUPS
Daniel Gorenstein
VOLUME 1: GROUPS OF NONCHARACTERISTIC 2 TYPE

ELLIPTIC DIFFERENTIAL EQUATIONS AND OBSTACLE PROBLEMS
Giovanni Maria Troianiello

FINITE SIMPLE GROUPS: An Introduction to Their Classification
Daniel Gorenstein

AN INTRODUCTION TO ALGEBRAIC NUMBER THEORY
Takashi Ono

INTRODUCTION TO PSEUDODIFFERENTIAL AND FOURIER INTEGRAL OPERATORS
François Treves
VOLUME 1: PSEUDODIFFERENTIAL OPERATORS
VOLUME 2: FOURIER INTEGRAL OPERATORS

MATRIX THEORY: A Second Course
James M. Ortega

A SCRAPBOOK OF COMPLEX CURVE THEORY
C. Herbert Clemens

TOPICS IN NUMBER THEORY
J. S. Chahal

An Introduction to Algebraic Number Theory

Takashi Ono

The Johns Hopkins University
Baltimore, Maryland

Springer Science+Business Media, LLC

Library of Congress Cataloging-in-Publication Data

Ono, Takashi.
[Sūron josetsu. English]
An introduction to algebraic number theory / Takashi Ono. -- 2nd ed.
p. cm. -- (The University series in mathematics)
Translation of: Sūron josetsu.
Includes bibliographical references.
ISBN 978-1-4612-7872-6 ISBN 978-1-4613-0573-6 (eBook)
DOI 10.1007/978-1-4613-0573-6
1. Algebraic number theory. I. Title. II. Series: University series in mathematics (Plenum Press)
QA247.056 1990
512'.74--dc20 90-30155
CIP

This volume is a translation (by the author) of the Second Edition of *Suron Josetsu (An Introduction to Algebraic Number Theory)*, originally published by Shokabo Publishing Co., Ltd., Tokyo, Japan. English translation rights were arranged with Shokabo Publishing Co., Ltd., through the Japan Foreign-Rights Centre.

Preface

This book is a translation of my book *Suron Josetsu* (*An Introduction to Number Theory*), Second Edition, published by Shokabo, Tokyo, in 1988. The translation is faithful to the original globally but, taking advantage of my being the translator of my own book, I felt completely free to reform or deform the original locally everywhere.

When I sent T. Tamagawa a copy of the First Edition of the original work two years ago, he immediately pointed out that I had skipped the discussion of the class numbers of real quadratic fields in terms of continued fractions and (in a letter dated 2/15/87) sketched his idea of treating continued fractions without writing explicitly continued fractions, an approach he had first presented in his number theory lectures at Yale some years ago. Although I did not follow his approach exactly, I added to this translation a section (Section 4.9), which nevertheless fills the gap pointed out by Tamagawa. With this addition, the present book covers at least T. Takagi's *Shoto Seisuron Kogi* (*Lectures on Elementary Number Theory*), First Edition (Kyoritsu, 1931), which, in turn, covered at least Dirichlet's *Vorlesungen*.

It is customary to assume basic concepts of algebra (up to, say, Galois theory) in writing a textbook of algebraic number theory. But I feel a little strange if I assume Galois theory and prove Gauss quadratic reciprocity. Since it does not occupy much space and time to introduce simple basic concepts of algebra and since number theory is a rich source of structures of algebra, such as groups, rings, and fields, I have included all definitions and proofs of propositions in algebra except those of basic facts in linear algebra such as linear independence, linear transformations, and determinants. In this way, the book covers the same basic facts of algebraic number theory contained in Hecke's *Vorlesungen* (except the quadratic reciprocity of an arbitrary number field) and in Part I of Takagi's *Daisuteki Seisuron* (*Algebraic Number Theory*), First Edition (Iwanami, 1948) (except the complete proof of Dedekind's discriminant theorem).

Part II of Takagi's book is an exposition of the class field theory based on the line of thought articulated in his fundamental paper (1920) on abelian field extensions, simplified in several places by the work of Artin, Hasse, and Chevalley in the early 1930s. A piece of good news for beginners is that the most important map in the whole theory (the Artin map) is now within their hands (Section 2.15 of this book). Guided by this map, they can reach sufficient height to appreciate the beautiful flowers discovered by those pioneers.

Takashi Ono

Baltimore

Contents

Notation and Conventions

Throughout this book, we use the standard symbols $\mathbb{N}$, $\mathbb{Z}$, $\mathbb{Q}$, $\mathbb{R}$, and $\mathbb{C}$ to represent natural numbers (i.e., positive integers), integers, rational numbers, real numbers, and complex numbers, respectively.

We assume that the reader is familiar with sets and the symbols $\cap$, $\cup$, $\supset$, $\subset$, and $\in$. If X, Y are sets, we write $X \subset Y$ to mean that X is contained in Y but may possibly be equal to Y. Similarly for $X \supset Y$. If $f: X \to Y$ is a mapping (or a map) from X to Y, we write $x \mapsto f(x)$ to denote the effect of f on $x \in X$. We say that f is injective if $x \neq y$ implies $f(x) \neq f(y)$. We say that f is surjective if for any $y \in Y$ there exists $x \in X$ such that $f(x) = y$. We say that f is bijective if it is both injective and surjective. We denote by $f(X)$ the set of all elements $f(x)$, with $x \in X$. In symbols, we have $f(X) = \{f(x); x \in X\}$. For a subset $Y' \subset Y$, we put $f^{-1}(Y') = \{x \in X; f(x) \in Y'\}$. We call $f(X)$ the image of f and $f^{-1}(Y')$ the inverse image of Y'. If g: $Y \to Z$ is another map, then we have a composite map $g \circ f: X \to Z$ such that $(g \circ f)(x) = g(f(x))$ for all $x \in X$.

We use occasionally the symbols $\forall$, $\exists$, and $\exists_1$ to mean "for any" (or "for all"), "there exists," and "there exists uniquely," respectively. If A and B are propositions, $A \Rightarrow B$ represents "A implies B" and $A \Leftrightarrow B$ represents "A if and only if B." When X is a finite set, we denote by $[X]$ the cardinality of X. If the equality sign "=" is used to define an object, we often write "$\stackrel{\text{def}}{=}$." We assume that the reader is familiar with the equivalence relation in a set X and the quotient set $X/\sim$ defined by the relation $\sim$.

1

To the Gauss Reciprocity Law

This chapter consists of *elementary number theory* and deals with the greatest common divisor, the euclidean algorithm, congruences, linear equations, primitive roots, and the quadratic reciprocity law. The material covered here corresponds to the first four chapters of Gauss's *Disquisitiones arithmeticae* (1801) and to the whole volume (70 pages) of Weil's *Number Theory for Beginners* (1985). Equally small (95 pages) Bakers *A Concise Introduction to the Theory of Numbers* (1984) contains, in addition to those standards, quadratic forms, diophantine approximation, Fermat primes, Mersenne primes, Goldbach's conjecture, twin primes, perfect numbers, the Riemann hypothesis, Euler's constant, $\zeta(2n+1)$, Fermat's conjecture, Catalan's conjecture, and so on. Following Weil, we regard this part of number theory as a rich source of structures of algebra such as groups, rings, and fields. Like computers in these days, the algebraic language is useful and economizes our thoughts. For example, group theory teaches us that the famous Fermat–Euler theorem on congruences is a special case of a simple theorem on finite groups and that the important existence theorem of primitive roots due to Gauss follows from the theorem stating that a finite subgroup of the multiplicative group of any field is cyclic. By doing so, number theory (elementary or advanced) is well ventilated. As for the proof of the reciprocity law, we followed the method of Gauss using Gauss sums because it is not only beautiful but also has a crucial influence on the later development of algebraic number theory.

1.1. Basic Facts

We denote by $\mathbb{Z}$ the set of all *integers*: $0, \pm 1, \pm 2, \pm 3, \ldots$ and by $\mathbb{Q}$ the set of all *rational numbers*. Basic rules in $\mathbb{Q}$ are:

1. $a \in \mathbb{Q}, b \in \mathbb{Q} \Rightarrow a + b \in \mathbb{Q}$ (addition).
2. $(a+b)+c = a+(b+c)$ (associative law).
3. $a+b=b+a$ (commutative law).
4. $a \in \mathbb{Q}, b \in \mathbb{Q} \Rightarrow \exists_1 x \in \mathbb{Q}$ such that $a+x=b$. x is written $b-a$ (subtraction).
5. $a \in \mathbb{Q}, b \in \mathbb{Q} \Rightarrow ab \in \mathbb{Q}$ (multiplication).
6. $(ab)c = a(bc)$ (associative law).
7. $ab = ba$ (commutative law).
8. $a \in \mathbb{Q}, a \neq 0, b \in \mathbb{Q} \Rightarrow \exists_1 x \in \mathbb{Q}$ such that $ax = b$. x is written $\dfrac{b}{a}$ or b/a (division).
9. $(a+b)c = ac + bc$ (distributive law).

Remark 1.1. There is a unique x such that $a+x=a$ by (4). Similarly, for $b \in \mathbb{Q}$, there is a unique y such that $b+y=b$. Since $(a+b)+x = a+(b+x) = a+(x+b) = (a+x)+b = a+b$ and $(a+b)+y = a+(b+y) = a+b$, we have $x=y$ by (4). Hence x such that $a+x=a$ is independent of a; this x is 0. Similarly, for the multiplication, the number x such that $ax=a$, $a \neq 0$, is independent of a; this x is 1.

Definition 1.1. Let K be a set which contains at least two elements. K is called a *field* if it has two operations $a+b$ and ab which satisfy (1)–(9) (with K instead of $\mathbb{Q}$).

Remark 1.2. As in Remark 1.1, K has 0_K and 1_K, or simply 0 and 1. It can be proved that $0 \neq 1$ (Exercise 1.2). There is a field containing 0, 1 only:

+	0	1
0	0	1
1	1	0

×	0	1
0	0	0
1	0	1

$\mathbb{Q}$, $\mathbb{R}$, and $\mathbb{C}$ are fields.

Definition 1.2. A set R is called a *ring* if it has two operations, $a+b$ and ab, which satisfy (1)–(9) except (8) (with R

instead of $\mathbb{Q}$). If R has an element 1 such that $1 \cdot a = a$, $a \in R$, then 1 is called the unit element of R. (Note that 1 is unique because $1 \cdot 1' = 1' \cdot 1 = 1 = 1'$.)

REMARK 1.3. $\mathbb{Z}$ is a ring with 1, so is any field. The set $2\mathbb{Z}$ of all even numbers is a ring but has no unit element. Usually when we speak of a ring we do not assume (7), but in this book by a ring we mean a commutative one unless otherwise stated.

EXERCISE 1.1. For a ring R, prove that $0 \cdot a = 0$ for all $a \in R$.

EXERCISE 1.2. For a field, prove that $0 \neq 1$.

EXERCISE 1.3. For a ring with 1, prove that $(-1)(-1) = 1$.

Since $\mathbb{R}$ may be identified with a straight line we can speak of the ordering of real numbers: $a < b \Leftrightarrow b > a \Leftrightarrow a$ lies to the left of b. a is positive (negative) if $a > 0$ ($a < 0$). If $a > 0$ and $b > 0$, then $a + b > 0$ and $ab > 0$. The ordering in $\mathbb{R}$ induces the one on $\mathbb{Q}$ and $\mathbb{Z}$. A positive integer is also called a *natural number*. We denote by $\mathbb{N}$ the set of all natural numbers: $\mathbb{N} = \{1, 2, 3, \ldots\}$. Any nonempty subset S of $\mathbb{N}$ contains the least integer. In fact, take any $n_0 \in S$; then the first integer among $1, 2, \ldots, n_0$ which is in S is the least one. From this follows the principle of mathematical induction.

PROPOSITION 1.1 (*Principle of Mathematical induction*). *Let $P(n)$ be a statement about a natural number n. We can deduce that $P(n)$ is true for every $n \in \mathbb{N}$ if we can prove both of the following conditions*:

1. *$P(1)$ is true.*
2. *For any natural number n, if $P(n)$ is true then $P(n+1)$ is true.*

PROOF. We must show that the set $S = \{n \in \mathbb{N};\ P(n)$ is not true$\}$ is empty. If not, let n_0 be the least element in S. Since $1 \notin S$, we have $n_0 - 1 \in \mathbb{N}$ and $n_0 - 1 \notin S$ and hence $P(n_0 - 1)$ is true. Then, by (2), $P(n_0)$ is true, or $n_0 \notin S$, a contradiction. Q.E.D.†

REMARK 1.4. Peano's axiom [V] in Note A (p. 210) is nothing but the principle of mathematical induction; the mapping φ is given by $\varphi(n) = n + 1$.

† Q.E.D. = quod erat demonstrandum (which was to be demonstrated). Stark says "This comes from English, Quite Easily Done" (See footnote on p. 224 of H. M. Stark: An Introduction to Number Theory, The MIT Press, 1978.)

Exercise 1.4. Prove that $(1^5+2^5+\cdots+n^5)+(1^7+2^7+\cdots+n^7)=2(1+2+\cdots+n)^4$.

Exercise 1.5. What is wrong with the following "proof" that everyone is bald? For any natural number n think of the statement: $P(n)$ = a person who has n hairs is bald. (1) $P(1)$ is obviously true; (2) if $P(n)$ is true then the person remains bald with one additional hair, i.e., $P(n+1)$ is true. Therefore, by the principle of mathematical induction $P(n)$ is true for any n.†

Definition 1.3. Let a and b be integers. If there is an integer x such that $b=ax$, b is said to be a *multiple* of a; a is said to be a *divisor* of b or to *divide* b. We write $a \mid b$ when a divides b.

Remark 1.5. Since $b=b\cdot 1$ for all $b \in \mathbb{Z}$, we have $1 \mid b$ for all b; since $0\cdot x=0$ for all x, $0 \mid a$ holds only when $a=0$; $a \mid a$ for all $a \neq 0$.

Definition 1.4. An integer $p>1$ is called a *prime* if it has no other positive divisor than 1 and p.

Exercise 1.6. Let a be a positive integer, $a>1$, which is not a prime. Prove that there is a divisor b of a such that $1<b\leq\sqrt{a}$. (If $a<100$ then $\sqrt{a}<10$; hence if a is not divisible by 2, 3, 5, and 7, then a is a prime.)

Example 1.6. The fundamental theorem of number theory states that every integer >1 can be written as a product of primes in only one way except for the order of the prime factors (Theorem 1.6). The uniqueness of the factorization in the theorem follows from the property of primes: $p \mid ab \Rightarrow p \mid a$ or $p \mid b$; and this property follows, in turn, from the properties of modules in $\mathbb{Z}$ discussed in the next section.

1.2. Modules in $\mathbb{Z}$

Proposition 1.2. *Let a and b be integers, $b>0$. Then there exist unique integers q and r such that $a=qb+r$, $0\leq r<b$. r is called the remainder of the division of a by b and q the quotient.*

† Borrowed from p. 6 of T. Kusaba: Suri to Hasso (Mathematical Principles and Conceptions, Kaitakusha, 1978).

PROOF. We can find $q \in \mathbb{Z}$ such that $bq \leq a < b(q+1)$. Then $r = a - bq$ satisfies the relation $0 \leq r < b$. If q' and r' are integers with properties $a = bq' + r'$, $0 \leq r' < b$, then $b(q - q') = r' - r$; so $b \mid (r' - r)$. Assuming, without loss of generality, that $r' \geq r$, we have $0 \leq r' - r < b$ and so $r' = r$. Therefore we have $q = q'$. Q.E.D.

DEFINITION 1.5. A nonempty subset M of $\mathbb{Z}$ is a *module* if it has the property $x, y \in M \Rightarrow x - y \in M$.

REMARK 1.7. For example, $M = 2\mathbb{Z}$ (the set of even numbers) and $3\mathbb{Z}$ (the set of all multiples of 3) are modules. More generally, for $m \in \mathbb{N}$, $M = m\mathbb{Z} = \{ma;\ a \in \mathbb{Z}\}$ is a module. $M = \{0\}$ is a module, too. The set of all odd numbers is not a module. For any module M and $x \in M$, we have $0 = x - x \in M$ and $-x = 0 - x \in M$; for $x, y \in M$, $x + y = x - (-y) \in M$. In other words, we may define a module M by the condition $x, y \in M \Rightarrow x \pm y \in M$.

THEOREM 1.1. *For a module $M \subset \mathbb{Z}$, there exists a unique $m \geq 0$ such that $M = m\mathbb{Z}$.*

PROOF. If $M = \{0\}$, $m = 0$ is the only choice. If $M \neq \{0\}$, take a nonzero number $a \in M$; by changing the sign if necessary, we can assume that $a > 0$. Call m the smallest natural number in M; clearly $m\mathbb{Z} \subset M$. Conversely take any $a \in M$ and divide it by m: $a = qm + r$, $0 \leq r < m$ (Proposition 1.2). Since $a, m \in M$, we have $r = a - qm \in M$, which implies $r = 0$ and so $a = qm$; or $M \subset m\mathbb{Z}$. Since m is the smallest number >0 in $M = m\mathbb{Z}$, it is uniquely determined by M. Q.E.D.

Let $a_1, \ldots, a_l$ be any integers and let

$$M = \{a_1x_1 + \cdots + a_lx_l;\ x_i \in \mathbb{Z},\ 1 \leq i \leq l\}.$$

As is easily verified, M is a module. By Theorem 1.1, there is a unique $d \geq 0$ such that $M = d\mathbb{Z}$.

THEOREM 1.2. *With the notation as above, the number d has the following properties:*

1. $d \mid a_i$, $1 \leq i \leq l$.
2. $e \mid a_i$, $1 \leq i \leq l$, $\Rightarrow e \mid d$.

PROOF.

1. Put $x_i = 1$, $x_j = 0$ $(j \neq i)$. Then $a_i \in M$ and hence $a_i = db_i$, i.e., $d \mid a_i$.
2. Since $d = d \cdot 1 \in M$, $d = a_1b_1 + \cdots + a_lb_l$, $b_i \in \mathbb{Z}$. Since $e \mid a_i$, we have $a_i = ec_i$ which implies that $d = ec_1b_1 + \cdots + ec_lb_l = e(c_1b_1 + \cdots + c_lb_l)$, i.e., $e \mid d$. Q.E.D.

REMARK 1.8. The number d determined by $a_1, \ldots, a_l$ as $M = d\mathbb{Z}$ has the properties (1) and (2) in Theorem 1.2. If $d' \geq 0$ also has the properties (1) and (2), then, by (2), $d \mid d'$ and $d' \mid d$, i.e., $d = d'$. The integer d is called the *greatest common divisor* (g.c.d.) of $a_1, \ldots, a_l$. We write $d = (a_1, \ldots, a_l)$. We often write $M = d\mathbb{Z} = a_1\mathbb{Z} + \cdots + a_l\mathbb{Z}$.

EXERCISE 1.7. For $a, b, c \in \mathbb{Z}$, prove that $(a, b, c) = ((a, b), c) = (a, (b, c))$.

EXERCISE 1.8. Define integers a_n, $n \geq 0$, by the recurrence relation $a_0 = 0$, $a_1 = 1$, $a_n = a_{n-2} + a_{n-1}$, $n \geq 2$. The first few terms of the sequence $\{a_n\}$ are 0, 1, 1, 2, 3, 5, 8, 13, 21, 34, 55, 89, 144, 233, 377, 610, 987, 1597, 2584, 4181, 6765, ... This is the Fibonacci (1170–1230) sequence. Prove that $(a_n, a_{n+1}) = 1$, $n \geq 0$. Prove also that $a_{n+1} \leq 2a_n$, $n \geq 1$.

EXERCISE 1.9. Consider the power series whose coefficients $\{a_n\}$ are the Fibonacci sequence

$$f(x) = \sum_{n=0}^{\infty} a_n \frac{x^n}{n!}.$$

Prove that

$$f(x) = \frac{1}{\sqrt{5}} \left[\exp\left(\frac{1+\sqrt{5}}{2}x\right) - \exp\left(\frac{1-\sqrt{5}}{2}x\right)\right]$$

and that

$$a_n = \frac{1}{\sqrt{5}} \left[\left(\frac{1+\sqrt{5}}{2}\right)^n - \left(\frac{1-\sqrt{5}}{2}\right)^n\right].$$

DEFINITION 1.6. Integers $a_1, \ldots, a_l$ are called *relatively prime* if $(a_1, \ldots, a_l) = 1$.

THEOREM 1.3. *Integers $a_1, \ldots, a_l$ are relatively prime if and only if the equation $a_1x_1 + \cdots + a_lx_l = 1$ has a solution $x_i \in \mathbb{Z}$, $1 \leq i \leq l$.*

PROOF. Let $d = (a_1, \ldots, a_l)$; $M = a_1\mathbb{Z} + \cdots + a_l\mathbb{Z} = d\mathbb{Z}$ (Theorem 1.1). Then $1 \in M \Leftrightarrow d = 1$. Q.E.D.

THEOREM 1.4. *Let a, b, and c be integers such that $(a, b) = 1$. Then $a \mid bc \Rightarrow a \mid c$.*

PROOF. By Theorem 1.3, there are $x, y \in \mathbb{Z}$ such that $ax + by = 1$. Then $c = c(ax + by) = a(cx) + (bc)y$. Since $a \mid bc$ we have $a \mid c$. Q.E.D.

THEOREM 1.5. *Let p be a prime. $p \mid ab \Rightarrow p \mid a$ or $p \mid b$.*

PROOF. Assume that $p \nmid a$ (p does not divide a). As p is a prime, we have $(p, a) = 1$. By Theorem 1.4, $p \mid ab$ implies $p \mid b$. Q.E.D.

THEOREM 1.6 (*Fundamental theorem of number theory*). *Every integer can be written as a product of primes; it can be so written in only one way except for the order of the factors.*

PROOF. Take $a > 1$; it has at leat one factor $b > 1$ (say, $b = a$). The smallest of such b is a prime p. If $a = p$, the theorem holds for a; if $a > p$, then $a/p > 1$ and $a > a/p$. If the first part (the existence of the factorization) is true for a/p, it is true for a. Therefore, the first part follows by induction on a. The second part (the uniqueness of the factorization) is more important. Assume that a is written in two ways as a product of primes: $a = p_1p_2 \cdots p_l = q_1q_2 \cdots q_k$. As p_1 divides a, Theorem 1.5 implies that p_1 must divide one of the primes $q_1, q_2, \ldots, q_k$, say q_1. Then $p_1 = q_1$; applying the second part of the theorem to a/p_1, we see that $q_2, \ldots, q_k$ must be the same as $p_2, \ldots, p_l$, except for the order. Q.E.D.

REMARK 1.9. Since there is no integer between 1 and 2, 2 is a prime; 2 is the only even prime. There are 25 primes < 100: 2, 3, 5, 7, 11, 13, 17, 19, 23, 29, 31, 37, 41, 43, 47, 53, 59, 61, 67, 71, 73, 79, 83, 89, 97.

EXERCISE 1.10. Verify that values of the quadratic polynomial $P(x) = x^2 + x + 41$ for $x = 0, 1, 2, \ldots, 39$ are all primes.

EXERCISE 1.11. Define a sequence $\{a_n\}$, $n \geq 1$, by $a_1 = 2$, $a_{n+1} = a_n^2 - a_n + 1$. Verify that $a_{n+1} = a_1 a_2 \cdots a_n + 1$, $n \geq 1$. Using this sequence, prove that there are infinitely many primes.

EXERCISE 1.12. Consider the Fermat (1601–1665) numbers $F_n = 2^{2^n} + 1$, $n \geq 0$. Verify that $F_{n+1} = F_0 F_1 \cdots F_n + 2$, $n \geq 0$. Using this sequence, prove that there are infinitely many primes.

1.3. Euclidean Algorithm and Continued Fractions

Let $a, b \in \mathbb{N}$, $a > b$. Repeat the division in Proposition 1.2 as follows until the remainder vanishes:

$$\text{(E)} \qquad \begin{cases} a = q_1 b + r_1, & 0 < r_1 < b; \\ b = q_2 r_1 + r_2, & 0 < r_2 < r_1; \\ r_1 = q_2 r_2 + r_3, & 0 < r_3 < r_2; \\ \vdots & \\ r_{n-3} = q_{n-1} r_{n-2} + r_{n-1}, & 0 < r_{n-1} < r_{n-2}; \\ r_{n-2} = q_n r_{n-1} + r_n, & 0 = r_n. \end{cases}$$

This process is the *euclidean algorithm*. As $a = q_1 b + r_1$ implies $r_1 = a - q_1 b$, we have $a\mathbb{Z} + b\mathbb{Z} = b\mathbb{Z} + r_1\mathbb{Z}$, i.e., $(a, b) = (b, r_1)$. Similarly, $(b, r_1) = (r_1, r_2) = \cdots = (r_{n-2}, r_{n-1}) = (r_{n-1}, 0) = r_{n-1}$; hence, $(a, b) = r_{n-1}$. In other words, we get the g.c.d. of a, b as the last remainder ($= r_{n-1}$) of the euclidean algorithm; we do not have to compare the prime factorizations of a, b.

EXERCISE 1.13. Find $(95, 432)$ in two different ways (prime factorization and euclidean algorithm).

Let us consider solving the following equation in integers:

$$ax + by = c, \qquad a, b \in \mathbb{N}, \qquad a > b, \qquad c \in \mathbb{Z}.$$

Since this is equivalent to saying that $c \in a\mathbb{Z} + b\mathbb{Z} = d\mathbb{Z}$, $d = (a, b)$ by Theorem 1.2, the latter equation has a solution if and only if $d \mid c$. Assuming $d \mid c$, let a pair x_0, y_0 be a solution: $ax_0 + by_0 = c$. Any other pair of solutions x, y satisfy $ax + by = c$, and so $a(x - x_0) = b(y_0 - y)$. Since $(a/d, b/d) = 1$, Theorem 1.4 implies that $(b/d) \mid (x - x_0)$ and $(a/d) \mid (y_0 - y)$, and so $x = x_0 + (b/d)t$ and $y = y_0 + (a/d)s$. We have $s = -t$ because

$$\frac{a}{d}(x - x_0) = \frac{b}{d}(y_0 - y).$$

Summarizing, we get

THEOREM 1.7. *Let $a, b \in \mathbb{N}$, $a \geq b$. A linear equation $ax + by = c$, $c \in \mathbb{Z}$, has a solution in $\mathbb{Z} \Leftrightarrow d \mid c$, $d = (a, b)$. When $d \mid c$, if x_0, y_0 is pair of solutions, any other pair x, y of solutions is given by $x = x_0 + (b/d)t$ and $y = y_0 - (a/d)t$, $t \in \mathbb{Z}$.*

Theorem 1.7 does not provide us with the method of finding a pair of initial solutions x_0, y_0. This is done by the euclidean algorithm. If we rewrite (E) in terms of matrices, we obtain the relations

$$\begin{pmatrix} a \\ b \end{pmatrix} = \begin{pmatrix} q_1 & 1 \\ 1 & 0 \end{pmatrix}\begin{pmatrix} b \\ r_1 \end{pmatrix}, \quad \begin{pmatrix} b \\ r_1 \end{pmatrix} = \begin{pmatrix} q_2 & 1 \\ 1 & 0 \end{pmatrix}\begin{pmatrix} r_1 \\ r_2 \end{pmatrix}, \quad \begin{pmatrix} r_1 \\ r_2 \end{pmatrix} = \begin{pmatrix} q_3 & 1 \\ 1 & 0 \end{pmatrix}\begin{pmatrix} r_2 \\ r_3 \end{pmatrix},$$

$$\text{(E*)} \quad \ldots, \quad \begin{pmatrix} r_{n-3} \\ r_{n-2} \end{pmatrix} = \begin{pmatrix} q_{n-1} & 1 \\ 1 & 0 \end{pmatrix}\begin{pmatrix} r_{n-2} \\ r_{n-1} \end{pmatrix}, \quad \begin{pmatrix} r_{n-2} \\ r_{n-1} \end{pmatrix} = \begin{pmatrix} q_n & 1 \\ 1 & 0 \end{pmatrix}\begin{pmatrix} r_{n-1} \\ 0 \end{pmatrix},$$

$$r_{n-1} = d = (a, b).$$

If we put

$$Q = \begin{pmatrix} q_1 & 1 \\ 1 & 0 \end{pmatrix}\begin{pmatrix} q_2 & 1 \\ 1 & 0 \end{pmatrix} \cdots \begin{pmatrix} q_n & 1 \\ 1 & 0 \end{pmatrix},$$

then (E*) implies that

$$\begin{pmatrix} a \\ b \end{pmatrix} = Q\begin{pmatrix} d \\ 0 \end{pmatrix}.$$

As Q is an integral matrix with determinant $(-1)^n$, the inverse Q^{-1} is also integral. A solution is given by

$$(x_0 \; y_0) = \begin{pmatrix} \frac{c}{d} & 0 \end{pmatrix} Q^{-1}.$$

In fact, we have

$$ax_0 + by_0 = (x_0\ y_0)\begin{pmatrix} a \\ b \end{pmatrix} = \begin{pmatrix} \frac{c}{d} & 0 \end{pmatrix} Q^{-1}\begin{pmatrix} a \\ b \end{pmatrix} = \begin{pmatrix} \frac{c}{d} & 0 \end{pmatrix} Q^{-1} Q \begin{pmatrix} d \\ 0 \end{pmatrix}$$

$$= \begin{pmatrix} \frac{c}{d} & 0 \end{pmatrix}\begin{pmatrix} d \\ 0 \end{pmatrix} = \frac{c}{d} \cdot d = c.$$

Exercise 1.14. Find an integral solution of $432x + 95y = 1$.

Exercise 1.15. Find an integral solution of $35x + 55y + 77z = 1$.

For $a, b \in \mathbb{N}$, $a > b$, let us rewrite the euclidean algorithm (E) as follows:

$$(\mathrm{E}^{\#}) \qquad \begin{cases} \dfrac{a}{b} = q_1 + \dfrac{r_1}{b}, & 0 < \dfrac{r_1}{b} < 1; \\[2mm] \dfrac{b}{r_1} = q_2 + \dfrac{r_2}{r_1}, & 0 < \dfrac{r_2}{r_1} < 1; \\ \vdots & \\ \dfrac{r_{n-3}}{r_{n-2}} = q_{n-1} + \dfrac{r_{n-1}}{r_{n-2}}, & 0 < \dfrac{r_{n-1}}{r_{n-2}} < 1; \\[2mm] \dfrac{r_{n-2}}{r_{n-1}} = q_n. & \end{cases}$$

As we have

$$\frac{a}{b} = q_1 + \frac{r_1}{b} = q_1 + \frac{1}{\frac{b}{r_1}} = q_1 + \frac{1}{q_2 + \frac{r_2}{r_1}} = q_1 + \cfrac{1}{q_2 + \cfrac{1}{q_3 + \frac{r_3}{r_2}}} = \cdots$$

$$= q_1 + \cfrac{1}{q_2 + \cfrac{1}{q_3 + \ddots + \cfrac{1}{q_{n-1} + \cfrac{1}{q_n}}}}$$

the rational number a/b can be expressed in terms of quotients $q_1, q_2, \ldots, q_n$ in euclidean algorithm. This oversized fraction is called a *continued fraction*.

If we put $\alpha = a/b$ in ($E^\#$) we have $q_1 = [\alpha]$, the integral part of α and $r_1/b = \langle\alpha\rangle$, the fractional part of α. Then equalities in ($E^\#$) are $\alpha = [\alpha] + \langle\alpha\rangle$, $\langle\alpha\rangle^{-1} = [\langle\alpha\rangle^{-1}] + \langle\langle\alpha\rangle^{-1}\rangle, \ldots$. Now definitions of $[\alpha]$ and $\langle\alpha\rangle$ make sense not only for rational numbers but also for *irrational numbers*, i.e., for real numbers which are not rational. The set of all irrational numbers can well be written as $\mathbb{R} - \mathbb{Q}$. For any real number α the integral part $[\alpha]$ and the fractional part $\langle\alpha\rangle$ equally make sense; we get the unique decomposition

$$\alpha = [\alpha] + \langle\alpha\rangle, \qquad [\alpha] \in \mathbb{Z}, \qquad 0 \le \langle\alpha\rangle < 1, \qquad \alpha \in \mathbb{R}.$$

For an irrational number α the algorithm ($E^\#$) never terminates and, conversely, if it terminates after a finite number of steps for a real number α, then α must be rational. For an irrational number α we shall denote by (C) the system of the infinite process:

$$\text{(C)} \begin{cases} \alpha = [\alpha] + \langle\alpha\rangle, & 0 < \langle\alpha\rangle < 1; \\ \langle\alpha\rangle^{-1} = [\langle\alpha\rangle^{-1}] + \langle\langle\alpha\rangle^{-1}\rangle, & 0 < \langle\langle\alpha\rangle^{-1}\rangle < 1; \\ \langle\langle\alpha\rangle^{-1}\rangle^{-1} = [\langle\langle\alpha\rangle^{-1}\rangle^{-1}] + \langle\langle\langle\alpha\rangle^{-1}\rangle^{-1}\rangle, & 0 < \langle\langle\langle\alpha\rangle^{-1}\rangle^{-1}\rangle < 1; \\ \ldots \quad \ldots & \end{cases}$$

Denoting by $q_1, q_2, q_3, \ldots$ the integral parts of α, $\langle\alpha\rangle^{-1}$, $\langle\langle\alpha\rangle^{-1}\rangle^{-1}, \ldots$ in (C), we obtain a mapping C:

$$C\colon \mathbb{R} - \mathbb{Q} \to \mathbb{Z} \times \prod_2^\infty \mathbb{N}$$

given by $C\alpha = \{q_n\}$, $n \ge 1$, where $\prod_2^\infty \mathbb{N}$ simply means the infinite product of countably many $\mathbb{N}$'s whose elements are $\{a_2, a_3, a_4, \ldots\}$, $a_i \in \mathbb{N}$. It will turn out that C is a bijection $\mathbb{R} - \mathbb{Q} \cong \mathbb{Z} \times \prod_2^\infty \mathbb{N}$ (Theorem 1.8).

Exercise 1.16. Verify that $C(\sqrt{2}) = \{1, 2, 2, 2, \ldots\}$.

1.4. Continued-Fraction Expansion of Irrational Numbers

For variables $\lambda_1, \ldots, \lambda_n$, put

$$[\lambda_1, \ldots, \lambda_n] = \lambda_1 + \cfrac{1}{\lambda_2 + \cfrac{1}{\lambda_3 + \cfrac{1}{\ddots \cfrac{1}{\lambda_{n-1} + \cfrac{1}{\lambda_n}}}}}.$$

Since the right-hand side is a rational function in $\lambda_1, \ldots, \lambda_n$, it is expressed as a ratio of polynomials:

$$[\lambda_1, \ldots, \lambda_n] = \frac{P_n(\lambda_1, \ldots, \lambda_n)}{Q_n(\lambda_1, \ldots, \lambda_n)}.$$

Since $[\lambda_1] = \lambda_1$, we have $P_1(\lambda_1) = \lambda_1$ and $Q_1(\lambda_1) = 1$.† Since $[\lambda_1, \lambda_2] = \lambda_1 + 1/\lambda_2 = (1 + \lambda_1\lambda_2)/\lambda_2$, we have $P_2(\lambda_1, \lambda_2) = 1 + \lambda_1\lambda_2$ and $Q_2 = (\lambda_1, \lambda_2) = \lambda_2$. Since

$$[\lambda_1, \lambda_2, \lambda_3] = \lambda_1 + \frac{1}{[\lambda_2, \lambda_3]} = \frac{\lambda_1 + \lambda_3 + \lambda_1\lambda_2\lambda_3}{1 + \lambda_2\lambda_3},$$

we have $P_3(\lambda_1, \lambda_2, \lambda_3) = \lambda_1 + \lambda_3 + \lambda_1\lambda_2\lambda_3$ and $Q_3(\lambda_1, \lambda_2, \lambda_3) = 1 + \lambda_2\lambda_3$, etc.

This calculation suggests that we define P_n and Q_n inductively as follows. First of all, for convenience, we put $P_0 = 1$, $Q_0 = 0$, $P_1 = \lambda_1$, and $Q_1 = 1$. Then, for $n \geq 2$, we put

$$P_n = \lambda_n P_{n-1} + P_{n-2}, \qquad Q_n = \lambda_n Q_{n-1} + Q_{n-2}, \qquad n \geq 2$$

† There is a clash of notation $[a]$. For a real number a, $[a]$ usually means the integral part of a, but if we substitute a for λ_1 we get $[a] = a$ according to the definition of $[\lambda_1]$. This ambiguity, however, should not confuse you because $[a]$ in the sense of continued fractions will seldom occur and will not then be important.

where $P_i = P_i(\lambda_1, \ldots, \lambda_n)$ and $Q_i = Q_i(\lambda_1, \ldots, \lambda_i)$, $i \geq 0$, with the above agreement for the case $i = 0, 1$. We leave here some formal properties of P_n and Q_n as exercises.

EXERCISE 1.17. Verify that $[\lambda_1, \ldots, \lambda_n] = P_n/Q_n$, $n \geq 1$.

EXERCISE 1.18. Verify that $P_nQ_{n-1} - P_{n-1}Q_n = (-1)^n$, $n \geq 1$.

EXERCISE 1.19. Verify that $P_nQ_{n-2} - P_{n-2}Q_n = (-1)^{n-1}\lambda_n$, $n \geq 2$.

Let us return to the mapping C defined near the end of the last section. The mapping C: $\mathbb{R} - \mathbb{Q} \to \mathbb{Z} \times \prod_2^\infty \mathbb{N}$ was defined by $C\alpha = \{q_n\}$, $n \geq 1$, where the sequence $\{q_n\}$ was determined by the following process (C):

$$(C) \qquad \begin{cases} \alpha = a_1 = q_1 + \dfrac{1}{\alpha_2}, & q_1 = [\alpha] = [\alpha_1]; \\ \alpha_2 = q_2 + \dfrac{1}{\alpha_3}, & q_2 = [\alpha_2]; \\ \alpha_3 = q_3 + \dfrac{1}{\alpha_4}, & q_3 = [\alpha_3]; \\ \ldots \quad \ldots & \\ \alpha_n = q_n + \dfrac{1}{\alpha_{n+1}}, & q_n = [\alpha_n]; \\ \ldots \quad \ldots & \end{cases}$$

THEOREM 1.8. *The mapping* C *is a bijection*: $\mathbb{R} - \mathbb{Q} \cong \mathbb{Z} \times \prod_2^\infty \mathbb{N}$.

PROOF. It is enough to find a mapping

$$D: \mathbb{Z} \times \prod_2^\infty \mathbb{N} \to \mathbb{R} - \mathbb{Q}$$

such that $DC\alpha = \alpha$ and $CDs = s$ for any $\alpha \in \mathbb{R} - \mathbb{Q}$ and any $s \in \mathbb{Z} \times \prod_2^\infty \mathbb{N}$. In this proof, for a sequence $\{q_n\}$, $n \geq 1$, we put $P_0 = 1$, $Q_0 = 0$, $P_n = P_n(q_1, \ldots, q_n)$, and $Q_n = Q_n(q_1, \ldots, q_n)$, $n \geq 1$. Then, for $n \geq 2$, we have $Q_n = q_nQ_{n-1} + Q_{n-2} > 0$. The inequalities

$$0 < Q_1(=1) \leq Q_2(=q_2) < Q_3 < Q_4 < \cdots < Q_n < Q_{n+1} < \cdots$$

imply that $Q_n \geq n-1$. Putting $a_n = P_n/Q_n$, we have, by Exercise 1.17,

$$a_n = \frac{P_n}{Q_n} = [q_1, \ldots, q_n].$$

Assuming $n \geq 3$, we have, by Exercise 1.19,

$$a_n - a_{n-2} = \frac{P_n}{Q_n} - \frac{P_{n-2}}{Q_{n-2}} = \frac{P_n Q_{n-2} - P_{n-2} Q_n}{Q_n Q_{n-2}} = \frac{(-1)^{n-1} q_n}{Q_n Q_{n-2}}.$$

Since q_n, Q_n, and Q_{n-2} are all >0, $a_n - a_{n-2} > 0$ if n is odd and $a_n - a_{n-2} < 0$ if n is even; in other words, the sequence $\{a_{2m+1}\}$ is increasing and the sequence $\{a_{2m}\}$ is decreasing. Next, by Exercise 1.18, we have

$$a_n - a_{n-1} = \frac{P_n}{Q_n} - \frac{P_{n-1}}{Q_{n-1}} = \frac{P_n Q_{n-1} - P_{n-1} Q_n}{Q_n Q_{n-1}} = \frac{(-1)^n}{Q_n Q_{n-1}},$$

which implies that $a_{2m-1} < a_{2m}$. Since $Q_n \geq n-1$, we have

$$|a_n - a_{n-1}| \leq \frac{1}{Q_n Q_{n-1}} \leq \frac{1}{(n-1)(n-2)} \to 0 \qquad (n \to \infty);$$

hence the real number $\alpha = \lim_{n\to\infty} a_n$ exists.

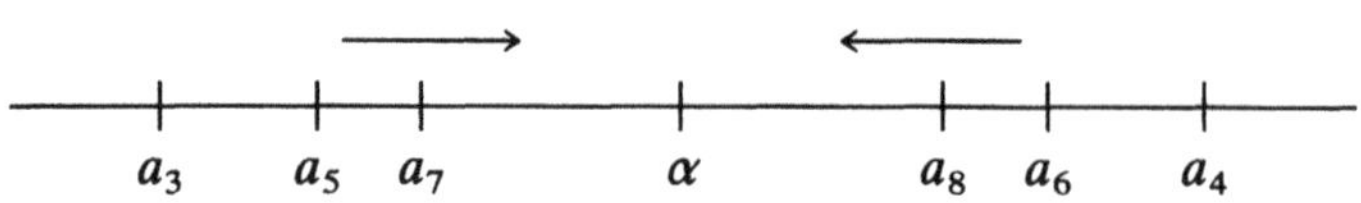

As $a_n = [q_1, \ldots, q_n]$, we can introduce the notation $[q_1, \ldots, q_n, \ldots]$ by

$$\alpha = \lim_{n\to\infty} a_n = \lim_{n\to\infty} [q_1, \ldots, q_n] \stackrel{\text{def}}{=} [q_1, \ldots, q_n, \ldots].$$

Applying what we did to the subsequence $\{q_n, q_{n+1}, \ldots\}$ of the full sequence $\{q_1, q_2, \ldots, q_n, \ldots\}$, we can determine the real number $\alpha_n = [q_n, q_{n+1}, \ldots]$. Then one proves easily that

$$\alpha = [q_1, \ldots, q_n, \ldots] = [q_1, \ldots, q_n, \alpha_n].$$

EXERCISE 1.20. Prove the last equality.

Since, for $N > n$, $[q_n, q_{n+1}, \ldots, q_N] > 1$ and $\{a_{2m+1}\}$ is increasing, by taking N $(N \to \infty)$ such that $N - n + 1$ is odd, we have

$$\alpha_n = [q_n, q_{n+1}, \ldots] > 1, \qquad n \geq 3.$$

Therefore, we have $0 < \alpha_n < 1$, $n \geq 3$. Furthermore,

$$\alpha_n = [q_n, q_{n+1}, \ldots] = q_n + \frac{1}{[q_{n+1}, \ldots]} = q_n + \frac{1}{\alpha_{n+1}}, \qquad n \geq 3,$$

which implies that $\alpha = [q_1, \ldots, q_{n-1}, \alpha_n]$ is irrational and that $C\alpha = \{q_1, q_2, \ldots, q_n, \ldots\}$. In other words, for a sequence $s = \{q_n\}$, $n \geq 1$, if we put

$$Ds \stackrel{\text{def}}{=} \alpha = \lim_{n\to\infty} [q_1, q_2, \ldots, q_n],$$

then what we observed above simply means that $CDs = s$ for all $s \in \mathbb{Z} \times \prod_2^\infty \mathbb{N}$. Hence, it remains to show that $DC\alpha = \alpha$ for all $\alpha \in \mathbb{R} - \mathbb{Q}$. To do this, put $C\alpha = \{q_1, \ldots, q_n, \ldots\}$. By the definition of the mapping (C), we have

$$\alpha = q_1 + \frac{1}{\alpha_2}, \qquad \alpha_2 = q_2 + \frac{1}{\alpha_3}, \qquad \ldots, \qquad \alpha_n = q_n + \frac{1}{\alpha_{n+1}}, \qquad \ldots,$$

i.e.,

$$\alpha = [q_1, \ldots, q_n, \alpha_{n+1}] = \frac{\alpha_{n+1}P_n + P_{n-1}}{\alpha_{n+1}Q_n + Q_{n-1}}.$$

On the other hand, we have

$$DC\alpha = \lim_{n\to\infty} [q_1, q_2, \ldots, q_n] = \lim_{n\to\infty} \frac{P_n}{Q_n}.$$

Since we have

$$\alpha - \frac{P_n}{Q_n} = \frac{\alpha_{n+1}P_n + P_{n-1}}{\alpha_{n+1}Q_n + Q_{n-1}} - \frac{P_n}{Q_n} = \frac{(\alpha_{n+1}P_n + P_{n-1})Q_n - (\alpha_{n+1}Q_n + Q_{n+1})P_n}{(\alpha_{n+1}Q_n + Q_{n-1})Q_n}$$
$$= \frac{P_{n-1}Q_n - P_nQ_{n-1}}{(\alpha_{n+1}Q_n + Q_{n-1})Q_n} = \frac{(-1)^{n-1}}{(\alpha_{n+1}Q_n + Q_{n-1})Q_n}$$

it follows that

$$\left|\alpha - \frac{P_n}{Q_n}\right| = \frac{1}{(\alpha_{n+1}Q_n + Q_{n-1})Q_n}.$$

In view of inequalities $\alpha_{n+1} > 1$ and $Q_n > Q_{n-1} \geq n-2$, we see that $|\alpha - P_n/Q_n| \to 0$ $(n \to \infty)$, i.e., $\alpha = \lim_{n\to\infty}(P_n/Q_n) = DC\alpha$. Q.E.D.

Exercise 1.21. Verify that $(\sqrt{5}+1)/2 = [1, 1, 1, \ldots]$. In this case, the sequence $\{Q_n\}$, $n \geq 0$, is the Fibonacci sequence (Exercise 1.8).

1.5. Concept of Groups

Definition 1.7. A group is a set G with two mappings $\pi: G \times G \to G$ and $\iota: G \to G$ satisfying the following conditions:

1. $\pi(\pi(a, b), c) = \pi(a, \pi(b, c))$ (associative law),
2. there is an element $e \in G$ such that $\pi(a, e) = \pi(e, a) = a$ for all $a \in G$,
3. for an element e in (2), $\pi(a, \iota(a)) = \pi(\iota(a), a) = e$ for all $a \in G$.

Remark 1.10. If e' is another element with (2), we have $\pi(e, e') = e = e'$; this unique e is called the unit element of G. Usually, we write $\pi(a, b) = ab$ and $\iota(a) = a^{-1}$ (the inverse of a). Therefore the group is also defined by the following conditions:

1'. $(ab)c = a(bc)$,
2'. $ae = ea = a$,
3'. $aa^{-1} = a^{-1}a = e$.

Definition 1.8. For a group G, if $ab = ba$ for all $a, b \in G$, G is called an *abelian* group or a *commutative* group; Abel (1802–1829).

REMARK 1.11. When G is abelian, we usually write $\pi(a, b) = a + b$; and write $e = 0$, $a^{-1} = -a$. An abelian group with this additive notation is also called a *module*. This general use of the word module is of course compatible with Definition 1.5.

Back to any group G, we define the power a^n, $a \in G$, $n \in \mathbb{Z}$, by $a^n = a \cdot a \cdot \cdots \cdot a$ (n times) when $n > 0$, by $a^n = (a^{-1})^{-n}$ when $n < 0$, and by $a^0 = e$. One verifies easily that $a^{n+m} = a^n a^m$ and $a^{nm} = (a^n)^m$. When the group operation is written additively, we write na instead of a^n and so we have $(n+m)a = na + ma$ and $(nm)a = n(ma)$.

For an element a of a group G, if there is a number $n \in \mathbb{N}$ such that $a^n = e$, a is said to be of finite order and the least of such an n is called the *order* of a: $n = \operatorname{ord} a$. An element $a \in G$ whose order is not finite is said to be of infinite order or *free*.

PROPOSITION 1.3. *Let m be the order of $a \in G$. If $a^n = e$, $n \in \mathbb{Z}$, then $m \mid n$.*

EXERCISE 1.22. Prove Proposition 1.3.

EXERCISE 1.23. Let R be a (not necessarily commutative) ring. Assume that R has the unit 1; $1 \cdot a = a \cdot 1 = a$ for all $a \in R$. An element $a \in R$ is said to be invertible if there is an element $a^* \in R$ such that $a^*a = aa^* = 1$. Prove that a^* when it exists is uniquely determined by a. Prove that the set $R^\times$ of all invertible elements of R forms a group.

REMARK 1.12. Let A be a (commutative) ring with 1 and $R = M_n(A)$ be the set of all $n \times n$ matrices of elements from A. For the matrix addition and multiplication, R becomes a ring (noncommutative when $n \geq 2$). We have $a \in R^\times \Leftrightarrow \det a \neq 0$. The group $R^\times$ is written $\mathrm{GL}_n(A)$ and is called the general linear group of degree n over A. When $A = \mathbb{Z}$, we have $\mathrm{GL}_n(\mathbb{Z}) = \{a \in M_n(\mathbb{Z});\ \det a = \pm 1\}$ (the unimodular group). Especially, the group $\mathrm{GL}_2(\mathbb{Z})$ is used to classify irrational numbers.

DEFINITION 1.9. Let G be a group and X be a set. G is said to be a *transformation group* on X if there is a mapping $T: G \times X \to X$ such that $T(e, x) = x$ and $T(\sigma\tau, x) = T(\sigma, T(\tau, x))$, for all $x \in X$, $\sigma, \tau \in G$.

Remark 1.13. When G is a transformation group on X, we often say that G *acts on* X and write $T(\sigma, x) = \sigma x$. Hence the action of G on X is described by $ex = x$ and $(\sigma\tau)x = \sigma(\tau x)$.

Exercise 1.24. When a group G acts on a set X, define a relation $x \sim y$, $x, y \in X$, by

$$x \sim y \overset{\text{def}}{\Leftrightarrow} y = \sigma x \qquad \text{for some } \sigma \in G.$$

Prove that $\sim$ is an equivalence relation in X.

As Exercise 1.24 shows, an action of G on X defines an equivalence relation in X and all equivalence classes make up the *quotient set* X/G.

As an application of the idea of transformation groups, let us reconsider the continued-fraction expansion of irrational numbers discussed in Section 1.4. Let $G = GL_2(\mathbb{Z})$ (Remark 1.12) and $X = \mathbb{R} - \mathbb{Q}$, the set of all irrational numbers. Define the action of G on X by

$$\sigma\alpha = \frac{a\alpha + b}{c\alpha + d}, \qquad \alpha \in X, \qquad \sigma = \begin{pmatrix} a & b \\ c & d \end{pmatrix} \in G.$$

Note that $\sigma \in G \Leftrightarrow a, b, c, d \in \mathbb{Z}$ and $\det \alpha = ad - bc = \pm 1$.

Exercise 1.25. Verify that the above really defines an action of G on X.

Now, by Theorem 1.8, there is a bijection

$$C: X = \mathbb{R} - \mathbb{Q} \xrightarrow{\sim} \mathbb{Z} \times \prod_{2}^{\infty} \mathbb{N}.$$

Therefore, if $C\alpha = \{q_1, q_2, \ldots, q_n, \ldots\}$, i.e., if $\alpha = [q_1, q_2, \ldots, q_n, \ldots]$, then the sequence $\{q'_1, q'_2, \ldots, q'_n, \ldots\}$ determined by $\sigma\alpha = [q'_1, q'_2, \ldots, q'_n, \ldots]$ must be described in terms of the sequence $\{q_1, q_2, \ldots, q_n, \ldots\}$ and $\sigma \in GL_2(\mathbb{Z})$. This is the first step in the theory of continued fractions.

Definition 1.10. In the set $X = \mathbb{R} - \mathbb{Q}$, define the following two equivalence relations. Let $\alpha = [q_1, q_2, \ldots, q_n, \ldots]$ and $\alpha' = [q'_1, q'_2, \ldots, q'_n, \ldots]$.

1. $\alpha \underset{G}{\sim} \alpha' \Leftrightarrow \alpha' = \sigma\alpha$ for some $\sigma \in G$,
2. $\alpha \underset{C}{\sim} \alpha' \Leftrightarrow \alpha_n = \alpha'_m$ for some $n, m \in \mathbb{N}$,

where α_n and α'_m are determined by $\alpha = [q_1, q_2, \ldots, q_{n-1}, \alpha_n]$ and $\alpha' = [q'_1, q'_2, \ldots, q'_{m-1}, \alpha'_m]$, respectively.

Exercise 1.26. Prove that $\alpha \underset{C}{\sim} \alpha'$ is an equivalence relation.

Theorem 1.9. *For $\alpha, \alpha' \in \mathbb{R} - \mathbb{Q}$, we have $\alpha \underset{G}{\sim} \alpha' \Leftrightarrow \alpha \underset{C}{\sim} \alpha'$.*

Proof. ($\Leftarrow$) By Exercise 1.17 we have, for $n \geq 2$,

$$\alpha = [q_1, q_2, \ldots, q_{n-1}, \alpha_n] = \frac{P_n}{Q_n} = \frac{\alpha_n P_{n-1} + P_{n-2}}{\alpha_n Q_{n-1} + Q_{n-2}} = \sigma_{n-1}\alpha_n$$

with

$$\sigma_{n-1} = \begin{pmatrix} P_{n-1} P_{n-2} \\ Q_{n-1} Q_{n-2} \end{pmatrix} \in \mathrm{GL}_2(\mathbb{Z})$$

(Exercise 1.18). Hence, we have $\alpha \underset{G}{\sim} \alpha_n$, $n \geq 2$. Since $\alpha = \alpha_1$, we have $\alpha \underset{G}{\sim} \alpha_n$, $n \in \mathbb{N}$. Therefore, $\alpha \underset{C}{\sim} \alpha'$ implies $\alpha \underset{G}{\sim} \alpha_n = \alpha'_m \underset{G}{\sim} \alpha'$ and so $\alpha \underset{G}{\sim} \alpha'$.

($\Rightarrow$) We first take care of a lemma:

Lemma. *Let β and γ be numbers in $\mathbb{R} - \mathbb{Q}$ with $\beta > 1$ and*

$$\sigma = \begin{pmatrix} a & b \\ c & d \end{pmatrix}$$

be a matrix in $\mathrm{GL}_2(\mathbb{Z})$ with $c > d > 0$. If $\gamma = \sigma\beta$ then $\beta = \gamma_m$ for some $m \in \mathbb{N}$.

Proof of Lemma. Let

$$\frac{a}{c} = [q_1, \ldots, q_{m-1}] = \frac{P_{m-1}}{Q_{m-1}}$$

be the continued-fraction expansion of the rational number a/c; $\varepsilon = ad - bc = \pm 1$ implies $(a, c) = 1$. Since the $c > 0$, we must have

$a = P_{m-1}$, $c = Q_{m-1}$. Since one can change the length of $a/c = [q_1, \ldots, q_{m-1}]$ by one by adjusting the last term as follows:

$$\frac{1}{q_{m-1}} = \frac{1}{(q_{m-1}-1)+\dfrac{1}{1}} \qquad \text{if } q_{m-1} > 1,$$

$$\frac{1}{q_{m-2}+\dfrac{1}{q_{m-1}}} = \frac{1}{q_{m-2}+1} \qquad \text{if } q_{m-1} = 1,$$

we may assume that $ad - bc = P_{m-1}Q_{m-2} - Q_{m-1}P_{m-2} = \varepsilon$. Then we have $P_{m-1}(d - Q_{m-2}) = Q_{m-1}(b - P_{m-2})$. Since $(P_{m-1}, Q_{m-1}) = 1$, we have $Q_{m-1} \mid (d - Q_{m-2})$. On the other hand, relations $Q_{m-2} \leq Q_{m-1}$ and $0 < d < c = Q_{m-1}$ imply $|d - Q_{m-2}| < Q_{m-1}$. Therefore we get $d = Q_{m-2}$, $b = P_{m-2}$. Since $\beta > 1$ by the assumption, we have

$$\gamma = \frac{a\beta + b}{c\beta + d} = \frac{\beta P_{m-1} + P_{m-2}}{\beta Q_{m-1} + Q_{m-2}} = [q_1, \ldots, q_{m-1}, \beta],$$

which means that $\beta = \gamma_m$. Q.E.D. of Lemma

Back to the proof of ($\Rightarrow$), assume that $\alpha \underset{\sigma}{\sim} \alpha'$, i.e.,

$$\alpha' = \sigma\alpha = \frac{a\alpha + b}{c\alpha + d}, \qquad \sigma = \begin{pmatrix} a & b \\ c & d \end{pmatrix} \in \mathrm{GL}_2(\mathbb{Z}).$$

Replacing σ by $-\sigma$ when necessary, we may assume that $c\alpha + d > 0$. Using σ_{n-1} in ($\Leftarrow$), we have $\alpha = \sigma_{n-1}\alpha_n$ and so $\alpha' = \sigma\sigma_{n-1}\alpha_n$. We shall show that we can choose n so that $c' > d' > 0$ for

$$\sigma' = \sigma\sigma_{n-1} = \begin{pmatrix} a' & b' \\ c' & d' \end{pmatrix}.$$

If that is so, applying Lemma with $\beta = \alpha_n$, $\gamma = \alpha'$, and $\sigma = \sigma'$, we get $\alpha_n = \alpha'_m$ for some m, i.e., $\alpha \underset{\sigma}{\sim} \alpha'$ and we are done. (Note that $\alpha_n = [q_n, q_{n+1}, \ldots] > 1$, $n \geq 3$, as shown in the proof of Theorem 1.8.)

Now we have

$$\sigma' = \sigma\sigma_{n-1} = \begin{pmatrix} a & b \\ c & d \end{pmatrix}\begin{pmatrix} P_{n-1} & P_{n-2} \\ Q_{n-1} & Q_{n-2} \end{pmatrix} = \begin{pmatrix} * & * \\ cP_{n-1} + dQ_{n-1} & cP_{n-2} + dQ_{n-2} \end{pmatrix}$$

and hence $c' = cP_{n-1} + dQ_{n-1} = Q_{n-1}(ca_{n-1} + d)$, $a_{n-1} = P_{n-1}/Q_{n-1}$, $d' = cP_{n-2} + dQ_{n-2} = Q_{n-2}(ca_{n-2} + d)$, and $a_{n-2} = P_{n-2}/Q_{n-2}$. Since $c\alpha + d > 0$ and a_n converges to α from both sides, we have, for large n, $ca_{n-1} + d > 0$ and $ca_{n-2} + d > 0$, which implies that $c', d' > 0$.

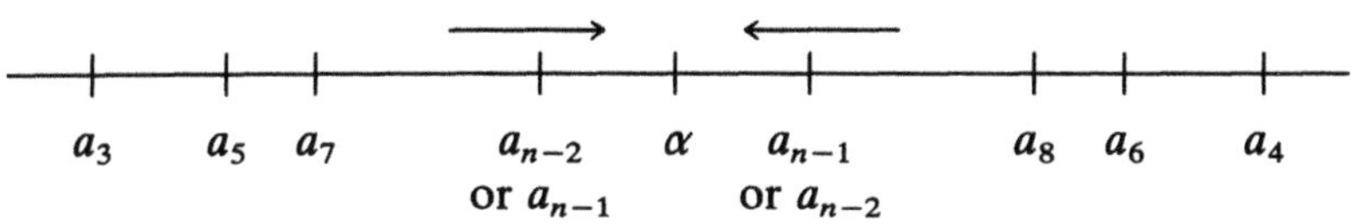

Finally, we have $c' - d' = Q_{n-1}(ca_{n-1} + d) - Q_{n-2}(ca_{n-2} + d) = Q_{n-2}(ca_{n-1}) + d) - Q_{n-2}(ca_{n-2} + d) = Q_{n-2}c(a_{n-1} - a_{n-2}) = c(-1)^{n-1}/Q_{n-1}$ (Exercise 1.18). By choosing a right parity of n, we have $c' > d'$ as we wanted. Q.E.D.

1.6. Subgroups and Quotient Groups

For subsets A and B of a group G, we put

$$AB \stackrel{\text{def}}{=} \{ab; a \in A, b \in B\} \qquad \text{and} \qquad A^{-1} \stackrel{\text{def}}{=} \{a^{-1}; a \in A\}.$$

When G is abelian and the operation is written additively, we shall use $A + B$ and $-A$ instead of AB and A^{-1}, respectively. One verifies easily that $AG = GA = G$, $G^{-1} = G$, $Ae = eA = A$, etc., for $A \neq \emptyset$.

Definition 1.11. A subset H of a group G which makes up a group under the same operation as G is called a *subgroup* of G.

Proposition 1.4. *A subset H of a group G is a subgroup* $\Leftrightarrow$ $HH^{-1} \subset H$.

Exercise 1.27. Prove Proposition 1.4.

Take any element a of a group G and put $H = \{a^n; n \in \mathbb{Z}\}$. It is easy to see that H is an abelian subgroup of G. This H is called a *cyclic* subgroup generated by a; we write $H = \langle a \rangle$. When the cardinality of a group G is finite, we denote it by $[G]$ and call it the *order* of G. When G is cyclic: $G = \langle a \rangle$, then $[G] = \operatorname{ord} a$ (Section 1.5).

Let H be a subgroup of a group G. By the mapping $T: H \times G \to G$ defined by $T(h, a) = ha$, H becomes a transformation group on the set G in the sense of Definition 1.9. Hence, by Exercise 1.24, an equivalence relation $a \sim b$ in G is defined by the condition $b = ha$, for some $h \in H$. The equivalence class containing a is then the set Ha which is called a *right coset* of G modulo H. Consider next the mapping $T': H \times G \to G$ defined by $T'(h, a) = ah^{-1}$. Since

$$T'(h, T'(k, a)) = (ak^{-1})h^{-1} = a(k^{-1}h^{-1}) = a(hk)^{-1} = T'(hk, a),$$

T' defines another action of H on the set G. The equivalence relation in G is $a \sim b \Leftrightarrow b = ah$, for some $h \in H$. The equivalence class containing a is this time the set aH which is called a *left coset* of G modulo H. If, in particular, H is such that $aH = Ha$ for all $a \in G$, then the distinction between right cosets and left cosets disappears; we call $aH = Ha$ a *coset* modulo H. Such a subgroup H is called a *normal* subgroup; usually a normal subgroup of G is written N. We shall denote by G/N the quotient set formed by cosets of G modulo N.

Proposition 1.5. *Let N be a normal subgroup of a group G. Then the quotient set G/N obtains naturally a group structure. This group G/N is called the quotient group of G modulo N.*

Exercise 1.28. Clarify what is meant by a natural group structure in G/N in Proposition 1.5 and verify the axiom of the group for G/N.

Every group $G \neq \{e\}$ has at least two normal subgroups: $N = G$ and $N = \{e\}$. G is said to be *simple* if those are the only normal subgroups. On the other hand, if G is abelian every subgroup of G is obviously normal.

Proposition 1.6. *Let G be a finite group, H a subgroup of G, and G/H be the quotient set consisting of right cosets of G modulo*

H. We have then $[G]=[H][G/H]$, *where* $[S]$ *denotes the cardinality of any finite set* S. *The same is true for left cosets of* G *modulo* H.

PROOF. Let Ha and Hb be any two right cosets. Since the mapping $xa \mapsto xb$, $x \in H$, is a bijection between Ha and Hb, we have $[Ha]=[Hb]=[H]$. As G is the disjoint union of $[G/H]$ right cosets, we get $[G]=[H][G/H]$. Q.E.D.

PROPOSITION 1.7. *Let* G *be a finite group of order* n. *For any element* $a \in G$, *we have* $a^n = e$.

PROOF. Let m be the order of the cyclic subgroup $H=\langle a \rangle$. By Proposition 1.6, we have $a^m = e$ and so $a^n = e$. Q.E.D.

1.7. Ideals and Quotient Rings

Let R be a (commutative) ring with 1. With respect to the addition, R is an abelian group, i.e., a module. If the set $R-\{0\}$ makes up a group, we have $R-\{0\}=R^\times$, the group of invertible elements of R, and R becomes a field (Definition 1.1).

DEFINITION 1.12. A subset I of a ring is an *ideal* if I is a subgroup of the additive group of R and $rI \subset I$ for all $r \in R$. (Here we do not have to assume that R has 1.)

REMARK 1.14. When $R=\mathbb{Z}$, since the condition $rI \subset I$ for any ideal is automatically satisfied whenever I is a submodule of $\mathbb{Z}$, any ideal of $\mathbb{Z}$ is $I = m\mathbb{Z}$, $m \geq 0$ (Theorem 1.1).

For any ring R, since I is a normal subgroup of the additive group of R, the quotient group R/I makes sense (Section 1.6). The sum of cosets $a+I$ and $b+I$ is defined by $(a+b)+I$ (Proposition 1.5 and Exercise 1.28); the condition $rI \subset I$, $r \in R$ guarantees the validity of defining the product $(a+I)(b+I)= ab+I$. We therefore get the following proposition.

PROPOSITION 1.8. *Let* I *be an ideal of a ring* R. *Then the quotient set* R/I *obtains naturally a ring structure. This ring* R/I *is called the quotient ring of* R *modulo* I.

EXERCISE 1.29. Prove Proposition 1.8.

In this section, we shall confine ourselves to the quotient ring $\mathbb{Z}/m\mathbb{Z}$, $m \in \mathbb{N}$. Since the set $\{0, 1, 2, \ldots, m-1\}$ forms a complete set of representatives of cosets of $\mathbb{Z}$ modulo $m\mathbb{Z}$, the ring $\mathbb{Z}/m\mathbb{Z}$ consists of m elements. For $a \in \mathbb{Z}$, we shall denote by $[a]$, by $[a]_m$, or by $a \bmod m$ the coset containing a. For $a, b \in \mathbb{Z}$, we shall write $a \equiv b(m)$ when a and b belong to the same coset, i.e., when $m \mid (a-b)$.

PROPOSITION 1.9. *For $a \in \mathbb{Z}$, $m \in \mathbb{N}$, we have $[a] \in (\mathbb{Z}/m\mathbb{Z})^\times \Leftrightarrow (a, m) = 1$.*

PROOF. ($\Rightarrow$) For $[a] \in (\mathbb{Z}/m\mathbb{Z})^\times$ there exists $[x] \in \mathbb{Z}/m\mathbb{Z}$ such that $[a][x] = [1]$, i.e., $ax \equiv 1(m)$ which implies $(a, m) = 1$.

($\Leftarrow$) If $(a, m) = 1$, then, by Theorem 1.3, there are $x, y \in \mathbb{Z}$ such that $ax + my = 1$ which implies that $ax \equiv 1(m)$, i.e, $[a][x] = [1]$ and so $[a] \in (\mathbb{Z}/m\mathbb{Z})^\times$. Q.E.D.

THEOREM 1.10. *Denote by $\varphi(m)$ the order of the group $(\mathbb{Z}/m\mathbb{Z})^\times$. Then we have $a^{\varphi(m)} \equiv 1(m)$ for all $a \in \mathbb{Z}$ with $(a, m) = 1$.*

PROOF. The assertion follows immediately from Proposition 1.7 and Proposition 1.9. Q.E.D.

DEFINITION 1.13. The mapping $\varphi: \mathbb{N} \to \mathbb{Z}$ defined in Theorem 1.10 is called Euler's function; Euler (1707–1783).

THEOREM 1.11. *Let p be a prime and a be an integer such that $p \nmid a$. Then we have $a^{p-1} \equiv 1(p)$.*

PROOF. The assertion follows from Theorem 1.10 because $\varphi(p) = p - 1$. Q.E.D.

REMARK 1.15. Thanks to the concept of groups and rings, Theorem 1.10 and Theorem 1.11 are merely corollaries to Proposition 1.7. Historically, Theorem 1.11 is Fermat's theorem and Theorem 1.10 is Euler's theorem. Theorem 1.10 is also called the Fermat–Euler theorem.

THEOREM 1.12. *The ring $\mathbb{Z}/m\mathbb{Z}$ is a field $\Leftrightarrow m = p$, a prime.*

PROOF. ($\Rightarrow$) Suppose m is not a prime. If $m = 1$, then $\mathbb{Z}/\mathbb{Z} = \{0\}$, i.e., $\mathbb{Z}/\mathbb{Z}$ contains only one element and so it is not a field by Definition 1.1. Next, suppose that $m > 1$ is not a prime. Then one can write $m = ab$, $a, b > 1$. Since $m \nmid a$ and $m \nmid b$, we have $[a] \neq 0$ and $[b] \neq 0$. Thus we have $0 = [m] = [ab] = [a][b]$, i.e., a product of two nonzero elements is 0, which implies that $\mathbb{Z}/m\mathbb{Z}$ is not a field.

($\Leftarrow$) Take any $[a] \neq 0$ from $\mathbb{Z}/p\mathbb{Z}$. Since $p \nmid a$, there are $x, y \in \mathbb{Z}$ such that $ax + by = 1$ (Theorem 1.3). In other words, we have $ax \equiv 1(p)$, or $[a][x] = 1$, i.e., $[a] \in (\mathbb{Z}/p\mathbb{Z})^{\times}$, which means that the ring $\mathbb{Z}/p\mathbb{Z}$ is a field. Q.E.D.

Definition 1.14. We denote by $\mathbb{F}_p$ the field $\mathbb{Z}/p\mathbb{Z}$ consisting of p (a prime) elements.

Exercise 1.30. Let $a \in \mathbb{N}$ and $p \neq 2$ be a prime. Prove that $p \mid (a^{2n} + 1) \Rightarrow 2^{n+1} \mid (p-1)$.

Exercise 1.31. Let $F_n = 2^{2^n} + 1$, $n \geq 0$, be Fermat numbers (Exercise 1.12). Verify that $F_0 = 3$, $F_1 = 5$, $F_2 = 17$, $F_3 = 257$, and $F_4 = 65537$ are all prime. Prove that $F_5 = 4294967297$ is not a prime. (It was Euler who pointed out that F_5 was not a prime, contradicting Fermat's guess that all integers F_n were prime.)

Exercise 1.32. By Theorem 1.11, $(2^{p-1} - 1)/p$ is an integer for any prime p. Prove that this number is a square if and only if $p = 3$ or $p = 7$.

Exercise 1.33. Let a be an integer ≥ 2 and p be an odd prime. Prove that $a^{p-1} - 1 = p^e$ for some $e \in \mathbb{N}$ if and only if $a = 2$ and $p = 3$.

1.8. Isomorphisms and Homomorphisms

Definition 1.15. A mapping $f: G \to G'$ of a group G to a group G' is a *homomorphism* if we have $f(ab) = f(a)f(b)$, $a, b \in G$. If, furthermore, f is a bijection between G and G', then f is called an *isomorphism*, written $f: G \xrightarrow{\sim} G'$. When there is an isomorphism of G with G', the two groups are said to be isomorphic, written $G \approx G'$.

Exercise 1.34. For a homomorphism $f: G \to G'$, prove that $f(e) = e'$, e and e' being the unit elements of G and G', respectively. Prove also that $f(a^{-1}) = f(a)^{-1}$, $a \in G$.

Theorem 1.13. *Let $f: G \to G'$ be a surjective homomorphism. Then the set $N = \{a \in G; f(a) = e'\}$ makes up a normal subgroup of G and f induces naturally an isomorphism $\bar{f}: G/N \xrightarrow{\sim} G'$. N is called the kernel of f: $N = \operatorname{Ker} f$.*

PROOF. Since $f(ab)=f(a)f(b)=e'\cdot e'=e'$ for $a, b \in N$, we have $ab \in N$; since $f(a^{-1})=f(a)^{-1}=(e')^{-1}=e'$, we have $a^{-1} \in N$ for $a \in N$, which shows that N is a subgroup of G. Next, for any $c \in G$ and $a \in N$,

$$f(cac^{-1})=f(c)f(a)f(c)^{-1}=f(c)e'f(c)^{-1}=f(c)f(c)^{-1}=e',$$

which means that $cac^{-1} \in N$. From this we find that $cN=Nc$, i.e., N is a normal subgroup of G. Now, for $Na \in G/N$, put $\bar{f}(Na)=f(a)$; as $f(N)=\{e'\}$, $\bar{f}$ is well-defined and one verifies easily that $\bar{f}$ is a surjective homomorphism $G/N \to G'$. Finally, if $\bar{f}(Na)=\bar{f}(Nb)$, then $f(a)=f(b)$, which implies $f(ab^{-1})=e'$ or $ab^{-1} \in N$, i.e., $Na=Nb$; therefore $\bar{f}$ is injective. Q.E.D.

We can do the same thing for rings.

DEFINITION 1.16. A mapping $f: R \to R'$ of a ring R to a ring R' is a *homomorphism* if we have $f(a+b)=f(a)+f(b)$ and $f(ab)=f(a)f(b)$, $a, b \in R$. If, furthermore, f is a bijection between R and R', then f is called an *isomorphism*, written $f: R \xrightarrow{\sim} R'$. When there is an isomorphism of R with R', the two rings are said to be isomorphic, written $R \approx R'$.

THEOREM 1.14. *Let $f: R \to R'$ be a surjective homomorphism. Then the set $I=\{a \in R;\ f(a)=0\}$ makes up an ideal of R and f induces naturaly an isomorphism $\bar{f}: R/I \xrightarrow{\sim} R'$. I is called the kernel of f: $I=\ker f$.*

PROOF. By Theorem 1.13 (applied to additive groups of R, R') we see that I is a subgroup of the additive group R. Now, for $a \in I$ and $b \in R$, we have $f(ab)=f(a)f(b)=0\cdot f(b)=0$ (Exercise 1.1), and so $ab \in I$, i.e., I is an ideal of R. Now, for $a+I \in R/I$, put $\bar{f}(a+I)=f(a)$; as $f(I)=\{0\}$, f is well defined and one verifies easily that $\bar{f}$ is a surjective homomorphism $R/I \to R'$. Finally, if $\bar{f}(a+I)=\bar{f}(b+I)$, then $f(a)=f(b)$, which implies $f(a-b)=0$ or $a-b \in I$, i.e., $a+I=b+I$, therefore $\bar{f}$ is injective. Q.E.D.

Let $m=m_1\cdots m_l$ be a decomposition of an integer $m>1$ such that each $m_i>1$ and $(m_i, m_j)=1$, $i \neq j$. We obtain rings $R=\mathbb{Z}/m\mathbb{Z}$, $R_i=\mathbb{Z}/m_i\mathbb{Z}$ and groups $R^\times$, $R_i^\times$, $1 \le i \le l$. To study relationships among these rings and groups, we need the notion of the direct product of groups and rings which will be introduced in the following exercise.

Exercise 1.35.

1. Let G_i, $1 \le i \le l$, be groups and put $G = G_1 \times \cdots \times G_l$, the product of sets; elements of G are l-tuples $x = (x_1, \ldots, x_l)$, $x_i \in G_i$. Prove that G becomes a group if we define the product of $x, y \in G$ by $xy = (x_1 y_1, \ldots, x_l y_l)$.
2. Let R_i, $1 \le i \le l$, be rings with 1 and put $R = R_1 \times \cdots \times R_l$, the product of sets. Prove that R becomes a ring with 1 if we define the sum and the product of $x, y \in R$ by $x + y = (x_1 + y_1, \ldots, x_l + y_l)$, $xy = (x_1 y_1, \ldots, x_l y_l)$.

Theorem 1.15. *For an integer $m = m_1 \cdots m_l$, $m_i > 1$, $(m_i, m_j) = 1$, $i \neq j$, put $R = \mathbb{Z}/m\mathbb{Z}$, $R_i = \mathbb{Z}/m_i\mathbb{Z}$. Then there are natural isomorphisms of rings and groups: $R \approx R_1 \times \cdots R_l$, $R^\times \approx R_1^\times \cdots \times R_l^\times$.*

Proof. For $a \in \mathbb{Z}$, write $[a]_n = a \bmod n \in \mathbb{Z}/n\mathbb{Z}$. Consider a mapping $f: \mathbb{Z} \to R_1 \times \cdots \times R_l$ defined by $f(a) = ([a]_{m_1}, \ldots, [a]_{m_l})$; f is clearly a ring homomorphism which is also surjective (Exercise 1.36). Therefore, by Theorem 1.14, f induces the isomorphism $f: \mathbb{Z}/I \xrightarrow{\sim} R_1 \times \cdots \times R_l$, $I = \operatorname{Ker} f$. Now, $a \in \operatorname{Ker} f \Leftrightarrow f(a) = 0 \Leftrightarrow [a]_{m_i} = 0$, $1 \le i \le l \Leftrightarrow m_i \mid a$, $1 \le i \le l \Leftrightarrow m \mid a \Leftrightarrow a \in m\mathbb{Z}$. Hence we have $I = \operatorname{Ker} f = m\mathbb{Z}$, i.e., $R \approx R_1 \times \cdots \times R_l$. Since the unit of R corresponds to the element $(1, \ldots, 1)$ of the product and $a = (a_1, \ldots, a_l) \in R^\times \Leftrightarrow a_i \in R_i^\times$, $1 \le i \le l$, the ring isomorphism induces the group isomorphism $R^\times \approx R_1^\times \times \cdots \times R_l^\times$. Q.E.D.

Exercise 1.36. Prove that the mapping f in the proof of Theorem 1.15 is surjective.

Proposition 1.10. *If $(n, m) = 1$ then $\varphi(nm) = \varphi(n)\varphi(m)$.*

Proof. Apply Theorem 1.15 with $l = 2$. Q.E.D.

Remark 1.16. When $m = p_1^{e_1} \cdots p_l^{e_l}$ is the factorization of an integer $m > 1$ into primes, we have $\varphi(m) = \varphi(p_1^{e_1}) \cdots \varphi(p_l^{e_l})$ by Theorem 1.15. Therefore the function φ is determined if we find $\varphi(p^e)$, $e \ge 1$. When $e = 1$, we find $\varphi(p) = p - 1$ because $\mathbb{Z}/p\mathbb{Z}$ is a field.

Proposition 1.11. *If* p *is a prime, we have* $\varphi(p^e)=p^{e-1}(p-1)$, $e\geq 1$.

Proof. Put $G=(\mathbb{Z}/p^{e+1}\mathbb{Z})$, $G'=(\mathbb{Z}/p^e\mathbb{Z})$, and consider a homomorphism $f\colon G\to G'$ defined by $(f[x]_{p^{e+1}})=[x]_{p^e}$. f is surjective, because if $[x]_{p^e}\in G'$ (hence $p\nmid x$) we have $f([x]_{p^{e+1}})=[x]_{p^e}$. Next, since $[x]_{p^{e+1}}\in \operatorname{Ker} f \Leftrightarrow f([x]_{p^{e+1}})=[1]_{p^e}\Leftrightarrow x\equiv 1 \pmod{p^e} \Leftrightarrow x\equiv 1, 1+p^e, 1+2p^e, \ldots, 1+(p-1)p^e \pmod{p^{e+1}}$, we have $G/\operatorname{Ker} f\approx G'$, which implies that $[G]/[\operatorname{Ker} f]=[G']$ or $\varphi(p^{e+1})=p\cdot\varphi(p^e)$. Therefore the assertion is proved by induction on e. Q.E.D.

1.9. Polynomial Rings

Let K be a field. For a variable X, we shall denote by $K[X]$ the set of all polynomials with coefficients in K. A polynomial $P=P(X)$ in $K[X]$ is an expression

$$P(X)=a_0+a_1X+\cdots+a_nX^n, \qquad a_i\in K, \qquad 0\leq i\leq n,$$

If $a_n\neq 0$, P is said to be of degree n; we write $n=\deg P$. $K[X]$ forms a ring with respect to addition and multiplication defined in a natural way; the field K is a subring of the ring $K[X]$ and $1\in K$ serves as the unit element for $K[X]$. For $P, Q\in K[X]$, with $P\neq 0$, $Q\neq 0$, we have $\deg(PQ)=\deg(P)+\deg(Q)$. The ring $K[X]$ resembles the ring $\mathbb{Z}$. In what follows, we shall asterisk the number for a statement on $K[X]$ corresponding to a similar statement on $\mathbb{Z}$ mentioned earlier.

Exercise 1.37. Prove that $(K[X])^\times=K^\times$ for a field K.

Definition 1.3*. Let A, B be polynomials in $K[X]$. If there is a polynomial $C\in K[X]$ such that $B=AC$, B is said to be a *multiple* of A; A is said to be a *divisior* of B or to *divide* B. We write $A\mid B$ when A divides B.

Definition 1.4*. Let A be a polynomial in $K[X]$ with $n=\deg A>0$. A is said to be *prime*, or *irreducible*, if it has no divisor of degree >0 and $<n$.

Remark 1.17. Every polynomial of degree 1 is irreducible. The property of a polynomial of being irreducible or reducible (not irreducible) depends on the field of coefficients; for example, x^2+1 is irreducible over $\mathbb{R}$, but not over $\mathbb{C}$, since $x^2+1=(x+i)(x-i)$, $i=\sqrt{-1}$.

REMARK 1.18. If $P=a_0+a_1x+\cdots+a_nx^n$, $a_n\neq 0$, is in $K[X]$, then $a_n^{-1}P$ becomes a polynomial in $K[X]$ with the highest coefficient 1; such a polynomial will be called *monic* or *normalized*. Since $(K[X])^\times = K^\times$ corresponds to $\mathbb{Z}^\times=\{\pm 1\}$, monic polynomials correspond to natural numbers.

PROPOSITION 1.2*. *Let A, B be polynomials in $K[X]$ with $B\neq 0$. Then there exist unique polynomials Q, R in $K[X]$ such that $A=QB+R$, where $R=0$ or $\deg R<\deg B$. R is called the remainder of the division of A by B and Q the quotient.*

EXERCISE 1.38. Prove Proposition 1.2*.

REMARK 1.19. When $B=X-a$, $a\in K$, we have $A=(X-a)Q+r$, $r\in K$. Therefore we have $A(a)=r$ (remainder theorem). Furthermore, $r=0 \Leftrightarrow A(a)=0$; thus A is a multiple of $X-a \Leftrightarrow a$ is a zero (or a root) of A.

THEOREM 1.1*. *For an ideal I of the ring $K[X]$, there exists a unique polynomial $P\in K[X]$, determined up to multiplication by elements of $K^\times$, such that $I=P\cdot K[X]$.*

EXERCISE 1.39. Prove Theorem 1.1*.

THEOREM 1.2*. *For $A_1, \ldots, A_l \in K[X]$, let*

$$I=\{P_1A_1+\cdots+P_lA_l;\ P_i\in K[X]\}.$$

I is an ideal of $K[X]$ and there is a unique monic polynomial D in $K[X]$ such that $I=D\cdot K[X]$. D has the following properties:

1. $D \mid A_i$, $1\leq i\leq l$,
2. $B \mid A_i$, $1\leq i\leq l \Rightarrow B\mid D$.

REMARK 1.8*. The monic polynomial D determined by $A_1, \ldots, A_l$ (Theorem 1.2*) is called the *greatest common divisor* (g.c.d.) of $A_1, \ldots, A_l$; we write $D=(A_1, \ldots, A_l)$.

DEFINITION 1.6*. Polynomials $A_1, \ldots, A_l$ are called *relatively prime* if $(A_1, \ldots, A_l)=1$.

Theorem 1.3*. *Polynomials* $A_1, \ldots, A_l$ *are relatively prime* $\Leftrightarrow$ *the equation* $A_1x_1 + \cdots + A_lx_l = 1$ *has a solution* $x_i \in K[X]$, $1 \le i \le l$.

Theorem 1.4*. *Let* A, B, C *be polynomials in* $K[X]$ *such that* $(A, B) = 1$. *Then* $A \mid BC \Rightarrow A \mid C$.

Theorem 1.5*. *Let* P *be irreducible.* $P \mid AB \Rightarrow P \mid A$ *or* $P \mid B$.

Theorem 1.6*. *Every polynomial in* $K[X]$ *can be written as a product of irreducible polynomials; it can be so written in one and only one way except for the order of factors and for multiplication by elements of* $K^\times$.

Remark 1.20. Proposition 1.2 and Proposition 1.2* show that the ring $R = \mathbb{Z}$ or $K[X]$ is "euclidean". An analogy between $\mathbb{Z}$ and $K[X]$ is striking, especially when K is finite, e.g., $K = \mathbb{F}_p = \mathbb{Z}/p\mathbb{Z}$.

Theorem 1.16. *Let* A *be a polynomial of degree* $n > 0$ *in* $K[X]$. *Then* A *can be written, uniquely up to the order of the factors, as*

$$A = (X - a_1) \cdots (X - a_m)Q, \qquad 0 \le m \le n, \qquad a_i \in K, \qquad 1 \le i \le m,$$

where Q *has no roots in* K.

Proof. If A has no roots in K, there is nothing to prove. Otherwise we prove the theorem by induction on n. If a is a root of A, we can write $A = (X - a)B$ with $\deg B = n - 1$ and we can apply the theorem to B. As for the uniqueness, let $A = (X - b_1) \cdots (X - b_r)R$ be another decomposition of A as above. Since $R(a) \ne 0$, $a = b_j$ for some j. Then we can use the theorem for $B = A(X - a)^{-1}$ to prove the uniqueness. Q.E.D.

Theorem 1.17. *A polynomial* A *in* $K[X]$ *of degree* $n > 0$ *has at most* n *roots in* K (*counted with multiplicities*).

Proof. The assertion follows immediately from Theorem 1.16. Q.E.D.

1.10. Primitive Roots

The main theorem (Theorem 1.19) asserts that for a prime p the group $(\mathbb{Z}/p\mathbb{Z})^\times$ is cyclic; this is the most basic and productive theorem in elementary number theory. To begin with, we shall prepare some facts on cyclic groups.

PROPOSITION 1.12. *Let $G = \langle x \rangle$ be a cyclic group of order m. For $a \in \mathbb{Z}$, we have* $\operatorname{ord} x^a = m/(a, m)$.

PROOF. If $a = 0$, we have $(0, m) = m$ and $\operatorname{ord} x^0 = \operatorname{ord} e = 1$, so the Proposition is true. For $a \neq 0$, put $d = (a, m)$, $t = \operatorname{ord} x^a$. Then $e = (x^a)^t = x^{at}$, which implies that $m \mid at$, or $(m/d) \mid (a/d)t$. Since $(m/d, a/d) = 1$, we have $(m/d) \mid t$. Conversely, $(x^a)^{m/d} = (x^m)^{a/d} = e$ implies that $t \mid (m/d)$; hence $t = m/d$. Q.E.D.

PROPOSITION 1.13. *Let $G = \langle x \rangle$ be a cyclic group of order m. For $a \in \mathbb{Z}$, we have $G = \langle x^a \rangle \Leftrightarrow (a, m) = 1$.*

PROOF. $G = \langle x^a \rangle \Leftrightarrow \operatorname{ord} x^a = m \xLeftrightarrow{\text{Prop. 1.12}} m/(a, m) = m \Leftrightarrow (a, m) = 1$. Q.E.D.

REMARK 1.21. By Proposition 1.13 we see that a cyclic group G of order m has $\varphi(m)$ generators. Note that such a group G is isomorphic to the additive group of the ring $\mathbb{Z}/m\mathbb{Z}$.

PROPOSITION 1.14. *Let $G = \langle x \rangle$ be a cyclic group of order m. For any $d \mid m$, $G_d = \langle x^{m/d} \rangle$ is a subgroup of G of order d. Conversely, any subgroup of G of order d must be the group G_d; in other words, there is a one to one correspondence between subgroups of G and divisors of m.*

PROOF. The first half of the assertion follows from the equalities $\operatorname{ord} x^{m/d} = m/(m/d, m) = m/(m/d) = d$ (Proposition 1.12). To prove the second half, let H be any subgroup of G of order d. If we put $M = \{n \in \mathbb{Z};\ x^n \in H\}$, M becomes a module in $\mathbb{Z}$ such that $M \neq \{0\}$ because $x^m = e \in H$, i.e., $m \in M$. By Theorem 1.1, we have $M = m_0\mathbb{Z}$, $m_0 \geq 1$; we have $m_0 \mid m$ since $m \in M$. Therefore we have $x^n \in H \Leftrightarrow m_0 \mid n$ and hence $H = \{x^{km_0};\ k \in \mathbb{Z}\} = \langle x^{m_0} \rangle$. Proposition 1.12 implies that $d = [H] = \operatorname{ord} x^{m_0} = m/(m_0, m) = m/m_0$, and so we have $H = \langle x^{m_0} \rangle = \langle x^{m/d} \rangle = D_d$. Q.E.D.

PROPOSITION 1.15. *For $m \in \mathbb{N}$, we have $m = \sum_{d \mid m} \varphi(d)$.*

PROOF. Let $G = \langle x \rangle$ be a cyclic group of order m, e.g., $G =$ the additive group of the ring $\mathbb{Z}/m\mathbb{Z}$. Put $G_d = \langle x^{m/d} \rangle$ and let

E_d denote the set of generators of G_d. We have $[E_d] = \varphi(d)$ by Proposition 1.13. Since $E_d \cap E_{d'} = \varnothing$, for $d \neq d'$, G is a disjoint union of E_d, $d \mid m$, and we get $m = \sum_{d \mid m} \varphi(d)$. Q.E.D.

PROPOSITION 1.16. *Let G be a finite group of order m. If, for each $d \mid m$, there are no more than d elements of G satisfying the equation $x^d = e$, then G is cyclic.*

PROOF. Denote by $\psi(d)$ the number of elements of G of order d; we must prove that $\psi(m) \geq 1$. Obviously, we have $m = \sum_{d \mid m} \psi(d)$. Now, if G has an element y of order d, then the subgroup $H = \langle y \rangle$ has order d. By the assumption, all $\psi(d)$ elements of G of order d are in H; by Remark 1.21, there are exactly $\psi(d)$ such elements. Therefore, we have proved that $\psi(d) = \varphi(d)$ whenever $\psi(d) \geq 1$. Since $m = \sum_{d \mid m} \varphi(d)$ (Proposition 1.15) and $m = \sum_{d \mid m} \psi(d)$, we have $\psi(d) = \varphi(d)$ for all d; in particular, $\psi(m) = \varphi(m) \geq 1$. Q.E.D.

THEOREM 1.18. *Let K be a field. Any finite subgroup of $K^\times$ is cyclic.*

PROOF. Let G be a subgroup of $K^\times$ of order m. For each divisor d of m, the equation $x^d = 1$ has at most d solutions (Theorem 1.17). The assertion then follows from Proposition 1.16. Q.E.D.

THEOREM 1.19. *For a prime p, the group $(\mathbb{Z}/p\mathbb{Z})^\times$ is cyclic.*

PROOF. The assertion follows immediately from Theorem 1.18. Q.E.D.

DEFINITION 1.17. Any generator of the cyclic group $(\mathbb{Z}/p\mathbb{Z})^\times$ is called a *primitive root* modulo p. (When $[a]_p \in (\mathbb{Z}/p\mathbb{Z})$ is a primitive root modulo p, we call a itself a primitive root modulo p.)

Let us fix, once and for all, a primitive root r modulo p. Hence the group $(\mathbb{Z}/p\mathbb{Z})^\times$ is a cyclic group of order $p-1$ generated by $[r]_p$ and, for any $a \in \mathbb{Z}$, we have

$$a \equiv r^x \quad (p)$$

where $x \in \mathbb{Z}$ is uniquely determined modulo $p-1$. This x is called the *index* of a to the base r and is denoted by

$$x = \operatorname{ind}_r a$$

For a ring R with 1, we denote by R^+ the additive group of R.

THEOREM 1.20. *The* ind_r *gives an isomorphism* $(\mathbb{Z}/p\mathbb{Z})^\times \xrightarrow{\sim} (\mathbb{Z}/(p-1)\mathbb{Z})^+$.

PROOF. If $x = \operatorname{ind}_r a$ and $y = \operatorname{ind}_r b$, then $a \equiv r^x \quad (p)$ and $b \equiv r^y \quad (p)$. As $ab \equiv r^x r^y = r^{x+y} \quad (p)$, we have $\operatorname{ind}_r(ab) = x + y = \operatorname{ind}_r a + \operatorname{ind}_r b$, i.e., ind_r is a homomorphism. As for the kernel, we have $\operatorname{ind}_r a = 0 \Leftrightarrow a \equiv r^0 = 1 \quad (p) \Leftrightarrow a \equiv 1 \quad (p)$, therefore ind_r is injective. Since the two groups have the same order $p-1$, ind_r is surjective, and hence an isomorphism. Q.E.D.

THEOREM 1.21. *Let p be a prime and a be an integer such that* $p \nmid a$. *Let* $e = (n, p-1)$, $n \in \mathbb{N}$, *and write* $p - 1 = ef$. *Then, we have*

$$x^n \equiv a \quad (p) \text{ has a solution} \Leftrightarrow a^f \equiv 1 \quad (p).$$

When the equation has a solution, the number of solutions (mod p) *is e.*

PROOF. By Theorem 1.20, to solve $x^n \equiv a \quad (p)$ is equivalent to solving $ny \equiv b \quad (p-1)$, a linear equation, where $y = \operatorname{ind}_r x$, $b = \operatorname{ind}_r a$. As for the linear equation, it has a solution if and only if $e \mid b$ (Remark 1.8). If that is so, dividing both sides of $ny \equiv b \quad (p-1)$ by (e), the equation $n'y \equiv b' \quad (f)$, $n = en'$, $b = eb'$, has a unique solution y_0 modulo f. Then e numbers $y_0, y_0 + f, \ldots, y_0 + (e-1)f$ make up solutions of $ny \equiv b \quad (p-1)$. Finally, we have $a^f \equiv 1 \quad (p) \Leftrightarrow r^{bf} \equiv 1 \quad (p) \Leftrightarrow (p-1) \mid bf \Leftrightarrow e \mid b$.
Q.E.D.

REMARK 1.22. An integer a, $p \nmid a$, is called an *nth power residue* or an *nth power nonresidue* according as the equation $x^n \equiv a \quad (p)$ has a solution or not. if p is an odd prime, an integer a, $p \nmid a$, is called a *quadratic residue* or a *quadratic nonresidue* according as the equation $x^2 \equiv a \quad (p)$ has a solution or not. (When $p = 2$, the matter is trivial.) Applying Theorem 1.21 with $n = 2$, $p \neq 2$, we see that, for $a \in \mathbb{Z}$, $p \nmid a$,

$$x^2 \equiv a \quad (p) \text{ has a solution} \Leftrightarrow a^f \equiv 1 \quad (p), \qquad 2f = p - 1.$$

In other words, we have the following proposition.

PROPOSITION 1.17. *Let a be an integer and p be an odd prime such that* $p \nmid a$. *Then we have*

$$a \text{ is a quadratic residue} \Leftrightarrow a^{(p-1)/2} \equiv 1 \quad (p).$$

EXERCISE 1.40. Find a primitive root r modulo $p=13$ and make a table of $\mathrm{ind}_r a$ for $1 \le a \le 12$. By your table find all primitive roots modulo 13. Furthermore, solve the following three equations: $7x \equiv 5 \quad (13)$, $x^2 \equiv -1 \quad (13)$, $x^3 \equiv 5 \quad (13)$.

1.11. Algebraic integers

DEFINITION 1.18. A complex number $\alpha \in \mathbb{C}$ is an *algebraic number* if there is a polynomial $f(X) \in \mathbb{Q}[X]$, $\deg f \ge 1$, such that $f(\alpha)=0$. We denote by $\bar{\mathbb{Q}}$ the set of all algebraic numbers.

REMARK 1.23. Since $\alpha \in \mathbb{Q}$ is a root of $f(X)=X-\alpha \in \mathbb{Q}[X]$ we see that $\mathbb{Q} \subset \bar{\mathbb{Q}}$. $\sqrt{2}$ is not in $\mathbb{Q}$ but is in $\bar{\mathbb{Q}}$ because it is a root of $f(X)=X^2-2 \in \mathbb{Q}[X]$. $\sqrt{-1}$ is a root of $f(X)=X^2+1 \in \mathbb{Q}[X]$, hence it is algebraic. For $m \ge 1$, $\alpha = e^{2\pi i/m}$ is algebraic because it is a root of $f(X)=X^m-1 \in \mathbb{Q}[X]$. The numbers e and π which appeared in the algebraic number $e^{2\pi i/m}$ are known to be not algebraic, i.e., $e, \pi \in \mathbb{C}-\bar{\mathbb{Q}}$; numbers in $\mathbb{C}-\bar{\mathbb{Q}}$ are called *transcendental numbers.*

REMARK 1.24. The polynomial $f(X)$ in Definition 1.18 can be normalized by multiplying some number in $\mathbb{Q}^\times$ (Remark 1.18). In other words,

$$\alpha \text{ is algebraic} \Leftrightarrow \alpha \text{ is a root of } f(X)=X^n+a_1X^{n-1}+\cdots+a_n, \qquad a_i \in \mathbb{Q}, \quad n \ge 1.$$

On restricting $a_i \in \mathbb{Q}$ in $\mathbb{Z}$, we reach the following definition.

DEFINITION 1.19. A complex number $\alpha \in \mathbb{C}$ is an *algebraic integer* if there is a polynomail $f(X)=X^n+a_1X^{n-1}+\cdots+a_n$, $a_i \in \mathbb{Z}$, $n \ge 1$, such that $f(\alpha)=0$. We denote by $\bar{\mathbb{Z}}$ the set of all algebraic integers. (Remark 1.24 implies that $\bar{\mathbb{Z}} \subset \bar{\mathbb{Q}}$.)

PROPOSITION 1.18. $\mathbb{Q} \cap \bar{\mathbb{Z}} = \mathbb{Z}$.

PROOF. Since $\alpha \in \mathbb{Z}$ is a root of $f(X)=X-\alpha$ we have $\alpha \in \bar{\mathbb{Z}}$ and so $\mathbb{Z} \subset \mathbb{Q} \subset \bar{\mathbb{Z}}$. Now take any $r \in \mathbb{Q} \cap \bar{\mathbb{Z}}$. Write $r=c/d$, $d \ge 1$, $(c, d)=1$. Since $r \in \bar{\mathbb{Z}}$, we find $f(X)=X^n+a_1X^{n-1}+\cdots+a_n$,

$a_i \in \mathbb{Z}$, $n \geq 1$, such that $f(r) = 0$, i.e., $(c/d)^n + a_1(c/d)^{n-1} + \cdots + a_n = 0$. Multiplying d^n on both sides, we have $c^n + a_1 c^{n-1} d + \cdots + a_n d^n = 0$, which implies that $d \mid c^n$. Then $d = 1$ because $(c, d) = 1$, i.e., $r \in \mathbb{Z}$. Therefore we have $\mathbb{Q} \cap \bar{\mathbb{Z}} = \mathbb{Z}$. Q.E.D.

PROPOSITION 1.19. *For any algebraic number $\alpha \in \bar{\mathbb{Q}}$, there is a natural number $a \in \mathbb{N}$ such that $a\alpha$ is an algebraic integer*: $a\alpha \in \bar{\mathbb{Z}}$.

PROOF. By the assumption, we have

$$c_0\alpha^n + c_1\alpha^{n-1} + \cdots + c_{n-1}\alpha + c_n = 0, \qquad c_i \in \mathbb{Q}, \qquad 0 \leq i \leq n.$$

Let d be a common denominator of c_i and write $c_i = b_i/d$, $b_i \in \mathbb{Z}$. Then we have $b_0\alpha^n + b_1\alpha^{n-1} + \cdots + b_{n-1}\alpha + b_n = 0$, where we can assume, without loss of generality, that $b_0 \geq 1$. Multiplying b_0^{n-1} on both sides, we have

$$\begin{aligned} 0 &= b_0^n\alpha^n + b_1 b_0^{n-1}\alpha^{n-1} + \cdots b_{n-1} b_0^{n-1}\alpha + b_n b_0^{n-1} \\ &= (b_0\alpha)^n + b_1(b_0\alpha)^{n-1} + \cdots + b_{n-1} b_0^{n-2}(b_0\alpha) + b_n b_0^{n-1}, \end{aligned}$$

which means that $b_0\alpha \in \bar{\mathbb{Z}}$. Q.E.D.

EXERCISE 1.41. Prove that $\mathbb{Q} \cdot \bar{\mathbb{Z}} \subset \bar{\mathbb{Q}}$. (Here $\mathbb{Q} \cdot \bar{\mathbb{Z}} = \{a\alpha;\ a \in \mathbb{Q},\ \alpha \in \bar{\mathbb{Z}}\}$. Since we have not yet proved that the set $\bar{\mathbb{Q}}$ is a field, the relations $\mathbb{Q} \subset \bar{\mathbb{Q}}$ and $\bar{\mathbb{Z}} \subset \bar{\mathbb{Q}}$ do not immediately imply $\mathbb{Q} \cdot \bar{\mathbb{Z}} \subset \bar{\mathbb{Q}}$.)

PROPOSITION 1.20. *Let $\gamma_1, \ldots, \gamma_l \in \mathbb{C}$ be complex numbers not all 0 and let M be the module in $\mathbb{C}$ generated by them*: $M = \{\gamma = \sum_{i=1}^{l} c_i\gamma_i;\ c_i \in \mathbb{Z}\}$. *If a complex number $\alpha \in \mathbb{C}$ has the property $\alpha M \subset M$, then α is an algebraic integer*: $\alpha \in \bar{\mathbb{Z}}$.

PROOF. By the assumption, we have $\alpha\gamma_i = \sum_{j=1}^{l} c_{ij}\gamma_j$, $c_{ij} \in \mathbb{Z}$, $1 \leq i \leq l$. Define a symbol δ_{ij} by

$$\delta_{ij} = \begin{cases} 1, & i = j, \\ 0, & i \neq j. \end{cases}$$

[This is called the Kronecker (1823–1891) delta.] Then we have $\alpha\gamma_i = \sum_{j=1}^{l} \alpha\delta_{ij}\gamma_j$ and so $\sum_{j=1}^{l} (\alpha\delta_{ij} - c_{ij})\gamma_j = 0$. Call A the matrix over $\mathbb{C}$ whose (i, j) component is $\alpha\delta_{ij} - c_{ij}$, $1 \leq i, j \leq l$. Since

$(\gamma_1, \ldots, \gamma_l) \neq (0, \ldots, 0)$ by the assumption, we must have $\det A = 0$. The definition of A then implies that

$$0 = \det A = \alpha^l + a_1\alpha^{l-1} + \cdots + a_{l-1}\alpha + a_l, \qquad a_i \in \mathbb{Z},$$

which means that $\alpha \in \bar{\mathbb{Z}}$. Q.E.D.

THEOREM 1.22. *$\bar{\mathbb{Z}}$ is a subring of the field $\mathbb{C}$. $\bar{\mathbb{Z}}$ is called the ring of algebraic integers.*

PROOF. We must verify that $\alpha, \beta \in \bar{\mathbb{Z}} \Rightarrow \alpha \pm \beta,\ \alpha\beta \in \bar{\mathbb{Z}}$. Obviously, we can assume that both of α and β are not zero. Suppose α satisfies

$$\alpha^n + a_1\alpha^{n-1} + \cdots + a_{n-1}\alpha + a_n = 0, \qquad a_i \in \mathbb{Z}, \qquad n \geq 1,$$

and β satisfies

$$\beta^m + b_1\beta^{m-1} + \cdots + b_{m-1}\beta + b_n = 0, \qquad b_i \in \mathbb{Z}, \quad m \geq 1.$$

Then consider the nonzero module M:

$$M = \left\{\gamma = \sum_{i,j} c_{ij}\alpha^i\beta^j;\, c_{ij} \in \mathbb{Z},\, 0 \leq i \leq n-1,\, 0 \leq j \leq m-1\right\}.$$

In view of the relations for α and β, we see that $\alpha M \subset M$ and $\beta M \subset M$. Since $(\alpha \pm \beta)M \subset \alpha M + \beta M \subset M + M \subset M$ and $\alpha\beta M = \alpha(\beta M) \subset \alpha M \subset M$, we have $\alpha \pm \beta \in \bar{\mathbb{Z}}$ and $\alpha\beta \in \bar{\mathbb{Z}}$ by Proposition 1.20. Q.E.D.

THEOREM 1.23. *$\bar{\mathbb{Q}}$ is a subfield of the field $\mathbb{C}$. $\bar{\mathbb{Q}}$ is called the field of algebraic numbers or the algebraic closure of $\mathbb{Q}$.*†

PROOF. For $\alpha, \beta \in \bar{\mathbb{Q}}$ there are $a, b \in \mathbb{N}$ such that $\alpha a, \beta b \in \bar{\mathbb{Z}}$ (Proposition 1.19). Since $\bar{\mathbb{Z}}$ is a ring (Theorem 1.22), we have $ab\alpha = ba\alpha \in \bar{\mathbb{Z}}$; similarly we have $ab\beta \in \bar{\mathbb{Z}}$. Therefore, $ab(\alpha \pm \beta) \in \bar{\mathbb{Z}}$. As $(ab)^{-1} \in \mathbb{Q}$, we have $\alpha \pm \beta \in \mathbb{Q}\bar{\mathbb{Z}} \subset \bar{\mathbb{Q}}$ (Exercise 1.41). Next, since $\bar{\mathbb{Z}}$ is a ring, we have $a\alpha b\beta = ab\alpha\beta \in \bar{\mathbb{Z}}$. As $(ab)^{-1} \in \mathbb{Q}$, we

† Compare Exercise 2.3.

have $\alpha\beta \in \mathbb{Q}\bar{\mathbb{Z}} \subset \bar{\mathbb{Q}}$, again. Finally, we must verify that $\alpha^{-1} \in \bar{\mathbb{Q}}$ if $\alpha \in \bar{\mathbb{Q}}$ and $\alpha \neq 0$. To do this, let $a_0\alpha^n + a_1\alpha^{n-1} + \cdots + a_{n-1}\alpha + a_n = 0$, $a_i \in \mathbb{Q}$, $n \geq 1$, $a_0 \neq 0$, be a relation for $\alpha \in \bar{\mathbb{Q}}$. Multiplying by α^{-n} on both sides, we get

$$a_0 + a_1\alpha^{-1} + \cdots + a_{n-1}(\alpha^{-1})^{n-1} + a_n(\alpha^{-1})^n = 0$$

Since at least one of $a_1, \ldots, a_n$ is not zero (as $a_0 \neq 0$), we see that $\alpha^{-1} \in \bar{\mathbb{Q}}$. Q.E.D.

For $\alpha, \beta, \mu \in \bar{\mathbb{Z}}$, $\mu \neq 0$, we can define an equivalence relation

$$\alpha \equiv \beta \quad (\mu) \overset{\text{def}}{\Leftrightarrow} \mu \mid (\alpha - \beta) \overset{\text{def}}{\Leftrightarrow} \alpha - \beta = \mu\gamma, \qquad \gamma \in \bar{\mathbb{Z}}.$$

When $\alpha, \beta, \mu \in \mathbb{Z}$, this relation induces the equivalence relation already defined in $\mathbb{Z}$ (use Proposition 1.18).

PROPOSITION 1.21. *Let p be a prime. For $\alpha, \beta \in \bar{\mathbb{Z}}$, we have $(\alpha + \beta)^p \equiv \alpha^p + \beta^p \quad (p)$.*

PROOF. We have

$$(\alpha + \beta)^p = \sum_{k=0}^{p} \binom{p}{j} \alpha^k \beta^{p-k}.$$

But $p \mid \binom{p}{k}$, $1 \leq k \leq p - 1$ (Exercise 1.42). Q.E.D.

EXERCISE 1.42. Prove that $p \mid \binom{p}{k}$, $1 \leq k \leq p - 1$.

1.12. Characters of Abelian Groups

Let G be a finite abelian group. A homomorphism $\chi: G \to \mathbb{C}^\times$ is called a *character*. We denote by $\hat{G}$ the set of all characters of G. If n is the order of the group G, we have $x^n = e$, $x \in G$, hence we have $1 = \chi(e) = \chi(x^n) = \chi(x)^n$, i.e., $\chi(x)$ is an nth root of 1. If we define the product of two characters $\chi, \chi' \in \hat{G}$ by $\chi\chi'(x) = \chi(x)\chi'(x)$, $x \in G$, then G becomes a group where $\chi^{-1} = \bar{\chi}$, with $\bar{\chi}(x) \overset{\text{def}}{=} \overline{\chi(x)}$. $\hat{G}$ is called the *character group* of G.

EXERCISE 1.43. Verify that $\hat{G}$ makes up a group and that $\chi^{-1} = \bar{\chi}$.

We shall denote by ϵ the identity element of the group $\hat{G}$: $\epsilon(x) = 1$, $x \in G$.

PROPOSITION 1.22. *Let n be the order of the group G. Then we have*

$$\sum_{x\in G} \chi(x) = \begin{cases} n, & \chi = \epsilon \\ 0, & \chi \neq \epsilon \end{cases} \qquad (\textit{orthogonality}).$$

PROOF. When $\chi = \epsilon$ we have $\sum_{x\in G}\epsilon(x) = \sum_{x\in G} 1 = n$. When $\chi \neq \epsilon$, we can find an element $x_0 \in G$ such that $\chi(x_0) \neq 1$. Put $S = \sum_{x\in G}\chi(x)$. Since $S = \sum_{x\in G}\chi(x) = \sum_{x\in G}\chi(x_0x) = \sum_{x\in G}\chi(x_0)\chi(x) = \chi(x_0)S$, we get $(1-\chi(x_0))S = 0$, which implies that $S = 0$ as $\chi(x_0) \neq 1$. Q.E.D.

Now let us assume that G is cyclic: $G = \langle g\rangle = \{1, g, \ldots, g^{n-1}\}$.† Then G is isomorphic to the additive group $(\mathbb{Z}/n\mathbb{Z})^+$. We shall determine the structure of the character group $\hat{G}$. First of all, $\chi(g^k) = \chi(g)^k$, $k \in \mathbb{Z}$, χ is determined by the value $\chi(g)$. Since $\chi(g)$ is an nth root of 1, we can write $\chi(g) = \zeta^a$, $a \in \mathbb{Z}$, where $\zeta = e^{2\pi i/n}$ and a is determined modulo n. Consider a mapping $\mathbb{Z} \to \hat{G}$ defined by $a \mapsto \chi_a$, $\chi_a(g) = \zeta^a$. In view of the equality $\zeta^{a+b} = \zeta^a\zeta^b$, the mapping is a homomorphism, i.e., we have $\chi_{a+b} = \chi_a\chi_b$. As we remarked above, any $\chi \in \hat{G}$ can be written as $\chi = \chi_a$ for some $a \in \mathbb{Z}$; hence the homomorphism is surjective. Now, we have $\chi_a = \epsilon \Leftrightarrow \chi_a(g) = \zeta^a = 1 \Leftrightarrow n \mid a$, which means that the kernel of the homomorphism is just $n\mathbb{Z}$. Therefore we have $(\mathbb{Z}/n\mathbb{Z})^+ \xrightarrow{\sim} \hat{G}$ (Theorem 1.13). Summing up, we get

$$G \approx \hat{G} \approx (\mathbb{Z}/n\mathbb{Z})^+, \qquad n = [G].$$

EXERCISE 1.44. Prove that

$$\sum_{\chi\in \hat{G}} \chi(x) = \begin{cases} n, & x = e, \\ 0, & x \neq e. \end{cases}$$

† For Gauss reciprocity, cyclic groups are enough. We shall discuss characters of any finite abelian groups in Section 3.7 on the occasion of introducing Dirichlet L-functions.

From now on, let p be an odd prime. Consider the cyclic group $G=(\mathbb{Z}/p\mathbb{Z})^\times=\langle [r]_p\rangle$ where $r\in\mathbb{Z}$ is a fixed primitive root modulo p (Theorem 1.19, Definition 1.17). By the above arguments, both groups are cyclic of order $p-1$ and so G and $\hat{G}$ have exactly one subgroup of order d for each factor d of $p-1$ (Proposition 1.14). Putting $d=2$, there is exactly one element of order 2 in G and $\hat{G}$. The character of order 2 in $\hat{G}$ is called the *Legendre* (1752–1833) *character*. We shall denote this character by λ_p. In other words, λ_p is a unique element in $\widehat{(\mathbb{Z}/p\mathbb{Z})^\times}$ such that $\lambda_p\neq\epsilon$, $\lambda_p^2=\epsilon$, $p\neq 2$. Since $\lambda_p^2(x)=\lambda_p(x)^2=1$, we have $\lambda_p(x)=\pm 1$. We regard λ_p as a function on the set of integers a such that $p\nmid a$ defined by $\lambda_p(a)=\lambda_p([a]_p)$.

Proposition 1.23. *Let p be an odd prime. For $a\in\mathbb{Z}$ such that $p\nmid a$, we have $\lambda_p(a)=(-1)^{\mathrm{ind}_r a}$ where r is a primitive root modulo p. Therefore, we have*

$$\lambda_p(a)=1\Leftrightarrow \exists x\in\mathbb{Z} \text{ such that } x^2\equiv a \quad (p).$$

The number of a $(\mathrm{mod}\, p)$ such that $\lambda_p(a)=1$ is $(p-1)/2$.

Proof. Like the mapping $a\mapsto\lambda_p(a)$, the mapping $a\mapsto(-1)^{\mathrm{ind}_r a}$ is also a character of the group $(\mathbb{Z}/p\mathbb{Z})^\times$ which we shall call ψ; the order of ψ is 2 because $\psi(r)=-1$ (note that $\mathrm{ind}_r(r)=1$). Since there is only one character of order 2, we must have $\psi=\lambda_p$. Next, we have

$$\lambda_p(a)=1\Leftrightarrow \mathrm{ind}_r\, a \text{ is even}\Leftrightarrow a\equiv r^{2k}\quad(p)\Leftrightarrow a\equiv x^2\quad(p),$$

which implies the second part of the Proposition. Finally, the mapping $\lambda_p\colon(\mathbb{Z}/p\mathbb{Z})^\times\to\{\pm1\}$ is a surjective homomorphism as $\lambda_p\neq\epsilon$. Hence, we have an isomorphism $(\mathbb{Z}/p\mathbb{Z})^\times/\mathrm{Ker}\,\lambda_p\approx\{\pm1\}$ (Theorem 1.13) and then $[\mathrm{Ker}\,\lambda_p]=(p-1)/2$ (Proposition 1.6).

Q.E.D.

Theorem 1.24. *Let p be an odd prime. For $a\in\mathbb{Z}$ such that $p\nmid a$, we have $\lambda_p(a)\equiv a^{(p-1)/2}(p)$.*

Proof. By Theorem 1.11, we have $r^{p-1}\equiv1\quad(p)$ and so $r^{(p-1)/2}\equiv\pm1\quad(p)$. Here we must have -1 because r is a primitive root modulo p, i.e., r has order $p-1$ in the group $(\mathbb{Z}/p\mathbb{Z})^\times$.

Then we have $(-1)^k \equiv r^{k(p-1)/2}(p)$ which means that $(-1)^{\mathrm{ind}_r a} \equiv a^{(p-1)/2}(p)$, $p \nmid a$. Our assertion then follows from Proposition 1.23. Q.E.D.

REMARK 1.25. Theorem 1.24 is called *Euler's criterion.* Usually, the Legendre character $\lambda_p(a)$ is written $\left(\frac{a}{p}\right)$ (the Legendre symbol). Since the latter notation occupies a good deal of space, we shall mainly use the former notation in this book. In terms of Remark 1.22, the middle part of Proposition 1.23 says that

$$\lambda_p(a) = \left(\frac{a}{p}\right) = 1 \Leftrightarrow a \text{ is a quadratic residue modulo } p.$$

Fixing an odd prime p, let $a = (-1)^\epsilon 2^\alpha q^\beta \cdots$ be the prime factorization of $a \in \mathbb{Z}$ such that $p \nmid a$, with $\epsilon = 0$ or 1, $\alpha, \beta, \ldots \in \mathbb{N} \cup \{0\}$, $q \neq p$. Since λ_p is a homomorphism, we have $\lambda_p(a) = \lambda_p(-1)^\epsilon(2)^\alpha \lambda_p(q)^\beta \cdots$ and so the determination of $\lambda_p(a)$ is reduced to that of $\lambda_p(-1)$, $\lambda_p(2)$, and $\lambda_p(q)$. The last one is the content of the Gauss reciprocity law.

THEOREM 1.25 (*The first supplementary law*).

$$\lambda_p(-1) = (-1)^{(p-1)/2}.$$

PROOF. Put $a = -1$ in Theorem 1.24. Since both sides of $\lambda_p(-1) \equiv (-1)^{(p-1)/2} \quad (p)$ are ± 1 and p is odd, we must have the equality. Q.E.D.

THEOREM 1.26 (*The second supplementary law*).

$$\lambda_p(2) = (-1)^{(p^2-1)/8}.$$

PROOF. Put $\zeta = e^{2\pi i/8}$ and $\eta = \zeta + \zeta^{-1}$. Since $\zeta^4 = -1$, we get $\zeta^2 + \zeta^{-2} = 0$ and $\eta^2 = (\zeta + \zeta^{-1})^2 = \zeta^2 + 2 + \zeta^{-2} = 2$. Obviously, ζ, η are algebraic integers: $\zeta, \eta \in \bar{\mathbb{Z}}$ (Definition 1.19). Putting $a = 2$ in Theorem 1.24, we have

$$\eta^{p-1} = (\eta^2)^{(p-1)/2} = 2^{(p-1)/2} \equiv \lambda_p(2) \quad (p).$$

In view of Proposition 1.21, we have $\lambda_p(2)\eta \equiv \eta^p = (\zeta + \zeta^{-1})^p \equiv \zeta^p + \zeta^{-p} \quad (p)$. Now, we have

$$p \equiv \pm 1 \quad (8) \Rightarrow \zeta^p + \zeta^{-p} = \zeta + \zeta^{-1} = \eta,$$
$$p \equiv \pm 3 \quad (8) \Rightarrow \zeta^p + \zeta^{-p} = \zeta^{-3} + \zeta^{-3} = -\zeta^{-1} - \zeta = -\eta,$$

and hence we have

$$\zeta^p + \zeta^{-p} = \begin{cases} \eta, & p \equiv \pm 1 \quad (8), \\ -\eta, & p \equiv \pm 3 \quad (8). \end{cases}$$

In other words, we get $(-1)^\omega \eta \equiv \lambda_p(2)\eta \quad (p)$, with $\omega = (p^2 - 1)/8$, and then $(-1)^\omega 2 \equiv \lambda_p(2)2 \quad (p)$ because $\eta^2 = 2$. Since $p \neq 2$, it follows that $(-1)^\omega \equiv \lambda_p(2) \quad (p)$ and so $(-1)^\omega = \lambda_p(2)$. Q.E.D.

1.13. The Gauss Reciprocity Law

In this section, p and q will denote distinct odd primes. We aim at the following theorem.

THEOREM 1.27. [*Gauss* (1777–1855) *reciprocity law*].

$$\lambda_p(q)\lambda_q(p) = (-1)^{(p-1)/2 \cdot (q-1)/2}.$$

REMARK 1.26. In this section, we put $p^* = (-1)^{(p-1)/2}p$. Theorem 1.27 is equivalent to the equality $\lambda_p(q)\lambda_q(p^*) = 1$ because $\lambda_q(p^*) = \lambda_q((-1)^{(p-1)/2}p) = (\lambda_q(-1))^{(p-1)/2}\lambda_q(p) = (-1)^{(q-1)/2 \cdot (p-1)/2}\lambda_q(p)$ by Theorem 1.25.

DEFINITION 1.20. Let p be an odd prime and let $\zeta = e^{2\pi i/p}$. For $a \in \mathbb{Z}$, put

$$\tau_a \overset{\text{def}}{=} \sum_{x \in \mathbb{F}_p^\times} \lambda_p(x)\zeta^{ax}, \qquad \mathbb{F}_p = \mathbb{Z}/p\mathbb{Z},$$

which is called a *Gauss sum*; we write $\tau_1 = \tau$ when $a = 1$.

REMARK 1.27. We shall extend the function λ_p on $\mathbb{F}_p^\times$ to a function on $\mathbb{F}_p$ (still written λ_p) by setting $\lambda_p(0) = 0$. Therefore, we can define τ_a as

$$\tau_a = \sum_{x \in \mathbb{F}_p} \lambda_p(x)\zeta^{ax}.$$

Such a modification is sometimes useful (e.g., the next Proposition).

PROPOSITION 1.24. $\tau_a = \lambda_p(a)\tau,\ a \in \mathbb{Z}$.

PROOF. Since both sides in the equality depend on a modulo p, we can consider a as an elemnt in $\mathbb{F}_p$. When $a = 0$, we have

$$\tau_0 = \sum_{x \in \mathbb{F}_p^\times} \lambda_p(x) = 0 = \lambda_p(0) \qquad \text{(Proposition 1.22).}$$

When $a \neq 0$, we have

$$\lambda_p(a)\tau_a = \sum_{x \in \mathbb{F}_p} \lambda_p(a)\lambda_p(x)\zeta^{ax} = \sum_{x \in \mathbb{F}_p} \lambda_p(ax)\zeta^{ax} = \sum_{y \in \mathbb{F}_p} \lambda_p(y)\zeta^y = \tau,$$

which implies $\tau_a = \lambda_p(a)\tau$. Q.E.D.

PROPOSITION 1.25. $\tau_a^2 = p^*,\ p \nmid a$.

PROOF. We compute the sum $S = \sum_{a \in \mathbb{F}_p} \tau_a \tau_{-a}$ in two ways. Since

$$\tau_a \tau_{-a} = \lambda_p(a)\lambda_p(-a)\tau^2 = \lambda_p(-a^2)\tau^2 = \lambda_p(-1)\tau^2,$$

we get $S = (-1)^{(p-1)/2}(p-1)\tau^2$ as $\tau_0 = 0$ (Proposition 1.24). On the other hand, since

$$\tau_a \tau_{-a} = \sum_{x \in \mathbb{F}_p} \sum_{y \in \mathbb{F}_p} \lambda_p(x)\lambda_p(y)\zeta^{ax}\zeta^{-ay},$$

we get

$$S = \sum_x \sum_y \lambda_p(x)\lambda_p(y) \sum_{a \in \mathbb{F}_p} \zeta^{a(x-y)}.$$

Applying Proposition 1.22 to the inner sum above, with $G = \mathbb{F}_p^+$, $\chi(a) = \zeta^{a(x-y)}$, we have [using $\chi = \epsilon \Leftrightarrow x - y \equiv 0 \quad (p)$]

$$S = \sum_x \sum_y \lambda_p(x)\lambda_p(y)p\delta_{x,y} = \sum_x \lambda_p(x)^2 p = (p-1)p.$$

Comparing the two results on S, we get $\tau_a^2 = \tau^2 = (-1)^{(p-1)/2} = p^*$. Q.E.D.

PROOF OF THEOREM 1.27. From Theorem 1.24 and Proposition 1.25, it follows that $\tau^{q-1} = (\tau^2)^{(q-1)/2} = p^{*(q-1)/2} \equiv \lambda_q(p^*) \quad (q)$. Therefore we get

$$\tau^q \equiv \lambda_q(p^*)\tau(q). \tag{1.1}$$

Next, from the definition of τ and Proposition 1.21, it follows that

$$\tau^q = \Big(\sum_{x\in\mathbb{F}_p} \lambda_p(x)\zeta^x\Big)^q \equiv \sum_{x\in\mathbb{F}_p} \lambda_p(x)^q\zeta^{qx} = \sum_{x\in\mathbb{F}_p} \lambda_p(x)\zeta^{qx} = \tau_q \quad (q),$$

i.e.,

$$\tau^q \equiv \tau_q \quad (q). \tag{1.2}$$

Since $\tau_q = \lambda_p(q)\tau$ by Proposition 1.24, we get, by (1.1) and (1.2),

$$\lambda_p(q)\tau \equiv \lambda_q(p^*)\tau(q). \tag{1.3}$$

Multiplying τ on both sides of (1.3), we have $\lambda_p(q)p^* \equiv \lambda_q(p^*)p^* \quad (q)$ by Proposition 1.25. From this, we have $\lambda_p(q) \equiv \lambda_q(p^*) \quad (q)$ and hence the equality $\lambda_p(q) = \lambda_q(p^*)$ holds because both sides of the congruence are ± 1. Q.E.D.

REMARK 1.27. Theorem 1.27 is certainly useful to determine whether the equation $x^2 \equiv a \quad (p)$ has a solution or not. We gain, however, a better understanding of the Gauss reciprocity law by developing the number theory in $\bar{\mathbb{Q}}$, i.e., the algebraic number theory. The use of the roots of 1 in Definition 1.20 already hint at this. (Compare Remark 2.20.)

EXERCISE 1.45. Let q be a prime such that $q \equiv 1 \quad (8)$ and a be an integer such that $p^2 \nmid a$ for any prime p and that $x^4 \equiv a \quad (q)$ has no solution in $\mathbb{Z}$. Prove that the equation $x^4 - qy^4 = az^2$ has no integral solution other than $x = y = z = 0$. Also find a pair $\{q, a\}$ which satisfies the above condition.

2

Basic Concepts of Algebraic Number Fields

In Chapter 2, we shall extend the results of Chapter 1 (number theory in $\mathbb{Q}$) to the case of an algebraic number field K. Such a field K is a subfield of the field $\bar{\mathbb{Q}}$ (the algebraic closure of $\mathbb{Q}$) which is of finite dimension over $\mathbb{Q}$. In particular, as a generalization of prime numbers in $\mathbb{Q}$, we shall introduce the concept of prime ideals in K and extend Theorem 1.6 (fundamental theorem of number theory) to the case of number fields. This is the ideal theory of Dedekind (1831–1916). Next, we shall introduce the theory of Hilbert (1862–1943) which combines the theory of Galois (1811–1832) on field extensions with the ideal theory of Dedekind. We shall give an alternative proof of the Gauss reciprocity law (Theorem 1.27) by considering, for an odd prime l, the quadratic subfield $\mathbb{Q}(\sqrt{l^*})$, $l^* = (-1)^{(l-1)/2}l$, of the lth cyclotomic field $K = \mathbb{Q}(e^{2\pi i/l})$ (Remark 2.20). This situation will turn out to be a special case of the reciprocity in the theory of relative abelian extensions, i.e., the class field theory, due to Takagi (1875–1960) and Artin (1898–1962).

2.1. Field Extensions

In Section 1.11 we proved that the set $\bar{\mathbb{Q}}$ of all algebraic numbers makes up a subfield of the field $\mathbb{C}$ of complex numbers (Theorem 1.23). Any subfield K of $\bar{\mathbb{Q}}$ will be called an *algebraic number field* or a *number field,* for short. Since K contains 1, it contains $\mathbb{Q}$ as a subfield. In general, for two fields K and k, K is said to be an *extension* of k if k is a subfield of K; we write K/k when K is an extension of k. A mapping $\sigma: K \to K$ of a field K is called an *automorphism* if it is bijective, i.e., injective and surjective, and it satisfies the following conditions: $\sigma(x+y) = \sigma(x) + \sigma(y)$ and $\sigma(xy) = \sigma(x)\sigma(y)$, for all $x, y \in K$. For an extension K/k, we shall denote by $G(K/k)$ the set of all automorphisms σ of K such that $\sigma(a) = a$ for all $a \in k$. For σ, τ in $G(K/k)$, define the product $\sigma\tau$ by $(\sigma\tau)(x) = \sigma(\tau(x))$, $x \in K$. We can verify that the set $G(K/k)$ makes up a group; the unit element e is the identity mapping $e(x) = x$, $x \in K$. This group $G(K/k)$ is the *automorphism group* of the extension K/k.

REMARK 2.1. We wrote $\sigma(x)$ for the action of $\sigma \in G(K/k)$. However, the exponential notation x^σ will often be used; in such a case, to make both notations compatible with each other, we use the convention $x^{\sigma\tau} = (x^\tau)^\sigma$. So we have, in general, $x^{\sigma\tau} \neq (x^\sigma)^\tau$ unless $G(K/k)$ is abelian. For a subset S of K, we put $\sigma S = S^\sigma \overset{\text{def}}{=} \{x^\sigma; x \in S\}$.

Now consider the automorphism group $G(\bar{\mathbb{Q}}/k)$ for an algebraic number field k. If K/k, $K \subset \bar{\mathbb{Q}}$, is an extension of k, we obtain another extension K^σ/k for each $\sigma \in G(\bar{\mathbb{Q}}/k)$; K^σ is called a *conjugate of K over k*.

DEFINITION 2.1. An extension K/k, $K \subset \bar{\mathbb{Q}}$, is called a *Galois extension* if $K^\sigma = K$ for all $\sigma \in G(\bar{\mathbb{Q}}/k)$. When that is so, $G(K/k)$ is called the *Galois group* of the extension K/k.

REMARK 2.2. When K/k is a Galois extension, $\sigma \in G(\bar{\mathbb{Q}}/k)$ induces an automorphism $\sigma_K \in G(K/k)$. The mapping $r_{K/k}: G(\bar{\mathbb{Q}}/k) \to G(K/k)$ given by $\sigma \mapsto \sigma_K$ is a homomorphism of groups such that $\operatorname{Ker} r_{K/k} = G(\bar{\mathbb{Q}}/\mathrm{K})$. We *wish* that $r_{K/k}$ would be surjective. We would then have a nice isomorphism $G(\bar{\mathbb{Q}}/k)/G(\bar{\mathbb{Q}}/K) \approx G(K/k)$. This section will end with good news (Theorem 2.1). An outline of the proof goes as follows. First of all, for fields K, L, an injective ring homomorphism $\psi: K \to L$ is called an embedding of K in L. [Verify that $\psi(1) = 1$.] Then we shall show that an embedding $\psi: K \to \bar{\mathbb{Q}}$ can be extended to an embedding $K(\alpha) \to \bar{\mathbb{Q}}$ of any simple extension $K(\alpha)$, $\alpha \in \bar{\mathbb{Q}}$ (Proposition 2.3). Repeating the process (or, strictly speaking, by Zorn's lemma), we will reach an embedding $\sigma: \bar{\mathbb{Q}} \to \bar{\mathbb{Q}}$ which will actually be an automorphism in $G(\bar{\mathbb{Q}}/\mathbb{Q})$ (Proposition 2.4). In particular, $\psi \in G(K/k)$ is an embedding $K \to \bar{\mathbb{Q}}$ and its extension σ is a desired automorphism in $G(\bar{\mathbb{Q}}/k)$ such that $r_{K/k}(\sigma) = \psi$.

Let us start with some elementary propositions.

PROPOSITION 2.1. *For an extension of fields K/k, K can be viewed as a vector space over k. When the dimension of the vector space is finite, the extension K/k is called finite and the dimension, written $[K: k]$, is called the degree of K/k. Let K, L, k be three fields such that $K \supset L \supset k$, then we have*

$$K/k \text{ is finite} \Leftrightarrow K/L \text{ and } L/K \text{ are both finite}$$

and, when this is the case, we have $[K: k] = [K: L][L: k]$.

EXERCISE 2.1. Prove Proposition 2.1.

For a number field K and an algebraic number $\alpha \in \bar{\mathbb{Q}}$, the set $K[\alpha] = \{f(\alpha); f \in K[X]\}$ makes up a ring and a mapping $\varphi: K[X] \to K[\alpha]$ given by $\varphi(f) = f(\alpha)$ becomes a surjective homomorphism of rings. By Theorem 1.1* we have $\operatorname{Ker} \varphi = (f_\alpha) \stackrel{\text{def}}{=} K[X]f_\alpha$ where f_α is a monic polynomial in $K[X]$; this f_α is the *minimal polynomial of α over K*. On the other hand, we denote by $K(\alpha)$ the smallest subfield of $\bar{\mathbb{Q}}$ containing K, i.e., the intersection of all subfields of $\bar{\mathbb{Q}}$ containing K.

PROPOSITION 2.2.

1. *f_α is an irreducible polynomial in $K[X]$.*
2. *f_α has no multiple roots in $\bar{\mathbb{Q}}$.*
3. *$[K(\alpha): K] = \deg f_\alpha$.*
4. *$\varphi: K[X] \to K[\alpha]$ induces an isomorphism $K[X]/(f_\alpha) \underset{K}{\approx} K(\alpha)$ of fields over K.*† *(Note that $K[\alpha] = K(\alpha)$.)*

EXERCISE 2.2. Prove Proposition 2.2.

EXERCISE 2.3. The algebraic number field $\bar{\mathbb{Q}}$ is algebraically closed, i.e., for any $f \in \bar{\mathbb{Q}}[X]$ there is an $\alpha \in \bar{\mathbb{Q}}$ such that $f(\alpha) = 0$.

PROPOSITION 2.3. *Let K be a number field and $\psi: K \to \bar{\mathbb{Q}}$ be an embedding of K in $\bar{\mathbb{Q}}$. Then, for any $\alpha \in \bar{\mathbb{Q}}$, ψ can be extended to*

† For extensions L, M over K, if an isomorphism $\psi: L \approx M$ is such that $\psi(a) = a$, $\forall a \in K$, then ψ is called an isomorphism over K and we write $\psi: L \underset{K}{\approx} M$.

an embedding φ of the field $K(\alpha)$ in $\bar{\mathbb{Q}}$.

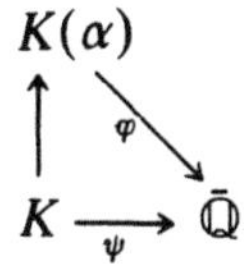

PROOF. $K^* = \psi(K)$ is a subfield of $\bar{\mathbb{Q}}$. We denote by $g^*(X)$ the polynomial in $K^*[X]$ which is obtained from $g(X) \in K[X]$ by applying the mapping ψ on each coefficient of $g(X)$. If $g(X)$ is irreducible, then so is $g^*(X)$. In particular, consider f^*_α for the minimal polynomial f_α of α over K and let α^* be a root of $f^*_\alpha(X) = 0$ in $\bar{\mathbb{Q}}$ (use Exercise 2.3). Then $f^*_\alpha = f_{\alpha^*}$, i.e., the minimal polynomial of α^* over K^*. Now, the isomorphism $\psi\colon K \xrightarrow{\sim} K^*$ can be extended naturally to an isomorphism $K[X]/(f_\alpha) \xrightarrow{\sim} K^*[X]/(f_{\alpha^*})$ and, by Proposition 2.2, this isomorphism gives an embedding $\varphi\colon K(\alpha) \xrightarrow{\sim} K^*(\alpha^*) \subset \bar{\mathbb{Q}}$, $\varphi(\alpha) = \alpha^*$, which extends ψ. Q.E.D.

PROPOSITION 2.4. *Any embedding $\psi\colon K \to \bar{\mathbb{Q}}$ can be extended to an automorphism $\varphi \in G(\bar{\mathbb{Q}}/\mathbb{Q})$.*

PROOF. In order to use Zorn's lemma (Note C), let us consider the set

$$X = \{x = (L, \theta);\ L/K,\ \theta\colon L \to \bar{\mathbb{Q}} \text{ is an embedding which is an extension of } \psi\}.$$

First of all, since $(K, \psi) \in X$, we have $X \neq \varnothing$. Next, we introduce an order $x < x'$ for $x, x' \in X$ by

$$x = (L, \theta) < x' = (L', \theta') \overset{\text{def}}{\Leftrightarrow} L'/L \text{ and } \theta' \text{ is an extension of } \theta;$$

X is easily seen to be an ordered set. we want to show that X is inductive. So let $Y = \{y_\lambda = (L_\lambda, \theta_\lambda), \lambda \in \Lambda\}$ be any totally ordered subset of X. Put $L = \bigcup_\lambda L_\lambda$ and define an embedding $\theta\colon L \to \bar{\mathbb{Q}}$ by $\theta(\alpha) = \theta_\lambda(\alpha)$, $\alpha \in L_\lambda$. Since Y is totally ordered, $b = (L, \theta)$ is well defined as an element of X, and one verifies easily that $y_\lambda < b$ for all $y_\lambda \in Y$; this shows that X is inductive. Therefore, by Zorn's lemma, X has a maximal element $m = (M, \varphi)$. We claim that $M = \bar{\mathbb{Q}}$. In fact, if $M \neq \bar{\mathbb{Q}}$, take a number $\alpha \in \bar{\mathbb{Q}}$ such that $\alpha \notin M$. By Proposition 2.3, φ can be extended to an embedding $\varphi^*\colon M^* = M(\alpha) \to \bar{\mathbb{Q}}$. Then we have $m^* = (M^*, \varphi^*) \in X$, $m < m^*$, $m \neq m^*$, which contradicts the maximality of m. Thus, we have $(\bar{\mathbb{Q}}, \varphi) \in X$, i.e., φ is an embedding of $\bar{\mathbb{Q}}$ to itself which extends ψ. It remains to show that $\varphi \in G(\bar{\mathbb{Q}}/\mathbb{Q})$; for this, it is enough to see that $\varphi(\bar{\mathbb{Q}}) = \bar{\mathbb{Q}}$. Now, since $\varphi(\bar{\mathbb{Q}}) \approx \bar{\mathbb{Q}}$, the field $\varphi(\bar{\mathbb{Q}})$ is algebraically closed, too. As any number $\alpha \in \bar{\mathbb{Q}}$ is algebraic over $\mathbb{Q}$, it is so over $\varphi(\bar{\mathbb{Q}})$ and so $\alpha \in \varphi(\bar{\mathbb{Q}})$ which means that $\varphi(\bar{\mathbb{Q}}) = \bar{\mathbb{Q}}$. Q.E.D.

THEOREM 2.1. *Let K/k be a Galois extension of algebraic number fields. Then, the restriction mapping $r_{K/k}: G(\bar{\mathbb{Q}}/k) \to G(K/k)$ is a surjective homomorphism. Therefore, we have an isomorphism: $G(\bar{\mathbb{Q}}/k)/G(\bar{\mathbb{Q}}/K) \approx G(K/k)$.*

PROOF. An automorphism ψ in $G(K/k)$ is an embedding $K \to \bar{\mathbb{Q}}$ and it can be extended to an automorphism σ in $G(\bar{\mathbb{Q}}/k)$ by Proposition 2.4; in other words, we have $r_{K/k}(\sigma) = \psi$, i.e., $r_{K/k}$ is surjective. Q.E.D.

2.2. Galois Theory

We begin with a fundamental proposition.

PROPOSITION 2.5. *Let K/k be a finite extension of number fields. Then there exists a number $\alpha \in \bar{\mathbb{Q}}$ such that $K = k(\alpha)$. A number such as α is called a primitive element for the extension K/k.*

PROOF. If $\alpha_1, \ldots, \alpha_n$, $n = [K: k]$, form a basis of the vector space K over k, we have $K = k(\alpha_1, \ldots, \alpha_n)$, i.e., the field K is generated by α_i's over k. Since we have $k(\alpha_1, \ldots, \alpha_n) = k(\alpha_1, \ldots, \alpha_{n-1})(\alpha_n)$, our assertion is reduced to proving the following proposition:

$$\text{For } \alpha, \beta \in \bar{\mathbb{Q}}, \text{ there is } \gamma \in \bar{\mathbb{Q}} \text{ such that } k(\alpha, \beta) = k(\gamma). \qquad (*)$$

To prove (*), let f_α and f_β be minimal polynomials of α and β over k, respectively. Decompose them over $\bar{\mathbb{Q}}$:

$$f_\alpha = (X - \alpha_1) \cdots (X - \alpha_n), \qquad \alpha_1 = \alpha,$$
$$f_\beta = (X - \beta_1) \cdots (X - \beta_m), \qquad \beta_1 = \beta.$$

If $m = 1$, then we can take $\gamma = \alpha$. So we may assume that $m \geq 2$. Since f_α and f_β have no multiple roots by Exercise 2.2(2) and k

contains infinitely many numbers, we can find a number c in k which is different from any number of the following finite set:

$$\left\{\frac{\alpha_i-\alpha_1}{\beta_1-\beta_j},\ 1\le i\le n,\ 2\le j\le m\right\}.$$

We shall show that the number $\gamma=\alpha+c\beta$ satisfies the required property (*). Since the relation $k(\alpha,\beta)\supset k(\gamma)$ is obvious, we have only to prove $k(\alpha,\beta)\subset k(\gamma)$. To do this, put $g(X)=f_\alpha(\gamma-cX)$; note that g and f_β are polynomials in $k(\gamma)[X]$. Now, we have $g(\beta)=f_\alpha(\gamma-c\beta)=f_\alpha(\alpha)=0$ and $f_\beta(\beta)=0$; this means that $X-\beta$ is a common divisor of g, f_β in $\bar{\mathbb{Q}}[X]$. However, $X-\beta_j$, $j\ge 2$, does not divide g because if so we must have $0=g(\beta_j)=f(\gamma-c\beta_j)$ which implies that $\gamma-c\beta_j=\alpha_i$ for some i, i.e., $c(\beta-\beta_j)=\alpha_i=\alpha$, contradicting the choice of c. Therefore we proved that $(g, f_\beta)=X-\beta$. Since g, f_β belong to $k(\gamma)[X]$ and the computation of common divisor (g, f_β) is carried out by the euclidean algorithm within the field $k(\gamma)$, we see that $X-\beta\in k(\gamma)[X]$, i.e., $\beta\in k(\gamma)$. Finally, we have $\alpha=\gamma-c\beta\in k(\gamma)$. Hence $k(\alpha,\beta)\subset k(\gamma)$.

Q.E.D.

PROPOSITION 2.6. *Let K/k be a finite extension of algebraic number fields. Then we have $[G(\bar{\mathbb{Q}}/k): G(\bar{\mathbb{Q}}/K)]=[K:k]$ where $[G:H]$, for any group G and its subgroup H, means the number of left (right) cosets of G modulo H.*

PROOF. By Proposition 2.5 there is a number $\alpha\in\bar{\mathbb{Q}}$ such that $K=k(\alpha)$. Call f_α the minimal polynomial of α over k. We have $\deg f_\alpha=[K:k]$ [Proposition 2.2(3)]. Let $R=\{\alpha_1,\ldots,\alpha_n\}$, $\alpha_1=\alpha$, $n=\deg f_\alpha$, be the set of all roots (in $\mathbb{C}$) of f_α. Since $f_\alpha(\beta)=0$ implies $f_\alpha(\sigma(\beta))=0$, $\sigma\in G(\bar{\mathbb{Q}}/k)$, the group $G(\bar{\mathbb{Q}}/k)$ acts naturally on R. Furthermore, this action is transitive. In fact, take any $\beta\in R$. Since $f_\beta=f_\alpha$, there is an isomorphism $\psi: k(\alpha)\underset{k}{\tilde{\rightarrow}} k(\beta)$ such that $\psi(\alpha)=\beta$ [Proposition 2.2(4)]. By Proposition 2.4, ψ can be extended to an automorphism $\sigma\in G(\bar{\mathbb{Q}}/k)$; hence the action is transitive. In view of the relation

$$\sigma(\alpha)=\tau(\alpha)\Leftrightarrow\tau\in\sigma G(\bar{\mathbb{Q}}/K)$$

we get a bijection between the set R and the set $G(\bar{\mathbb{Q}}/k)/G(\bar{\mathbb{Q}}/K)$ of left cosets of $G(\bar{\mathbb{Q}}/k)$ modulo $G(\bar{\mathbb{Q}}/K)$ and so $[G(\bar{\mathbb{Q}}/k): G(\bar{\mathbb{Q}}/K)]=[R]=n=[K:k]$. Q.E.D.

THEOREM 2.2. *If K/k is a finite Galois extension of number fields, then we have $[G(K/k)] = [K: k]$.*

PROOF. If K/k is a Galois extension, the mapping $r_{K/k}: G(\bar{\mathbb{Q}}/k) \to G(K/k)$ is a surjective homomorphism with $\operatorname{Ker} r_{K/k} = G(\bar{\mathbb{Q}}/K)$ by Theorem 2.1 and hence we have an isomorphism $G(\bar{\mathbb{Q}}/k)/G(\bar{\mathbb{Q}}/K) \approx G(K/k)$. On the other hand, since K/k is finite, we have $[G(\bar{\mathbb{Q}}/k): G(\bar{\mathbb{Q}}/K)] = [K: k]$ by Proposition 2.6. Therefore, we get $[G(K/k)] = [K: k]$. Q.E.D.

Now we shall introduce some notations for a finite Galois extension K/k. We denote by $\mathscr{F}$ the set of all fields F such that $K \supset F \supset k$, and by $\mathscr{G}$ the set of all subgroups G of $G(K/k)$. Next, we define a mapping $\#: \mathscr{F} \to \mathscr{G}$ and a mapping $\flat: \mathscr{G} \to \mathscr{F}$ as follows:

$$F^{\#} = \{\sigma \in G(K/k); \sigma(a) = a, \forall a \in F\}, \qquad \text{for } F \in \mathscr{F},$$

$$G^{\flat} = \{a \in K; \sigma(a) = a, \forall \sigma \in G\}, \qquad \text{for } G \in \mathscr{G}.$$

One checks easily that $F^{\#}$ is a subgroup of $G(K/k)$ and $G^{\flat}$ is a subfield of K containing k. It follows immediately by definition that

$$F \subset F^{\#\flat}, \qquad G \subset G^{\flat\#} \tag{2.1}$$

and

$$F_1 \subset F_2 \Rightarrow F_1^{\#} \supset F_2^{\#}, \tag{2.2}$$

$$G_1 \subset G_2 \Rightarrow G_1^{\flat} \supset G_2^{\flat}. \tag{2.3}$$

THEOREM 2.3. *Let K/k be a finite Galois extension of number fields. Let $\mathscr{F}$ be the set of all fields in between K and k, and $\mathscr{G}$ be the set of all subgroups of $G(K/k)$. Then the mapping $\#: \mathscr{F} \to \mathscr{G}$, $\flat: \mathscr{G} \to \mathscr{F}$ are both bijective, order reversing, and inverse of each other*: $\flat \cdot \# = \mathrm{id}_{\mathscr{F}}$, $\# \cdot \flat = \mathrm{id}_{\mathscr{G}}$.

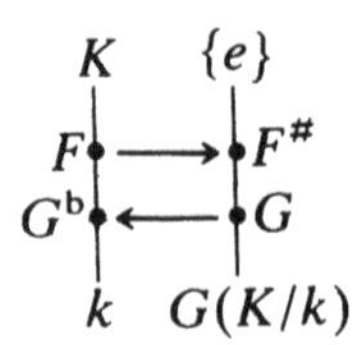

Proof. In view of (2.2) and (2.3), it is enough to prove that

$$G = G^{\flat\#} \tag{2.4}$$

and

$$F = F^{\#\flat}. \tag{2.5}$$

We shall first show that

$$[K: F] = [F^{\#}], \qquad F \in \mathscr{F} \tag{2.6}$$

and

$$[K: G^{\flat}] = [G], \qquad G \in \mathscr{G}. \tag{2.7}$$

As for (2.6), since K/k is a Galois extension, we have $K^{\sigma} = K$ for all $\sigma \in G(\bar{\mathbb{Q}}/k)$ and so $K^{\sigma} = K$ for all $\sigma \in G(\bar{\mathbb{Q}}/F)$, which means that K/F is also a Galois extension. By the definition of the mapping $\#$, we have $F^{\#} = G(K/F)$ from which follows (2.6) using Theorem 2.2. To verify (2.7), let α be a number such that $K = K(\alpha)$ (Proposition 2.5) and let f_{α} be the minimal polynomial of α over $G^{\flat}$: $f_{\alpha} \in G^{\flat}[X]$. By Proposition 2.2(3), we have $\deg f_{\alpha} = [K: G^{\flat}]$. For $\sigma \in G$, by the definition of $G^{\flat}$, we have $f_{\alpha}(\alpha^{\sigma}) = 0$. As α is a generator of K over k, we have $\alpha^{\sigma} \neq \alpha^{\tau}$ when $\sigma \neq \tau$, $\sigma, \tau \in G$, which implies that $\deg f_{\alpha} \geq [G]$. On the other hand, since the polynomial $g(X) = \prod_{\sigma \in G} (X - \alpha^{\sigma})$ belongs to $G^{\flat}[X]$ and has α as a root, we must have $f_{\alpha} \mid g$; therefore we find that $\deg f_{\alpha} \leq [G]$. Thus we get $[K: G^{\flat}] = \deg f_{\alpha} = [G]$, which proves (2.7). If we put $F = G^{\flat}$ in (2.6), we get $[K: G^{\flat}] = [G^{\flat\#}]$; it follows from (2.7) and (2.1) ($G \subset G^{\flat\#}$) that $G = G^{\flat\#}$ (Proposition 1.6), which proves (2.4). Finally put $G = F^{\#}$ in (2.7). Then we get $[K: F^{\#\flat}] = [F^{\#}]$ and it follows from (2.6) and (2.1) ($F \subset F^{\#\flat}$) that $F = F^{\#\flat}$ (Proposition 2.1), which proves (2.5). Q.E.D.

Theorem 2.4. *For $F_1, F_2 \in \mathscr{F}$, we have*

F_1 and F_2 are conjugate over $k \Leftrightarrow F_1^{\#}$ and $F_2^{\#}$ are conjugate in $G(K/k)$.†

In particular, we have

F/k is a Galois extension $\Leftrightarrow F^{\#}$ is a normal subgroup of $G(K/k)$.

† Two subgroups H and H' of a group G are conjugate if $H' = gHg^{-1}$ for some $g \in G$.

PROOF. Suppose first that F_1 and F_2 are conjugate over k, i.e., there is an automorphism $\sigma \in G(\bar{\mathbb{Q}}/k)$ such that $F_2 = \sigma(F_1)$.† Then we have

$$\begin{aligned} F_2^\# \ni \tau \Leftrightarrow \tau(b) = b, \quad \forall b \in F_2 \Leftrightarrow \tau\sigma(a) = \sigma(a), \quad \forall a \in F_1 \\ \Leftrightarrow \sigma^{-1}\tau\sigma(a) = a, \quad \forall a \in F_1 \Leftrightarrow \sigma^{-1}\tau\sigma \in F_1^\# \end{aligned}$$

which implies that $F_2^\# = \sigma F_1^\# \sigma^{-1}$. Conversely, suppose that $F_2^\# = \sigma F_1^\# \sigma^{-1}$. Taking $\flat$ of both sides, we get $F_2 = (\sigma F_1^\# \sigma^{-1})^\flat$ and we are reduced to verifying that $(\sigma F_1^\# \sigma^{-1}) = \sigma(F_1)$. Now we have

$$\begin{aligned} a \in (\sigma F_1^\# \sigma^{-1})^\flat \Leftrightarrow \sigma\tau\sigma^{-1}(a) = a, \quad \forall \tau \in F_1^\# \\ \Leftrightarrow \tau(\sigma^{-1}(a)) = \sigma^{-1}(a), \quad \forall \tau \in F_1^\# \Leftrightarrow \sigma^{-1}(a) \in F_1 \\ \Leftrightarrow a \in \sigma(F_1) \end{aligned}$$

which proves our assertion. Q.E.D.

EXERCISE 2.4. Let $K/\mathbb{Q}$ be a quadratic extension: $[K:\mathbb{Q}] = 2$. Then $K = \mathbb{Q}(\sqrt{m})$, where m is an integer which has no square factors. Conversely, let m be an integer ($m \neq 0, 1$) which has no square factors. Then $K = \mathbb{Q}(\sqrt{m})$ is a quadratic extension. For such *square free* integers m_1 and m_2, we have $\mathbb{Q}(\sqrt{m_1}) = \mathbb{Q}(\sqrt{m_2}) \Leftrightarrow m_1 = m_2$. Therefore the set of all quadratic extensions are in 1–1 correspondence with the set of all square free integers: $m = -1, \pm 2, \pm 3, \pm 5, \pm 6, \pm 7, \pm 10, \ldots$. Any quadratic extension $K/\mathbb{Q}$ is a Galois extension. The Galois group $G(K/\mathbb{Q})$ is a cyclic group of order 2 generated by the automorphism σ defined by $\sigma(\sqrt{m}) = -\sqrt{m}$.

EXERCISE 2.5. Let l be an odd prime and let $\zeta = e^{2\pi i/l}$. The field $K = \mathbb{Q}(\zeta)$ is the lth *cyclotomic field*.

1. $K/\mathbb{Q}$ is a Galois extension.
2. The minimal polynomial of ζ is $f_\zeta(X) = 1 + X + X^2 + \cdots + X^{l-1} \in \mathbb{Q}[X]$.
3. Define an integer $j(\sigma)$ for each $\sigma \in G(K/\mathbb{Q})$ by $\zeta^\sigma = \zeta^{j(\sigma)}$. Then we get an isomorphism $G(K/\mathbb{Q}) \xrightarrow{\sim} (\mathbb{Z}/l\mathbb{Z})^\times = \mathbb{F}_l^\times$ given by $\sigma \mapsto j(\sigma) \bmod l$.
4. Put $l^* = (-1)^{(l-1)/2}$. Show that the quadratic field $\mathbb{Q}(\sqrt{l^*})$ is a subfield of $\mathbb{Q}(\zeta)$.
5. The subfield k of $K = \mathbb{Q}(\zeta)$ such that $[K; k] = 2$ is given by $k = \mathbb{Q}(\zeta + \zeta^{-1})$. (This k is called the maximal real subfield of K. Why?)

† As F_1 and F_2 are contained in K, we regard this σ as $r_{K/K}(\sigma)$.

2.3. Norm, Trace, and Discriminant

Let K/k be a finite extension. By Proposition 2.6 we have $[G(\bar{\mathbb{Q}}/k): G(\bar{\mathbb{Q}}/K)] = [K: k] = n$; this is, as the proof of that proposition shows, a consequence of the observation that the group $G(\bar{\mathbb{Q}}/k)$ acts on the set $R = \{\alpha_1, \dots, \alpha_n\}$ of all roots of the minimal polynomial $f_\alpha \in k[X]$ of $\alpha = \alpha_1$ such that $K = k(\alpha)$. For $\sigma \in G(\bar{\mathbb{Q}}/k)$ we denoted by K^σ and α^σ the conjugates of K and α, respectively. When $\sigma_1, \sigma_2, \dots, \sigma_n$, $\sigma_1 = e$, are representatives of left cosets of $G(\bar{\mathbb{Q}}/k)/G(\bar{\mathbb{Q}}/K)$, we put $K^{\sigma_i} = K^{(i)}$, $\alpha^{\sigma_i} = \alpha^{(i)}$, $1 \leq i \leq n$; for a number $\gamma \in K$, we put $\gamma^{\sigma_i} = \gamma^{(i)}$; therefore, we have $a^{(i)} = a$, $1 \leq i \leq n$, for all $a \in k$.

Definition 2.2. For a finite extension K/k, the *norm* $N_{K/k}$ and the *trace* $t_{K/k}$ of a number $\gamma \in K$ is defined by

$$N_{K/k}\gamma = \gamma^{(1)} \cdots \gamma^{(n)}, \qquad t_{K/k}\gamma = \gamma^{(1)} + \cdots + \gamma^{(n)}.$$

Remark 2.3. When K/k is a Galois extension, we have

$$N_{K/k}\gamma = \prod_{\sigma \in G(K/k)} \gamma^\sigma, \qquad t_{K/k}\gamma = \sum_{\sigma \in G(K/k)} \gamma^\sigma,$$

where we used the same notation σ for $r_{K/k}(\sigma)$, the restriction of $G(\bar{\mathbb{Q}}/k)$ on the subfield K of $\bar{\mathbb{Q}}$.

Proposition 2.7. *Let K/k be a finite extension. The trace $t_{K/k}: K \to k$ is a k-linear mapping of vector space K, k over k. The norm $N_{K/k}: K \to k$ induces a homomorphism $K^\times \to k^\times$ of the multiplicative groups of fields K, k. In particular, $N_{K/k}a = a^n$, $a \in k$, $n = [K: k]$.*

Proof. If $\sigma_1, \dots, \sigma_n$ are representatives of left cosets of $G(\bar{\mathbb{Q}}/k)/G(\bar{\mathbb{Q}}/K)$, then so are $\sigma\sigma_1, \dots, \sigma\sigma_n$ for any $\sigma \in G(\bar{\mathbb{Q}}/k)$. Therefore,

$$\begin{aligned} N_{K/k}\gamma &= \gamma^{\sigma_1} \cdots \gamma^{\sigma_n} = \gamma^{\sigma\sigma_1} \cdots \gamma^{\sigma\sigma_n} = (\gamma^{\sigma_1})^\sigma \cdots (\gamma^{\sigma_n})^\sigma \\ &= (\gamma^{\sigma_1} \cdots \gamma^{\sigma_n})^\sigma = (N_{K/k}\gamma)^\sigma, \end{aligned}$$

for any $\gamma \in K$. We claim that $\beta = N_{K/k}\gamma \in k$. In fact, if $\beta \notin k$, put $L = k(\beta)$. Then the minimal polynomial $f_\beta \in k[X]$ of β over k has at least two roots $\beta, \beta^*, \beta \neq \beta^*$. Call ψ an embedding $L \to \bar{\mathbb{Q}}$ defined by $\psi(\beta) = \beta^*$ and $\sigma \in G(\bar{\mathbb{Q}}/k)$ be an extension of ψ

(Proposition 2.4). We have then $\beta^\sigma = \beta^* \neq \beta$ which contradicts our observation above. Similarly, one proves that $t_{K/k}\gamma \in k$, $\gamma \in K$. When a is a number in k, we have

$$t_{K/k}(a\gamma) = (a\gamma)^{(1)} + \cdots + (a\gamma)^{(n)} = a\gamma^{(1)} + \cdots + a\gamma^{(n)}$$
$$= a(\gamma^{(1)} + \cdots + \gamma^{(n)}) = at_{K/k}\gamma.$$

Furthermore, the following relations are obvious: $t_{K/k}(\gamma_1 + \gamma_2) = t_{K/k}\gamma_1 + t_{K/k}\gamma_2$, $N_{K/k}(\gamma_1\gamma_2) = (N_{K/k}\gamma_1)(N_{K/k}\gamma_2)$, and $N_{K/k}\gamma = 0 \Leftrightarrow \gamma = 0$.

Q.E.D.

EXERCISE 2.6. Let $K \supset L \supset k$ be fields such that K/k is finite. Prove that $N_{K/k}\gamma = N_{L/k}(N_{K/L}\gamma)$ and $t_{K/k}\gamma = t_{L/k}(t_{K/L}\gamma)$ for $\gamma \in K$.

Now we shall consider the discriminant for a finite extension K/k of number fields. For $\gamma_1, \ldots, \gamma_n \in K$, $n = [K: k]$, put

$$d_{K/k}(\gamma_1, \ldots, \gamma_n) \stackrel{\text{def}}{=} (\det(\gamma_j^{(i)}))^2, \qquad 1 \le i, j \le n.$$

Since this equality can also be written as

$$d_{K/k}(\gamma_1, \ldots, \gamma_n) = \det(t_{K/k}(\gamma_i\gamma_j)),$$

it follows that $d_{K/k}(\gamma_1, \ldots, \gamma_n) \in k$. In particular, for a primitive element α for K/k: $K = k(\alpha)$, we have

$$D_{K/k}(\alpha) \stackrel{\text{def}}{=} d_{K/k}(1, \alpha, \ldots, \alpha^{n-1}) = \prod_{1 \le i < j \le n} (\alpha^{(i)} - \alpha^{(j)})^2.$$

Since f_α has no multiple roots [Proposition 2.2(2)], we have $D_{K/k}(\alpha) \neq 0$. We call $D_{K/k}(\alpha)$ the *discriminant* of α. Since each of the numbers $\gamma_1, \ldots, \gamma_n$ is a linear combination of $1, \alpha, \ldots, \alpha^{n-1}$ over k, we have

$$(\gamma_1, \ldots, \gamma_n) = (1, \alpha, \ldots, \alpha^{n-1})A, \qquad A \in M_n(k)$$

and hence

$$d_{K/k}(\gamma_1, \ldots, \gamma_n) = D_{K/k}(\alpha)(\det A)^2$$

from which we see that

$$\{\gamma_1, \ldots, \gamma_n\} \text{ is a basis for } K/k \Leftrightarrow d_{K/k}(\gamma_1, \ldots, \gamma_n) \neq 0.$$

EXERCISE 2.7. Let K/k be a finite extension of degree n and let $K = k(\alpha)$. Let f_α be the minimal polynomial of α over k. Prove that $D_{K/k}(\alpha) = (-1)^{n(n-1)/2} N_{K/k} f'(\alpha)$ where f'_α denotes the derivative of f_α.

EXERCISE 2.8. Let $K/\mathbb{Q}$ be a quadratic field: $K = \mathbb{Q}(\sqrt{m})$ (Exercise 2.4). Verify that $D_{K/\mathbb{Q}}(\sqrt{m}) = 4m$.

EXERCISE 2.9. Let l be an odd prime and let $K = \mathbb{Q}(\zeta)$ be the lth cyclotomic field (Exercise 2.5). Verify that $D_{K/\mathbb{Q}}(\zeta) = (-1)^{(l-1)/2} l^{l-2}$.

2.4. Gauss Sum and Jacobi Sum

In Section 1.13 we used the Gauss sum (for the Legendre character) to prove the Gauss reciprocity law. For an odd prime l, let $\mathbb{F}_l = \mathbb{Z}/l\mathbb{Z}$ be the finite field with l elements and χ be a character of the multiplicative group $\mathbb{F}_l^\times$. We shall denote by ϵ the identity character: $\epsilon(x) = 1$, $\forall x \in \mathbb{F}_l^\times$. According to the notation in Section 1.12, the Legendre character λ_l is the unique character of order 2 (in the group $\hat{\mathbb{F}}_l^\times$ of all characters of $\mathbb{F}_l^\times$). We shall put $\chi(0) = 0$ for any χ and occasionally consider χ as a function defined on the whole set $\mathbb{F}_l$. Since $\mathbb{F}_l^\times$ is a cyclic group of order $l-1$, the values of χ are either 0 or $(l-1)$th roots of 1. If r is a primitive root modulo l, i.e., if $\mathbb{F}_l^\times = \langle r \rangle$, then any character can be written $\chi = \chi_a$ where $\chi_a(r) = \xi^a$ and $\xi = e^{2\pi i/(l-1)}$, for some $a \in \mathbb{Z}$. In particular, the character χ_1 which has order $l-1$ generates the character group $\hat{\mathbb{F}}_l^\times$; we have $\chi_1(x) \neq 1$ if $x \neq 1 \in \mathbb{F}_l^\times$. By Exercise 1.44, we see easily that

$$\sum_{\chi \in \hat{\mathbb{F}}_l^\times} \chi(x) = \begin{cases} l-1, & x = 1, \\ 0, & x \neq 1. \end{cases}$$

DEFINITION 2.3. For an odd prime l, $\zeta = e^{2\pi i/l}$, $\chi \in \hat{\mathbb{F}}_l^\times$, $a \in \mathbb{F}_l$, put $\tau_a(\chi) = \sum_{x \in \mathbb{F}_l} \chi(x) \zeta^{ax}$ and $\tau_1(\chi) = \tau(x)$. We call $\tau_a(\chi)$ the *Gauss sum* with respect to the character χ.

PROPOSITION 2.8.

$$\tau_a(\chi) = \begin{cases} \overline{\chi(a)}\tau(\chi), & a \neq 0, \quad \chi \neq \epsilon, \\ -1, & a \neq 0, \quad \chi = \epsilon, \\ 0, & a = 0, \quad \chi \neq \epsilon, \\ l-1, & a = 0, \quad \chi = \epsilon. \end{cases}$$

PROOF. If $a \neq 0$ and $\chi \neq \epsilon$, we have $\chi(a)\tau_a(\chi) = \chi(a)\Sigma_x \chi(x)\zeta^{ax} = \Sigma_x \chi(ax)\zeta^{ax} = \Sigma_y \chi(y)\zeta^y = \tau(\chi)$ and the equality follows because $\chi(a)^{-1} = \overline{\chi(a)}$. If $a \neq 0$ and $\chi = \epsilon$, then $\tau_a(\epsilon) = \Sigma_x \epsilon(x)\zeta^{ax} = \Sigma_x \zeta^{ax} = 0$ (Proposition 1.22). Finally, we have $\tau_0(\chi) = \Sigma_x \chi(x) = 0$ if $\chi \neq \varepsilon$ and $= l - 1$ if $\chi = \epsilon$. Q.E.D.

PROPOSITION 2.9. *If $\chi \neq \epsilon$, we have $|\tau(\chi)| = \sqrt{l}$.*

PROOF. We shall compute the sum $S = \Sigma_{a \in \mathbb{F}_l} |\tau_a(\chi)|^2$ in two ways. First of all, when $a \neq 0$, the first equality in Proposition 2.8 implies $|\tau_a(\chi)|^2 = \tau_a(\chi)\overline{\tau_a(\chi)} = \chi(a)\tau(x)\overline{\chi(a)\tau(x)} = |\tau(\chi)|^2$. Next, when $a = 0$, we have $\tau_0(\chi) = 0$ (Proposition 2.8). Therefore $S = (l-1)\,|\tau(\chi)|^2$. On the other hand, we have

$$
\begin{aligned}
|\tau_a(\chi)|^2 = \tau_a(\chi)\overline{\tau_a(\chi)} &= \Big(\sum_x \chi(x)\zeta^{ax}\Big)\Big(\sum_y \overline{\chi(y)}\zeta^{-ay}\Big) \\
&= \sum_x \sum_y \chi(x)\overline{\chi(y)}\zeta^{a(x-y)}
\end{aligned}
$$

and so

$$
\begin{aligned}
S = \sum_a |\tau_a(\chi)|^2 &= \sum_{x,y} \chi(x)\overline{\chi(y)} \sum_a \zeta^{a(x-y)} \\
&= \sum_{x,y} \chi(x)\overline{\chi(y)}\, l\delta_{x,y} = l \sum_x |\chi(x)|^2 = l(l-1).
\end{aligned}
$$

Therefore, we get $|\tau(\chi)|^2 = l$. Q.E.D.

REMARK 2.4. Since $\overline{\tau(\chi)} = \Sigma_x \overline{\chi(x)}\zeta^{-x} = \Sigma_x \overline{\chi(-x)}\,\zeta^x = \chi(-1)\,\Sigma_x \bar{\chi}(x)\zeta^x = \chi(-1)\tau(\bar{\chi})$, Proposition 2.9 may be written $\tau(\chi)\tau(\bar{\chi}) = \chi(-1)l$. If, in particular, $\chi = \lambda_l$ (the Legendre character), then the last equality is a special case of Proposition 1.25 ($a = 1$) because $\bar{\lambda}_l = \lambda_l$.

DEFINITION 2.4. Let χ and ψ be characters of the group $\mathbb{F}_l^\times$. we call the sum $J(\chi, \psi) \stackrel{\text{def}}{=} \Sigma_{x+y=1} \chi(x)\psi(y)$ the *Jacobi* (1804–1851) *sum.*

PROPOSITION 2.10. *Assume that $\chi \neq \epsilon$, $\psi \neq \epsilon$, and $\chi\psi \neq \epsilon$. Then we have*

$$
J(\chi, \psi) = \frac{\tau(\chi)\tau(\psi)}{\tau(\chi\psi)}, \qquad |J(\chi, \psi)| = \sqrt{l}.
$$

PROOF. We have

$$\tau(\chi)\tau(\psi)=\sum_x \chi(x)\zeta^x \sum_y \psi(y)\zeta^y=\sum_{x,y}\chi(x)\psi(y)\zeta^{x+y}$$
$$=\sum_t\Big(\sum_{x+y=t}\chi(x)\psi(y)\Big)\zeta^t.$$

If $t=0$, then

$$\sum_{x+y=0}\chi(x)\psi(y)=\sum_x \chi(x)\psi(-x)=\psi(-1)\sum_x(\chi\psi)(x)=0$$

by the assumption $\chi\psi\neq\epsilon$. If $t\neq 0$, put $x=tu$ and $y=tv$. Then we have

$$\sum_{x+y=t}\chi(x)\psi(y)=(\chi\psi)(t)\sum_{u+v=1}\chi(u)\psi(v)=(\chi\psi)(t)J(\chi,\psi)$$

and so $\tau(\chi)\tau(\psi)=\sum_t(\chi\psi)(t)J(\chi,\psi)\zeta^t=J(\chi,\psi)\tau(\chi\psi)$. Finally, applying Proposition 2.9 to χ, ψ, and $\chi\psi$ we obtain $|J(\chi,\psi)|=\sqrt{l}$.

Q.E.D.

REMARK 2.5. The Gauss sum and Jacobi sum (for the field $\mathbb{F}_l$) bear a resemblance to the gamma function and beta function (for the field $\mathbb{R}$).

Gauss sum	Jacobi sum
$\tau(\chi)=\sum_x \chi(x)\zeta^x$	$J(\chi,\psi)=\sum_x \chi(x)\psi(1-x)=\dfrac{\tau(\chi)\tau(\psi)}{\tau(\chi\psi)}$
Gamma function	Beta function
$\Gamma(s)=\int_0^\infty x^{s-1}e^{-x}\,dx$	$B(p,q)=\int_0^1 x^{p-1}(1-x)^{q-1}\,dx=\dfrac{\Gamma(p)\Gamma(q)}{\Gamma(p+q)}$

PROPOSITION 2.11. *Let $m\geq 3$ and l be a prime such that $l\equiv 1(m)$. If χ is a character of $\mathbb{F}_l^\times$ of order m, then we have*

$$\tau(\chi)^m=\chi(-1)lJ(\chi,\chi)J(\chi,\chi^2)\cdots J(\chi,\chi^{m-2}).$$

Proof. Proposition 2.10 implies $\tau(\chi)^2 = J(\chi, \chi)\tau(\chi^2)$. Multiplying $\tau(\chi)$ on both sides, we have

$$\tau(\chi)^3 = J(\chi, \chi)\tau(\chi)\tau(\chi^2) = J(\chi, \chi)J(\chi, \chi^2)\tau(\chi^3).$$

Continuing this way, we arrive at

$$\tau(\chi)^{m-1} = J(\chi, \chi)J(\chi, \chi^2) \cdots J(\chi, \chi^{m-2})\tau(\chi^{m-1}).$$

Since $\chi^{m-1} = \chi^{-1} = \bar{\chi}$ and $\tau(\chi)\tau(\bar{\chi}) = \chi(-1)l$ (Remark 2.4), the above equality yields the equality in the proposition. Q.E.D.

2.5. Construction of a Regular l-gon

Let l be an odd prime. Let us try to draw a regular l-gon (a regular polygon with l sides) by ruler and compass. Choose two points on a paper; call one 0 and the other 1. Two points 0 and 1 determine the x-axis. The y-axis is obtained by joining intersections of two circles, one passing through -1 with center at 1 and the other passing through 1 with center at -1. The imaginary unit i is the intersection of the y-axis with the circle passing through 1 with center 0. Starting with points $\{0, 1\}$, we can construct all points in $\mathbb{Q}$ in the cartesian plane, i.e., in $\mathbb{C}$, by finite applications of ruler and compass. Call F any field between $\mathbb{C}$ and $\mathbb{Q}$. For points $a, b \in F$, the (infinite) line joining a and b is the *F-line* and the circle passing through b with center a is the *F-circle* (Figure 2.1). One sees easily that the points of intersections of an F-line

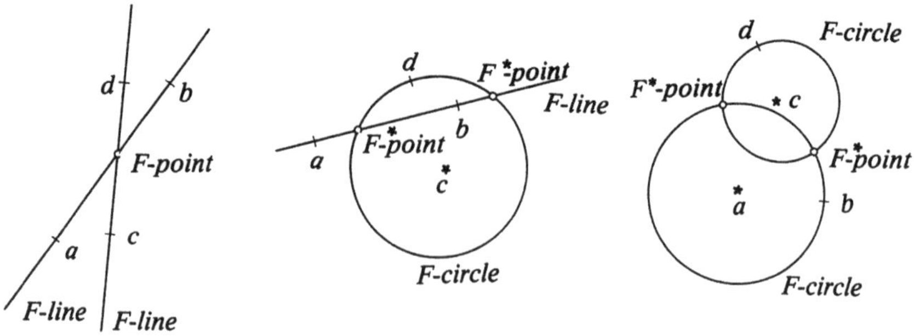

Figure 2.1

and another F-line, an F-line and an F-circle, or an F-circle and another F-circle belong to an extension F^* of F such that $[F^*: F] \leq 2$.

Therefore, the *constructible* number $\alpha \in \mathbb{C}$ may be defined as follows:

DEFINITION 2.5. A complex number $\alpha \in \mathbb{C}$ is constructible if there exists a finite sequence of fields $\mathbb{Q} = F_0 \subset F_1 \subset F_2 \subset \cdots \subset F_n$ such that $[F_i: F_{i-1}] \leq 2$, $1 \leq i \leq n$, and $\alpha \in F_n$.

In what follows, we put $\zeta_k = e^{2\pi i/k}$, for $k \in \mathbb{N}$.

PROPOSITION 2.12. *Numbers* ζ_{2^k}, $k \in \mathbb{N}$ *are constructible.*

PROOF. If $k=1$, $\zeta_2 = -1$ is trivially constructible. Suppose now that ζ_{2^k} is constructible; call F_n the last field such that $\zeta_{2^k} \in F_n$ (Definition 2.5). Since $(\zeta_{2^{k+1}})^2 = \zeta_{2^k}$, we have $\zeta_{2^{k+1}} \in F_{n+1} = F_n(\sqrt{\zeta_{2^k}})$ which proves that $\zeta_{2^{k+1}}$ is constructible. Q.E.D.

THEOREM 2.5. ζ_l *is constructible* $\Leftrightarrow l$ *is a Fermat prime, i.e., a prime of the form* $l = 2^{2^n} + 1$, $n \geq 0$.†

PROOF. ($\Rightarrow$) If ζ_l is constructible, then ζ_l will belong to the last field F_n of a finite sequence of fields (Definition 2.5). Since $[F_n: \mathbb{Q}]$ is a power of 2 and $[\mathbb{Q}(\zeta_l): \mathbb{Q}] \mid [F_n: \mathbb{Q}]$, $[\mathbb{Q}(\zeta_l): \mathbb{Q}] = \varphi(l) = l-1$ is a power of 2 and so $l = 1 + 2^k$. If k contains an odd factor, then l cannot be a prime. Therefore one must have $l = 2^{2^n} + 1$.

($\Leftarrow$) This direction is more important. We shall reduce the constructibility of ζ_l to that of a Gauss sum: $\tau(\chi) = \sum_{x \in \mathbb{F}_l} \chi(x) \zeta_l^x$, $\chi \in \hat{\mathbb{F}}_l^\times$. Since

$$\sum_{x \in \mathbb{F}_l^\times} \chi(x) = \begin{cases} l-1, & x = 1, \\ 0, & x \neq 1, \end{cases}$$

as we remarked before Definition 2.3, we have

$$\sum_{\chi} \tau(\chi) = \sum_{x} \Big(\sum_{\chi} \chi(x) \Big) \zeta_l^x = (l-1)\zeta_l$$

which shows that ζ_l is constructible if each $\tau(\chi)$ is constructible.

† Compare Exercise 1.12 and Exercise 1.31. Theorem 2.5 appears in the last chapter of Gauss's *Disquisitiones*. the constructibility of ζ_{17} was discovered by him on 3/30/1796 when he was 18 years old.

Now, by the assumption, l is of the form 2^k+1 and since the group $\hat{\mathbb{F}}_l^\times$ is cyclic of order $l-1$, each character χ has the order which is a power of 2: $\operatorname{ord}\chi = 2^r$, $r \le k$. Proposition 2.11 (with $m=2^r$) implies then

$$\tau(\chi)^{2^r} = \chi(-1)lJ(\chi,\chi)J(\chi,\chi^2)\cdots J(\chi,\chi^{2^r-2}).\dagger$$

As only multiplicative characters are used to define $J(\chi,\chi^j)$, we have $J(\chi,\chi^j) \in \mathbb{Q}(\zeta_{l-1}) = \mathbb{Q}(\zeta_{2^k})$ and we see that $J(\chi,\chi^j)$ are all constructible (Proposition 2.12). Then $\tau(\chi)^{2^r}$ is constructible and so is $\tau(\chi)$. Q.E.D.

2.6. Subfields of the lth Cyclotomic Field

Let l be an odd prime, $\zeta = e^{2\pi i/l}$, and $K=\mathbb{Q}(\zeta)$, the lth cyclotomic field. Putting $l^* = (-1)^{(l-1)/2}l$, we obtained the quadratic subfield $\mathbb{Q}(\sqrt{l^*})$ of $\mathbb{Q}(\zeta)$ [Exercise 2.5(4)]. We know also that $k=\mathbb{Q}(\zeta+\zeta^{-1}) = \mathbb{Q}(\zeta+\bar{\zeta})$ is the maximal real subfield of K with $[K:k]=2$ [Exercise 2.5(5)]. Since $K/\mathbb{Q}$ is a cyclic field, i.e., a Galois extension whose Galois group is cyclic, for each factor n of $l-1=[K:\mathbb{Q}]$ there is one and only one subfield k_n of K such that $[k_n:\mathbb{Q}]=n$ (Proposition 1.14). Since the group $G(K/\mathbb{Q})$ is cyclic, every $k_n/\mathbb{Q}$ is a Galois extension (Theorem 2.4). In particular, we have $\mathbb{Q}=k_1$, $K=k_{l-1}$, $\mathbb{Q}(\sqrt{l^*})=k_2$, and $\mathbb{Q}(\zeta+\bar{\zeta})=k_{(l-1)/2}$.

In Exercise 2.5(2), we obtained an isomorphism $G(K/\mathbb{Q}) \xrightarrow{\sim} \mathbb{F}_l^\times$ defined by the mapping $\sigma \mapsto j(\sigma) \bmod l$ with $\zeta^\sigma = \zeta^{j(\sigma)}$. Here, we consider rather the inverse mapping $a \mapsto \sigma_a$ defined by $\zeta^{\sigma_a} = \zeta^a$, $a \in \mathbb{F}_l^\times$. To begin with, we shall explain some of the notation in the diagram below:

$$\begin{array}{cccc}
K & \{e\} & \{1\} & \hat{\mathbb{F}}_l^\times \\
| & | & | & | \\
k_n & k_n^\# & H_n = (\mathbb{F}_l^\times)^n & H_n^* \\
| & | & | & | \\
k' & k'^\# & H' & H'^* \\
| & | & | & | \\
\mathbb{Q} & G(K/\mathbb{Q}) & \xleftarrow{\sim} \mathbb{F}_l^\times & \{\epsilon\} \\
 & \sigma_a & \longleftarrow\!\shortmid\; a &
\end{array}$$

† When $r=0$ or 1, one cannot use Proposition 2.11 as $m<3$. But, $r=0 \Rightarrow \chi=\epsilon$, $\tau(\epsilon)=0$; $r=1 \Rightarrow \chi=\lambda_l$, $\tau(\lambda_l)^2 = l^*$ and so there are no problems in these cases.

As we said before, k_n, $n \mid l-1$, is the subfield with $[k_n: \mathbb{Q}] = n$. Next, $k_n^{\#}$ is the subgroup of $G(K/\mathbb{Q})$ corresponding to k_n in the sense of Galois theory (Theorem 2.3) with $[k_n^{\#}] = [K: k_n]$. The third column is the copy of the second column by the isomorphism $a \mapsto \sigma_a$. H_n denotes the subgroup of $\mathbb{F}_l^\times$ corresponding to the subgroup k_n of $G(K/\mathbb{Q})$. Call r a generator of $\mathbb{F}_l^\times$, i.e., a primitive root mod l. Since $[\mathbb{F}_l^\times: H_n] = n$, we have $H_n = \langle r^n \rangle = (\mathbb{F}_l^\times)^n$. Hence we obtain a coset decomposition:

$$\mathbb{F}_l^\times = H_n \cup rH_n \cup \cdots \cup r^{n-1}H_n.$$

The fourth column shows subgroups of the character group $\hat{\mathbb{F}}_l^\times$ of $\mathbb{F}_l^\times$. To H_n corresponds the group H_n^* (the annihilator of H_n):

$$H_n^* = \{\chi \in \hat{\mathbb{F}}_l^\times ; \chi(a) = 1, \forall a \in H_n\}.$$

As $H_n = \langle r^n \rangle$, we have

$$H_n^* = \{\chi \in \hat{\mathbb{F}}_l^\times ; \chi(r)^n = 1\}.$$

On the other hand, since any character χ can be written $\chi = \chi_a$, with $\chi_a(r) = \xi^a$ and $\xi = e^{2\pi i/(l-1)}$, $a \in \mathbb{Z}$, we have

$$\chi = \chi_a \in H_n^* \Leftrightarrow \chi_a(r)^n = \xi^{an} = 1 \Leftrightarrow (l-1) \mid an \Leftrightarrow (l-1)/n \mid a.$$

Such a's are, taken $\bmod (l-1)$, 0, $(l-1)/n$, $2(l-1)/n, \cdots,$ $(n-1)(l-1)/n$ and so we have $[H_n^*] = n$.

PROPOSITION 2.13. *Let n and m be factors of $l-1$. We have*

$$k_n \subset k_m \Leftrightarrow n \mid m.$$

PROOF. $k_n \subset k_m \Leftrightarrow k_n^{\#} \supset k_m^{\#} \Leftrightarrow H_n \supset H_m \Leftrightarrow \langle r^n \rangle \supset \langle r^m \rangle \Leftrightarrow n \mid m.$ Q.E.D.

THEOREM 2.6. *For a factor n of $l-1$, we have $k_n = \mathbb{Q}(t_{K/k_n}\zeta)$.*

PROOF. Since n is fixed, we put $k = k_n$, $H = H_n$, and $H^* = H_n^*$ in the proof. We also put $\eta = t_{K/k}\zeta$. As we have $\eta \in k$, it is enough to prove that $k \subset \mathbb{Q}(\eta)$. From now on, we put $k' = \mathbb{Q}(\eta)$. In view of 1–1 correspondence of subfields and subgroups, to prove $k' \supset k$, it is enough to see that $H^* \subset H'^*$. Thus, take any $\chi \in H^*$.

If $\chi=\epsilon$, the matter is trivial. For $\chi\neq\epsilon$, consider the Gauss sum

$$\tau_a(\chi)=\sum_{x\in\mathbb{F}_l}\chi(x)\zeta^{ax}, \qquad a\in\mathbb{F}_l^\times.$$

Since $\tau_a(\chi)=\bar{\chi}(a)\tau(\chi)$ (Proposition 2.8) and $|\tau(\chi)|=\sqrt{l}$ (Proposition 2.9), we have $\tau(\chi)\neq 0$. Using the coset decomposition of $\mathbb{F}_l^\times/H$, we have

$$\tau_a(\chi)=\sum_x\chi(x)\zeta^{ax}=\sum_{i=0}^{n-1}\sum_{h\in H}\chi(r^ih)\zeta^{ar^ih}=\sum_{i=0}^{n-1}\chi(r)^i\sum_{h\in H}\zeta^{ar^ih}$$
$$=\sum_{i=0}^{n-1}\chi(r)^i\Big(\sum_{h\in H}\zeta^{\sigma_h}\Big)^{\sigma_a\sigma_r^i}=\sum_{i=0}^{n-1}\chi(r)^i\eta^{\sigma_a\sigma_r^i}.$$

In particular, we have $\sigma_a\in k'^{\#}$ for any $a\in H'$ and so we have $\eta^{\sigma_a}=\eta$ because $\eta\in k'$. Therefore we have

$$\bar{\chi}(a)\tau(\chi)=\tau_a(\chi)=\sum_{i=0}^{n-1}\chi(r)^i\eta^{\sigma_r^i}=\tau(\chi).$$

Since $\tau(\chi)\neq 0$, we have $\chi(a)=1$ for all $a\in H'$, i.e., $\chi\in H'^*$.

Q.E.D.

REMARK 2.6. Let $n=2$. In this case, $\mathbb{F}_l^\times=\langle r\rangle=H_2\cup rH_2$, $H_2=\langle r^2\rangle$, and $H_2^*=\langle\lambda_l\rangle$, λ_l being the Legendre character, $\lambda_l(r)=-1$. Then, we have $\eta=t_{K/k_2}\zeta=\sum_{h\in H_2}\zeta^h$ and

$$\tau(\lambda_l)=\sum_{x\in\mathbb{F}_l}\lambda_l(x)\zeta^x=\sum_{h\in H_2}\lambda_l(h)\zeta^h+\sum_{h\in H_2}\lambda_l(rh)\zeta^{rh}$$
$$=\sum_h\zeta^h-\sum_h\zeta^{hr}=\eta-\eta^{\sigma_r}.$$

On the other hand, $\eta+\eta^{\sigma_r}=\sum_x\zeta^x=0$. Therefore we have $-\eta^{\sigma_r}=\eta$ and $\tau(\lambda_l)=2\eta$. Since $\tau(\lambda_l)=\sqrt{l^*}$, we get $k_2=\mathbb{Q}(\eta)=\mathbb{Q}(\sqrt{l^*})$ [Exercise 2.5(4)]. Next, let $n=(l-1)/2$. Then, we have $H_2=\langle r^n\rangle=\langle -1\rangle$, $\eta=t_{K/k_n}\zeta=\zeta+\zeta^{-1}=\zeta+\bar{\zeta}$, and $k_{(l-1)/2}=\mathbb{Q}(\zeta+\zeta^{-1})$ [Exercise 2.5(5)].

EXERCISE 2.10. Let $l=5$. Prove that $k_2=\mathbb{Q}(\zeta+\zeta^{-1})=\mathbb{Q}(\sqrt{5})$. [Do not use the fact that $k_2=\mathbb{Q}(\sqrt{l^*})$ with $l^*=5$.]

2.7. Cohomology of Cyclic Groups

Let K/k be a finite Galois extension. The Galois group $G(K/k)$ acts on K as a group of automorphisms of the field K. However, this group acts not only on the field K but, as you will see later, on the group $K^\times$ of nonzero elements of K, on the group $\mathfrak{o}_K^\times$ of units of K, on the group H_K of ideal classes of K, etc. Some important results in algebraic number theory are described as propositions about the action of the Galois group on certain objects. Since, in this book, we have in mind an application of cohomology only for quadratic fields (Section 4.6), we shall restrict ourselves to the case of cyclic groups.

Definition 2.6. When a group G acts on a module A (Definition 1.9) in such a way that $\sigma(a+b)=\sigma a+\sigma b$, $a, b \in A$, then A is called a *G-module.*

Remark 2.7. When the law of composition in A is written multiplicatively, the above condition for a G-module means $\sigma(ab)=\sigma a \cdot \sigma b$. A $\mathbb{Z}$-module is nothing but a module (or an abelian group).

From now on let $G=\langle\sigma\rangle$ be a cyclic group of order n generated by σ. We introduce symbols Δ and N:

$$\Delta = 1-\sigma, \qquad N = 1+\sigma+\cdots+\sigma^{n-1}.$$

For a G-module A, we define

$$\Delta(a)=a-\sigma(a), \qquad N(a)=a+\sigma(a)+\cdots+\sigma^{n-1}(a), \qquad a\in A.$$

For a module A, a homomorphism $f\colon A \to A$ is called an *endomorphism* of A. The set $\operatorname{End} A$ of all endomorphisms of A forms naturally a (usually noncommutative) ring: the *endomorphism ring.*

Exercise 2.11. Verify that the set $\operatorname{End} A$ forms a ring.

Obviously, the symbols Δ and N, viewed as mappings of A into itself, belong to $\operatorname{End} A$. They satisfy the relations $\Delta N = N\Delta = 0$. In fact,

$$\begin{aligned}\Delta N(a) &= \Delta(a+\sigma(a)+\cdots+\sigma^{n-1}(a)) = \Delta(a)+\Delta\sigma(a)+\cdots+\Delta\sigma^{n-1}(a)\\ &= (a-\sigma(a))+(\sigma(a)-\sigma^2(a))+\cdots+(\sigma^{n-1}(a)-\sigma^n(a))=0,\\ N\Delta(a) &= N(a-\sigma(a)) = Na - N\sigma(a)\\ &= (a-\sigma(a)+\cdots+\sigma^{n-1}(a))-(\sigma(a)+\sigma^2(a)+\cdots+\sigma^n(a))=0.\end{aligned}$$

In general, when $f: A \to B$ is a homomorphism (of groups or rings) we put

$$\operatorname{Ker} f = \{a \in A; f(a) = 0\}, \qquad \operatorname{Im} f = \{f(a) \in B; a \in A\}.$$

Therefore the relations $\Delta N = N\Delta = 0$ can also be written

$$\operatorname{Im} N \subset \operatorname{Ker} \Delta, \qquad \operatorname{Im} \Delta \subset \operatorname{Ker} N. \tag{2.8}$$

For homomorphisms $f: A \to B$ and $g: B \to C$, the sequence

$$A \xrightarrow{f} B \xrightarrow{g} C$$

is said to be *exact* at B if $\operatorname{Im} f = \operatorname{Ker} g$. More generally, a sequence of homomorphisms

$$\cdots \to A_{n-1} \xrightarrow{f_{n-1}} A_n \xrightarrow{f_n} A_{n+1} \to \cdots$$

is called an *exact sequence* if $\operatorname{Im} f_{n-1} = \operatorname{Ker} f_n$ for all n. Since the sequences $A \xrightarrow{N} A \xrightarrow{\Delta} A$ and $A \xrightarrow{\Delta} A \xrightarrow{N} A$ are not necessarily exact at the middle A, one wants to know the deviation from being exact; this motivates the cohomology. In view of (2.8), we introduce the following definition.

Definition 2.7. Let $G = \langle \sigma \rangle$ be a finite cyclic group. For a G-module A, the groups

$$H^0(A) = \operatorname{Ker} \Delta / \operatorname{Im} N, \qquad H^1(A) = \operatorname{Ker} N / \operatorname{Im} \Delta$$

are called the *cohomology groups* of A.

Remark 2.8. Frequently, we write A^G for $\operatorname{Ker} \Delta$; A^G is the set of fixed points, i.e., $A^G = \{a \in A; \sigma a = a\}$. It is, of course, natural to write NA for $\operatorname{Im} N$. We write sometimes ${}_NA$ for $\operatorname{Ker} N$. Finally, $\operatorname{Im} \Delta$ can be written $(1 - \sigma)A$ (or $A^{1-\sigma}$).

Proposition 2.14. *Let A and B be G-modules and $f: A \to B$ be a G-homomorphism, i.e., a homomorphism such that $f(\tau a) = \tau f(a)$, $a \in A$, $\tau \in G$. Then, f induces homomorphisms f_i: $H^i(A) \to H^i(B)$, $i = 0, 1$.*

Proof. Since f is a G-homomorphism, it commutes with σ, Δ, and N: $\sigma f(a) = f(\sigma a)$, $f(Na) = Nf(a)$, and $f(\Delta a) = \Delta f(a)$.

Thus $f(A^G) \subset B^G$, $f(NA) \subset NB$, and the map f_0: $H^0(A) \to H^0(B)$ is defined by $f_0(a + NA) = f(a) + NB$. Next, since $f({}_NA) \subset {}_NB$ and $f(\Delta A) \subset \Delta B$, the map f_1: $H^1(A) \to H^1(B)$ is defined by $f_1(a + \Delta A) = f(a) + \Delta B$. Q.E.D.

THEOREM 2.7. *Let G be a finite cyclic group. For an exact sequence*

$$1 \to A \xrightarrow{f} B \xrightarrow{g} C \to 1$$

of G-modules A, B, and C, there exist homomorphisms δ_0 and δ_1 so that the following hexagonal sequence is exact:

$$\begin{array}{ccccccc} & & H^0(A) & \xrightarrow{f_0} & H^0(B) & & \\ & \nearrow^{\delta_1} & & & & \searrow^{g_0} & \\ H^1(C) & & & & & & H^0(C) \\ & \nwarrow_{g_1} & & & & \swarrow_{\delta_0} & \\ & & H^1(B) & \xleftarrow{f_1} & H^1(A) & & \end{array}$$

PROOF. We shall only give definitions of δ_0 and δ_1 and leave the rest as an exercise (Exercise 2.12). First, we define δ_0: $H^0(C) = C^G/NC \to H^1(A) = {}_NA/\Delta A$. For $c \in C^G$, since g is surjective, one finds $b \in B$ such that $c = g(b)$. We have $0 = \Delta c = \Delta g(b) = g(\Delta b)$ and so $\Delta b \in \operatorname{Ker} g = \operatorname{im} f$, i.e., $\Delta b = f(a)$, for some $a \in A$. For this a, since $f(Na) = Nf(a) = N\,\Delta b = 0$, we have $Na \in \operatorname{Ker} f = 0$, i.e., $a \in {}_NA$. We now put $\delta_0(c + NC) = a + \Delta A$. Here, we must, of course, verify that if $c + NC = c' + NC$, $c' \in C^G$, then $a + \Delta A = a' + \Delta A$ when a' is chosen by c' in the same way as a is chosen by c. Thus, call b' an element of B such that $g(b') = c'$ and $b' = f(a')$. Since $c - c' = Ng(b'')$, $b'' \in B$, we have $g(b - b' - Nb'') = 0$ and so there is an $a'' \in A$ such that $b - b' - Nb'' = f(a'')$. Applying Δ on both sides of the last equality, $\Delta b - \Delta b' = f(a - a') = f(\Delta a'')$. Since f is injective, we get $a - a' = \Delta a'' \in \Delta A$, i.e., $a + \Delta A = a' + \Delta A$ which shows that the map δ_0 is well defined. As all maps (f, g, Δ, N, etc.) are homomorphisms, one verifies easily that δ_0 is also a homomorphism.

Next, we define δ_1: $H^1(C) = {}_NC/\Delta C \to H^0(A) = A^G/NA$. For $c \in {}_NC$, one finds $b \in B$ such that $c = g(b)$. Since $0 = Nc = Ng(b) = g(Nb)$, we have $Nb \in \operatorname{Ker} g = \operatorname{Im} f$. Call a an element of A such that $Nb = f(a)$. We put $\delta_1(c + \Delta C) = a + NA$. If $c + \Delta C = c' + \Delta C$, $c' \in {}_NC$, we must verify that $a + NA = a' + NA$ where $g(b') =$

c' and $f(a') = Nb'$. Since $g(b - b') = c - c' - \Delta g(b'') = g(\Delta b'')$, we have $b - b' - \Delta b'' = f(a'')$, $a'' \in A$, and, applying N on both sides of the last equality, we have $f(a) - f(a') = Nb - Nb' = f(Na'')$. As f is injective, we get $a - a' \in NA$ as we wanted. One verifies easily that the map δ_1 is a homomorphism. not quite Q.E.D.

Exercise 2.12. Complete the proof of Theorem 2.7. We remind you here about the definitions of six maps:

$$f_0(a + NA) = f(a) + NB, \qquad g_0(b + NB) = g(b) + NC,$$

$$f_1(a + \Delta A) = f(a) + \Delta B, \qquad g_1(b + \Delta B) = g(b) + \Delta C,$$

$$\delta_0(c + NC) = a + \Delta A, \qquad c = g(b), \qquad \Delta b = f(a),$$

$$\delta_1(c + \Delta C) = a + NA, \qquad c = g(b), \qquad Nb = f(a).$$

Definition 2.8. Let G be a finite cyclic group and A be a G-module. When both $H^0(A)$ and $H^1(A)$ are finite groups, the rational number $Q(A) = [H^1(A)]/[H^0(A)]$ is called the *Herbrand* (1908–1932) *quotient* of A.

Proposition 2.15. *Let* $1 \to A \to B \to C \to 1$ *be an exact sequence of* G*-modules. If any two of three Herbrand quotients* $Q(A)$, $Q(B)$, *and* $Q(C)$ *make sense* (*in the sense of Definition* 2.8), *then the third one also makes sense and we have* $Q(A)Q(C) = Q(B)$.

Proof. Since $H^0(C)/\operatorname{Ker}\delta_0 \approx \operatorname{Im}\delta_0$ and $\operatorname{Ker}\delta_0 = \operatorname{Im} g_0$, we have $H^0(C)/\operatorname{Im} g_0 \approx \operatorname{Im}\delta_0$. If $H^i(A)$ and $H^i(B)$, $i = 0, 1$, are finite, then $H^0(C)$ is finite. In view of the symmetry of the hexagonal diagram (Theorem 2.7), the remaining two cases are settled in the same way. Now, from the hexagonal diagram, we have

$$[H^0(A)] = [\operatorname{Im} f_0][\operatorname{Ker} f_0], \qquad [H^1(A)] = [\operatorname{Im} f_1][\operatorname{Ker} f_1],$$

$$[H^0(C)] = [\operatorname{Im}\delta_0][\operatorname{Ker}\delta_0], \qquad [H^1(C)] = [\operatorname{Im}\delta_1][\operatorname{Ker}\delta_1],$$

$$[H^1(B)] = [\operatorname{Im} g_1][\operatorname{Ker} g_1], \qquad [H^0(B)] = [\operatorname{Im} g_0][\operatorname{Ker} g_0].$$

Since $[\operatorname{Im} f_1] = [\operatorname{Ker} g_1]$, $[\operatorname{Ker} f_1] = [\operatorname{Im}\delta_0]$, $[\operatorname{Im}\delta_1] = [\operatorname{Ker} f_0]$, $[\operatorname{Ker}\delta_1] = [\operatorname{Im} g_1]$, $[\operatorname{Im} g_0] = [\operatorname{Ker}\delta_0]$, and $[\operatorname{Ker} g_0] = [\operatorname{Im} f_0]$, we get

$$[H^0(A)][H^0(C)][H^1(B)] = [H^1(A)][H^1(C)][H^0(B)].$$

i.e., $Q(A)Q(C) = Q(B)$. Q.E.D.

PROPOSITION 2.16 *Let B be a G-module and A be a G-submodule of B such that $C = B/A$ is finite. If one of $Q(A)$, $Q(B)$ makes sense, then both of them make sense and we have $Q(A) = Q(B)$.*

PROOF. Since $H^0(C) = \operatorname{Ker}\Delta/\operatorname{Im} N$ and $H^1(C) = \operatorname{Ker} N/\operatorname{Im}\Delta$, we have

$$Q(C) = [H^1(C)]/[H^0(C)] = ([\operatorname{Ker} N]/[\operatorname{Im}\Delta])([\operatorname{Im} N]/[\operatorname{Ker}\Delta])$$
$$= ([\operatorname{Ker} N][\operatorname{Im} N])/([\operatorname{Im}\Delta][\operatorname{Ker}\Delta]) = [C]/[C] = 1,$$

and the assertion follows from Proposition 2.15. Q.E.D.

THEOREM 2.8 (*Hilbert Theorem* 90).† *Let K/k be a finite Galois extension of algebraic number fields and $G(K/k)$ be the Galois group. If a set $\{x_\sigma \in K^\times;\ \sigma \in G(K/k)\}$ of nonzero numbers of K satisfies the condition $x_{\sigma\tau} = x_\sigma x_\tau^\sigma$, σ, $\tau \in G(K/k)$, then there exists an $a \in K^\times$ such that $x_\sigma = a^{1-\sigma}$.*

PROOF. For any $b \in K$, put $a = \sum_\tau b^\tau x_\tau$. Then, we have

$$a^\sigma x_\sigma = \sum_\tau (b^\tau)^\sigma x_\tau^\sigma x_\sigma = \sum_\tau b^{\tau\sigma} x_\sigma x_\tau^\sigma = \sum_\tau b^{\sigma\tau} x_{\sigma\tau} = \sum_\sigma b^\sigma x_\sigma = a.$$

Hence, if one could choose b so that $a \neq 0$, then the assertion is proved. Suppose, on the contrary, that $\sum_\sigma b^\sigma x_\sigma = 0$ for all $b \in K$. Thus, for any basis $\{\gamma_1, \ldots, \gamma_n\}$ of the vector space K/k, we have $\sum_\sigma \gamma_j^\sigma x_\sigma = 0$, $1 \leq j \leq n$. However, since $d_{K/k}(\gamma_1, \ldots, \gamma_n) = (\det(\gamma_j^\sigma))^2 \neq 0$ (Section 2.3), we must have $x_\sigma = 0$ for all $\sigma \in G(K/k)$, which contradicts the assumption that $x_\sigma \in K^\times$. Q.E.D.

THEOREM 2.9. *Let K/k be a cyclic extension with the Galois group $G(K/k) = \langle\sigma\rangle$. If we regard the multiplicative group $K^\times$ as a $G(K/k)$-module, then we have $H^1(K^\times) = 1$.*

PROOF. Since $H^1(K^\times) = {}_N K^\times/\Delta K^\times$, it is enough to prove that for any $a \in K^\times$ such that $N_{K/k}a = 1$ one finds $b \in K^\times$ with $a = b^{1-\sigma}$. Let $n = [K:k]$ and $x_e = 1$, $x_\sigma = a$, $x_{\sigma^2} = aa^\sigma, \ldots, x_{\sigma^{n-1}} = aa^\sigma \cdots a^{\sigma^{n-2}}$. Then one verifies that $x_{\tau\rho} = x_\tau x_\rho^\tau$, τ, $\rho \in G(K/k)$: in

† Satz 90 of Hilbert's Bericht (1897).

fact, put $\tau = \sigma^i$, $\rho = \sigma^j$, $0 \le i,j \le n-1$. Then we have

1. $i+j \le n-1$: $x_{\sigma^i} x_{\sigma^j}^{\sigma^i} = (aa^\sigma \cdots a^{\sigma^{i-1}})(aa^\sigma \cdots a^{\sigma^{j-1}})^{\sigma^i}$
$$= aa^\sigma \cdots \alpha^{\sigma^{i-1}} a^{\sigma^i} \cdots \alpha^{\sigma^{i+j-1}} = x_{\sigma^{i+j}},$$
2. $i+j \ge n$: put $i+j = n+k$, $x_{\sigma^i} x_{\sigma^j}^{\sigma^i}$
$$= (aa^\sigma \cdots \alpha^{\sigma^{i-1}})(aa^\sigma \cdots a^{\sigma^{j-1}})^{\sigma^i}$$
$$= (aa^\sigma \cdots a^{\sigma^{n-1}})(aa^\sigma \cdots a^{\sigma^{k-1}})$$
$$= aa^\sigma \cdots a^{\sigma^{k-1}} = x_{\sigma^k} = x_{\sigma^{i+j}}.$$

Hence $x_\sigma = a = b^{1-\sigma}$ by Theorem 2.8. Q.E.D.

2.8. Finite Fields

Let p be a prime and $\mathbb{F}_p = \mathbb{Z}/p\mathbb{Z}$ be the field of integers modulo p. The fact that the group $\mathbb{F}_p^\times$ is cyclic (Theorem 1.19, Definition 1.17) is the core of the elementary number theory as we have seen in Chapter 1. Since an algebraic number field naturally provides many finite fields as the field $\mathbb{Q}$ does with $\mathbb{F}_p$, we shall consider here any finite field.

Let K be a field with finitely many elements. Since 0 and $1\,(0 \ne 1)$ are in K (Exercise 1.2), K contains $n \cdot 1$ for all $n \in \mathbb{N}$. On the other hand, since K contains only finitely many elements, we must have $n \cdot 1 = m \cdot 1$, i.e., $(n-m) \cdot 1 = 0$ for some positive integers n and m such that $n > m$. So let us call n_0 the smallest $n \in \mathbb{N}$ such that $n \cdot 1 = 0$. As any field has no zero divisor (i.e., $ab = 0 \Rightarrow a = 0$ or $b = 0$), we have $n_0 = p$, a prime number. This p is called the *characteristic* of K. Thus K is an extension of $\mathbb{F}_p = \mathbb{Z}/_p\mathbb{Z}$. If $n = [K: \mathbb{F}_p]$, then K contains $q = p^n$ elements. The multiplicative group $K^\times$ is a cyclic group of order $q-1$: $K^\times = \langle \theta \rangle$ (Theorem 1.18). We have $K = \{0, 1, \theta, \ldots, \theta^{q-2}\}$ as sets. Hence $K = \mathbb{F}_p(\theta)$. We have $f_\theta \mid X^q - X$ where $f_\theta(X)$ is the minimal polynomial of θ over $\mathbb{F}_p$. Let K' be another extension of $\mathbb{F}_p$ of degree n. As in the case of K, K' consists of all roots of $X^q - X = 0$, and so there is an element $\theta' \in K'$ such that $f_\theta(\theta') = 0$. Therefore we have $[\mathbb{F}_p(\theta'): \mathbb{F}_p] = \deg f_\theta = n$ and hence $K' = \mathbb{F}_p(\theta')$. One proves that there is an isomorphism

$$K \xrightarrow[\mathbb{F}_p]{\sim} K'$$

such that $\theta \mapsto \theta'$.† Thus we shall denote by $\mathbb{F}_q$ the unique finite field with $q = p^n$ elements. Conversely, one constructs, for any n, a finite field $\mathbb{F}_q$ with $q = p^n$ using all roots of $X^q - X = 0$. (We shall not prove the last statement because all finite fields considered in this book are those obtained naturally from algebraic number fields.)

Let $\mathbb{F}_{q^f}/\mathbb{F}_q$ be an extension of degree f. As in the case of number fields we denote by $G(\mathbb{F}_{q^f}/\mathbb{F}_q)$ the group of all automorphisms of $\mathbb{F}_{q^f}$ over $\mathbb{F}_q$.

THEOREM 2.10. *The group $G(\mathbb{F}_{q^f}/\mathbb{F}_q)$ is cyclic of order f generated by the mapping $\sigma_F x = x^q$, $x \in \mathbb{F}_{q^f}$. This mapping σ_F is called the Frobenius* (1849–1917) *automorphism.*

PROOF. Put $K = \mathbb{F}_{q^f}$, $k = \mathbb{F}_q$; $[K: k] = f$. In a field of characteristic p, we have $(x+y)^p = x^p + y^p$ (Exercise 1.42) and hence $(x+y)^{p^n} = x^{p^n} + y^{p^n}$, $n \in \mathbb{N}$. Since K is finite, the mapping σ_F which is an injective endomorphism $K \to K$ is actually an automorphism. Furthermore, all elements of $\mathbb{F}_q$ are roots of $X^q - X = 0$ and so $\sigma_F \in G(K/k)$. From the equalities $\sigma_F^2 x = x^{q^2}, \ldots, \sigma_F^{f-1} x = x^{q^{f-1}}$, we see that σ_F is of order f in $G(K/k)$ and that $[G(K/k)] \geq f = [K: k]$. Let $K = \mathbb{F}_p(\theta)$, and f_θ be the minimal polynomial of θ over k. Then $K = k(\theta)$ and $[K: k] = \deg f_\theta = f$. For any $\sigma \in G(K/k)$, we have $f_\theta(\theta^\sigma) = 0$; this implies that $[G(K/k)] \leq \deg f_\theta = f$ because $\theta^\sigma \neq \theta^\tau$ if $\sigma \neq \tau$. Therefore we get $[G(K/k)] = f$ and $G(K/k) = \langle \sigma_F \rangle$. Q.E.D.

EXERCISE 2.13. For $K = \mathbb{F}_{q^f}$ and $k = \mathbb{F}_q$, prove that the norm $N_{K/k}: K^\times \to k^\times$ is a surjection.

EXERCISE 2.14. Let K and k be as above. Consider $K^\times$ as a $G(K/k)$-module and determine cohomology groups $H^0(K^\times)$ and $H^1(K^\times)$.

2.9. Ring of Integers, Ideals, and Discriminant

In Section 1.11 we introduced the ring $\bar{\mathbb{Z}}$ of all algebraic integers. For a finite algebraic number field k, we call the ring

$$\mathfrak{o}_k \overset{\text{def}}{=} k \cap \bar{\mathbb{Z}}$$

† Proposition 2.2 holds not only for number fields but also for finite extensions of a finite field. That f_θ has no multiple roots follows from the fact that $X^q - X$ has q distinct roots $0, 1, \theta, \ldots, \theta^{q-2}$.

the *ring of integers* in k. Clearly, we have $\mathfrak{o}_k \supset \mathbb{Z}$. When we speak of ideals $\mathfrak{a}$ of $\mathfrak{o}_k$ we always assume that $\mathfrak{a} \neq 0$.

PROPOSITION 2.17. *Any ideal $\mathfrak{a}$ of $\mathfrak{o}_k$ contains a basis of the vector space k over $\mathbb{Q}$.*

PROOF. Let $u_1, \ldots, u_n$ be any basis of k over $\mathbb{Q}$. One finds $a \in \mathbb{N}$ such that $au_1, \ldots, au_n \in \mathfrak{o}_k$ (Proposition 1.19). Take any $\alpha \in \mathfrak{a}$, $\alpha \neq 0$. Then $a\alpha u_1, \ldots, a\alpha u_n$ all belong to $\mathfrak{a}$ and make up a basis of k over $\mathbb{Q}$. Q.E.D.

Now remember the norm and trace in k over $\mathbb{Q}$ (Section 2.3). For $\alpha \in k$, the conjugates of α over $\mathbb{Q}$ are numbers $\alpha^\sigma \in k^\sigma$, $\sigma \in G(\bar{\mathbb{Q}}/\mathbb{Q})$. Since $\alpha^\sigma \in \bar{\mathbb{Z}}$ if $\alpha \in \bar{\mathbb{Z}}$, we have $\alpha^\sigma \in \mathfrak{o}_{k^\sigma}$ for $\alpha \in \mathfrak{o}_k$. Let $\sigma_1, \ldots, \sigma_n$, $\sigma_1 = e$, be a complete set of representatives of left cosets in $G(\bar{\mathbb{Q}}/\mathbb{Q})/G(\bar{\mathbb{Q}}/k)$ and let $k^{\sigma_i} = k^{(i)}$, $\alpha^{\sigma_i} = \alpha^{(i)}$, $1 \leq i \leq n$. Then norms and traces were $\mathbb{Q}$-valued functions defined by

$$N_{k/\mathbb{Q}}\alpha = \alpha^{(1)} \cdots \alpha^{(n)}, \qquad t_{k/\mathbb{Q}}\alpha = \alpha^{(1)} + \cdots + \alpha^{(n)}.$$

If, in particular, $\alpha \in \mathfrak{o}_k$, then these become $\mathbb{Z}$-valued functions as we have $\mathbb{Q} \cap \bar{\mathbb{Z}} = \mathbb{Z}$ (Proposition 1.18). Similarly, if $\alpha_1, \ldots, \alpha_n \in \mathfrak{o}_k$, then

$$d_{k/\mathbb{Q}}(\alpha_1, \ldots, \alpha_n) = (\det(\alpha_j^{(i)})^2 = \det(t_{k/\mathbb{Q}}(\alpha_i\alpha_j)) \in \mathbb{Z}.$$

THEOREM 2.11. *Let $\mathfrak{a}$ be an ideal in $\mathfrak{o}_k$ and let $\alpha_1, \ldots, \alpha_n$ be numbers in $\mathfrak{a}$, linearly independent over $\mathbb{Q}$, such that $|d_{k/\mathbb{Q}}(\alpha_1, \ldots, \alpha_n)| \in \mathbb{N}$ is minimal. Then we have $\mathfrak{a} = \mathbb{Z}\alpha_1 + \cdots + \mathbb{Z}\alpha_n$. (In general, a set of n numbers $\alpha_1, \ldots, \alpha_n$ is called an integral basis of an ideal $\mathfrak{a}$ if we have $\mathfrak{a} = \mathbb{Z}\alpha_1 + \cdots + \mathbb{Z}\alpha_n$. If that is so, we write $\mathfrak{a} = [\alpha_1, \ldots, \alpha_n]$.)*

PROOF. Any $\alpha \in \mathfrak{a}$ can be written $\alpha = a_1\alpha_1 + \cdots + a_n\alpha_n$ with $a_i \in \mathbb{Q}$, $1 \leq i \leq n$. It is enough to show that all $a_i \in \mathbb{Z}$. Suppose, on the contrary, that some $a_i \notin \mathbb{Z}$; we may assume that $a_1 \notin \mathbb{Z}$ without loss of generality. Write $a_1 = b + \theta$, $b \in \mathbb{Z}$, $0 < \theta < 1$ and put

$$\beta_1 = \alpha - b\alpha_1, \qquad \beta_2 = \alpha_2, \ldots, \beta_n = \alpha_n.$$

Since $\beta_1 = \theta\alpha_1 + a_2\alpha_2 + \cdots + a_n\alpha_n$, we have

$$(\beta_1, \beta_2, \ldots, \beta_n) = (\alpha_1, \alpha_2, \ldots, \alpha_n)\begin{pmatrix} \theta & 0 & 0 & \cdots & 0 \\ a_2 & 1 & 0 & \cdots & 0 \\ \vdots & \vdots & \vdots & \ddots & \vdots \\ a_n & 0 & 0 & \cdots & 1 \end{pmatrix}$$

and hence

$$|d_{k/\mathbb{Q}}(\beta_1, \ldots, \beta_n)| = \theta^2 \, |d_{k/\mathbb{Q}}(\alpha_1, \ldots, \alpha_n)| < |d_{k/\mathbb{Q}}(\alpha_1, \ldots, \alpha_n)|.$$

Since $\beta_i \in \mathfrak{a}$ and $\beta_1, \ldots, \beta_n$ are independent over $\mathbb{Q}$, the above inequality contradicts the minimality of $\alpha_1, \ldots, \alpha_n$. Q.E.D.

For two integral bases of an ideal $\mathfrak{a}$: $\mathfrak{a} = [\alpha_1, \ldots, \alpha_n] = [\beta_1, \ldots, \beta_n]$ (Theorem 2.11), the α_i's and β_j's are related as $(\alpha_1, \ldots, \alpha_n) = (\beta_1, \ldots, \beta_n)A$, $A \in GL_n(\mathbb{Z})$ (Remark 1.12) and we find that $d_{k/\mathbb{Q}}(\alpha_1, \ldots, \alpha_n)$ depends only on the ideal $\mathfrak{a}$; we denote it by $d_{k/\mathbb{Q}}(\mathfrak{a})$. In particular, the *discriminant* of k is defined by

$$\Delta_k \overset{\text{def}}{=} d_{k/\mathbb{Q}}(\mathfrak{o}_k)$$

This is an important invariant of k. An integral basis of $\mathfrak{o}_k$ is also called an integral basis of k.

PROPOSITION 2.18. *Let k be a number field and let $\omega_1, \ldots, \omega_n$ be an integral basis of k. Let $\alpha_1, \ldots, \alpha_n$ be a basis of $k/\mathbb{Q}$ with $\alpha_i \in \mathfrak{o}_k$, $1 \le i \le n$. Call A a matrix in $M_n(\mathbb{Z})$* (Remark 1.12) *such that*

$$(\alpha_1, \ldots, \alpha_n) = (\omega_1, \ldots, \omega_n)A$$

and put $M = \mathbb{Z}\alpha_1 + \cdots + \mathbb{Z}\alpha_n$, this being a module in $\mathfrak{o}_k$. Then we have $[\mathfrak{o}_k : M] = |\det A|$ and $d_{k/\mathbb{Q}}(\alpha_1, \ldots, \alpha_n) = \Delta_k [\mathfrak{o}_k : M]^2$.

PROOF. We must show that $[\mathfrak{o}_k : M] = |\det A|$ (the last equality follows immediately from this). Now, by Exercise 2.15, there are matrices $U, V \in GL_n(\mathbb{Z})$ such that

$$UAV = D = \begin{bmatrix} d_1 & & 0 \\ & \ddots & \\ 0 & & d_n \end{bmatrix}, \qquad d_i \in N, \qquad d_i \mid d_{i+1}.$$

Thus we have $(\alpha_1, \ldots, \alpha_n)V = (\omega_1, \ldots, \omega_n)U^{-1}D$ and if we use new bases $(\beta_1, \ldots, \beta_n) = (\alpha_1, \ldots, \alpha_n)V$ and $(\eta_1, \ldots, \eta_n) = (\omega_1, \ldots, \omega_n)U^{-1}$ for M and $\mathfrak{o}_k$, respectively, we get $(\beta_1, \ldots, \beta_n) = (\eta_1, \ldots, \eta_n)D$, i.e., $\beta_i = d_i \eta_i$, $1 \le i \le n$. Therefore we have

$$\mathfrak{o}_k / M \approx \mathbb{Z}/d_1\mathbb{Z} \times \cdots \times \mathbb{Z}/d_n\mathbb{Z}$$

and so $[\mathfrak{o}_k : M] = d_1 \cdots d_n = \det D = |\det A|$. Q.E.D.

EXERCISE 2.15. Prove that for $A \in M_n(\mathbb{Z})$ there are $U, V \in GL_n(\mathbb{Z})$ such that

$$UAV = \begin{bmatrix} d_1 & & & & & 0 \\ & \ddots & & & & \\ & & d_r & & & \\ & & & 0 & & \\ & & & & \ddots & \\ 0 & & & & & 0 \end{bmatrix}, \qquad r = \operatorname{rank} A, \quad d_i \in \mathbb{N}, \quad d_i \mid d_{i+1}.$$

PROPOSITION 2.19. *Let k be a number field of degree n, α be an integer in $\mathfrak{o}_k$ such that $k = \mathbb{Q}(\alpha)$, and f_α be the minimal polynomial of α over $\mathbb{Q}$. Then we have $f_\alpha(X) \in \mathbb{Z}[X]$, $\mathbb{Z}[\alpha] = \mathbb{Z} + \mathbb{Z}\alpha + \cdots + \mathbb{Z}\alpha^{n-1}$, and $d_{k/\mathbb{Q}}(1, \alpha, \ldots, \alpha^{n-1}) = D_{k/\mathbb{Q}}(\alpha) = \Delta_k[\mathfrak{o}_k: \mathbb{Z}[\alpha]]^2$* (Proposition 2.18). *In particular, if $d_{k/\mathbb{Q}}(1, \alpha, \ldots, \alpha^{n-1}) = D_{k/\mathbb{Q}}(\alpha)$ has no square factors, then $\mathfrak{o}_k = \mathbb{Z}[\alpha] = [1, \alpha, \ldots, \alpha^{n-1}]$.*

PROOF. For $\alpha \in \bar{\mathbb{Q}}$ one finds $a \in \mathbb{N}$ such that $a\alpha \in \bar{\mathbb{Z}}$ (Proposition 1.19), hence there is an $\alpha \in \mathfrak{o}_k$ such that $k = \mathbb{Q}(\alpha)$. Since all conjugates $\alpha^{(i)}$, $1 \le i \le n$, belong to $\bar{\mathbb{Z}}$ and $f_\alpha(X) = \prod_{i=1}^n (X - \alpha^{(i)})$, all coefficients of $f_\alpha(X)$ belong to $\mathbb{Q} \cap \bar{\mathbb{Z}} = \mathbb{Z}$ (Proposition 1.18). The rest of the assertion is easy to see. Q.E.D.

EXERCISE 2.16. Let $k/\mathbb{Q}$ be a quadratic field and let $\alpha \in K$. Prove that $\alpha \in \mathfrak{o}_k \Leftrightarrow N_{k/\mathbb{Q}}\alpha$ and $t_{k/\mathbb{Q}}\alpha \in \mathbb{Z}$.

EXERCISE 2.17. Let $k = \mathbb{Q}(\sqrt{m})$ be a quadratic field (Exercise 2.4). Prove that

$$\mathfrak{o}_k = \begin{cases} [1, \sqrt{m}], & m \equiv 2, 3 \quad (4) \\ \left[1, \dfrac{1 + \sqrt{m}}{2}\right], & m \equiv 1 \quad (4) \end{cases}$$

and

$$\Delta_k = \begin{cases} 4m, & m \equiv 2, 3 \quad (4) \\ m, & m \equiv 1 \quad (4). \end{cases}$$

EXERCISE 2.18. Let l be an odd prime and $k = \mathbb{Q}(\zeta)$ the lth cyclotomic field with $\zeta = e^{2\pi i/l}$. Prove that $\mathfrak{o}_k = \mathbb{Z}[\zeta] = [1, \zeta, \ldots, \zeta^{l-2}]$ and $\Delta_k = (-1)^{(l-1)/2} l^{l-2}$.

DEFINITION 2.9. Let $k/\mathbb{Q}$ be a finite Galois extension and $G = G(k/\mathbb{Q})$ be the Galois group. A set of numbers $\{\omega^\sigma\}_{\sigma\in G}$ such that $\mathfrak{o}_k = \sum_{\sigma\in G} \mathbb{Z}\omega^\sigma = [\omega^\sigma, \sigma \in G]$ is called a *normal basis* of $\mathfrak{o}_k$.

REMARK 2.9. There are Galois extensions $k/\mathbb{Q}$ with no normal bases for $\mathfrak{o}_k$. For example, the Gauss field $k = \mathbb{Q}(i)$ has no normal bases. In fact, suppose $\omega = a + ib$, $a, b \in \mathbb{Z}$, provided one. Then, we must have

$$(\omega, \omega^\sigma) = (1, i)\begin{pmatrix} a & a \\ b & -b \end{pmatrix} \qquad \text{with } A = \begin{pmatrix} a & a \\ b & -b \end{pmatrix} \in \mathrm{GL}_2(\mathbb{Z}),$$

but this is impossible because $\det A = -2ab \neq \pm 1$. On the other hand, for the lth cyclotomic field $k = \mathbb{Q}(\zeta)$ (Exercise 2.18), we have $\mathfrak{o}_k = \mathbb{Z}[\zeta] = [1, \zeta, \ldots, \zeta^{l-2}] = [\zeta, \zeta^2, \ldots, \zeta^{l-1}]$ as $\zeta \in \mathfrak{o}_k^\times$ and this shows that $\mathfrak{o}_k$ has a normal basis.

EXERCISE 2.19. Let $k = \mathbb{Q}(\sqrt{m})$ be a quadratic field. Prove that

$$\mathfrak{o}_k \text{ has a normal basis} \Leftrightarrow m \equiv 1 \quad (4).$$

Let $k/\mathbb{Q}$ be a finite Galois extension with Galois group $G = G(k/\mathbb{Q})$. Let H be a subgroup of G, $L =$ the subfield of k corresponding to H, and $G = H\sigma_1 \cup \cdots \cup H\sigma_m$, $\sigma_1 = e$, $m = [L: \mathbb{Q}]$, be the decomposition of G with respect to right cosets modulo H. Suppose that $\mathfrak{o}_k$ has a normal basis: $\mathfrak{o}_k = [\omega^\sigma]$, $\sigma \in G$. When we write $\alpha \in \mathfrak{o}_k$ as $\alpha = \sum_\sigma c_\sigma \omega^\sigma$, we have

$$\alpha \in \mathfrak{o}_L \Leftrightarrow \alpha^\tau = \alpha, \qquad \forall \tau \in H$$
$$\Leftrightarrow \alpha^\tau = \sum_\sigma c_\sigma (\omega^\sigma)^\tau = \sum_\sigma c_\sigma \omega^{\tau\sigma} = \sum_\sigma c_{\tau^{-1}\sigma} \omega^\sigma = \alpha$$
$$\Leftrightarrow c_{\tau^{-1}\sigma} = c_\sigma \Leftrightarrow c_\sigma = c_{\tau\sigma}.$$

Hence, for $\alpha \in \mathfrak{o}_L$, we get

$$\alpha = \sum_\sigma c_\sigma \omega^\sigma = \sum_{i=1}^m \sum_{\tau\in H} c_{\tau\sigma_i} \omega^{\tau\sigma_i} = \sum_{i=1}^m c_{\sigma_i} \sum_{\tau\in H} (\omega^{\sigma_i})^\tau$$

which means that $\mathfrak{o}_L = [\sum_{\tau\in H} (\omega^{\sigma_i})^\tau]$. If, in particular, $k/\mathbb{Q}$ is abelian, i.e., if the group $G = G(k/\mathbb{Q})$ is abelian, then we have $(\omega^{\sigma_i})^\tau = \omega^{\tau\sigma_i} = \omega^{\sigma_i\tau} = (\omega^\tau)^{\sigma_i}$, and so $\mathfrak{o}_L = [\eta^s]$, $s \in G/H = G(L/\mathbb{Q})$ with $\eta = \sum_{\tau\in H} \omega^\tau$. Thus, we have proved the following theorem.

THEOREM 2.12. *Let $k/\mathbb{Q}$ be an abelian extension with the Galois group $G = G(k/\mathbb{Q})$. Suppose that $\mathfrak{o}_k$ has a normal basis: $\mathfrak{o}_k = [\omega^\sigma]$, $\sigma \in G$. Then, for any subfield L of k, $\mathfrak{o}_L$ has a normal basis: $\mathfrak{o}_L = [\eta^s]$, $s \in G(L/\mathbb{Q})$, where $\eta = \sum_{\tau \in L^\#} \omega^\tau = t_{k/L}\omega$.*

REMARK 2.10. We know that the lth cyclotomic field $k = \mathbb{Q}(\zeta) = k_{l-1}$ has a normal basis for $\mathfrak{o}_k$: $\mathfrak{o}_k = [\zeta, \zeta^2, \ldots, \zeta^{l-1}]$. Hence, by Theorem 2.12, every subfield k_n, $n \mid l-1$, of k, has a normal basis $\mathfrak{o}_{k_n} = [\eta^s]$, $s \in G(k_n/\mathbb{Q})$, $\eta = t_{k/k_n\zeta}$; this provides an alternative proof of Theorem 2.6.

2.10. Fundamental Theorem of Ideal Theory

Let $k/\mathbb{Q}$ be an algebraic extension of degree $n = [k: \mathbb{Q}]$. For α ($\neq 0$), $(\alpha) = \mathfrak{o}_k\alpha$ is an ideal in $\mathfrak{o}_k$ which is called the *principal ideal* generated by α. For two ideals $\mathfrak{a}$, $\mathfrak{b}$ in $\mathfrak{o}_k$, the *product* is defined by

$$\mathfrak{a}\mathfrak{b} = \left\{\sum \alpha_i\beta_i \text{ (finite sum)},\ \alpha_i \in \mathfrak{a},\ \beta_i \in \mathfrak{b}\right\}.$$

An ideal $\mathfrak{p}$ ($\neq \mathfrak{o}_k$) is called a *prime ideal* if it has the property: $\alpha\beta \in \mathfrak{p} \Rightarrow \alpha \in \mathfrak{p}$ or $\beta \in \mathfrak{p}$, $\alpha, \beta \in \mathfrak{o}_k$.

PROPOSITION 2.20. *For an ideal $\mathfrak{a}$ in $\mathfrak{o}_k$, the quotient ring $\mathfrak{o}_k/\mathfrak{a}$ is finite.*

PROOF. For basis expressions $\mathfrak{o}_k = [\omega_1, \ldots, \omega_n]$ and $\mathfrak{a} = [\alpha_1, \ldots, \alpha_n]$, one finds a matrix $A \in M_n(\mathbb{Z})$ such that $(\alpha_1, \ldots, \alpha_n) = (\omega_1, \ldots, \omega_n)A$. Then $[\mathfrak{o}_k: \mathfrak{a}] = |\det A|$ (Proposition 2.18) which is enough to show the finiteness of $\mathfrak{o}_k/\mathfrak{a}$. Q.E.D.

PROPOSITION 2.21. *The ring $\mathfrak{o}_k$ is noetherian, i.e., for an ascending chain of ideals in $\mathfrak{o}_k$: $\mathfrak{a}_1 \subset \mathfrak{a}_2 \subset \cdots \subset \mathfrak{a}_n \subset \cdots$, there exists a natural number N such that $\mathfrak{a}_m = \mathfrak{a}_{m+1}$ for all $m \geq N$.*

PROOF. Since the ring $\mathfrak{o}_k/\mathfrak{a}$ is finite (Proposition 2.20), there are only finitely many ideals containing $\mathfrak{a}_1$. Q.E.D.

PROPOSITION 2.22. *For a prime ideal $\mathfrak{p}$, the ring $\mathfrak{o}_k/\mathfrak{p}$ is a field.*

PROOF. Let $[\alpha_1], \ldots, [\alpha_l]$, $[\alpha_i] = \alpha_i \bmod \mathfrak{p}$, be all elements of the finite ring $\mathfrak{o}_k/\mathfrak{p}$ other than zero. Take any number $\xi \notin \mathfrak{p}$ and consider elements $[\xi][\alpha_i]$, $1 \leq i \leq l$. We claim that all these elements are not zero. In fact, if, for some i, $[\xi][\alpha_i] = [0]$, then we have

$$[\xi][\alpha_i] = [\xi\alpha_i] = [0] \Rightarrow \xi\alpha_i \in \mathfrak{p} \Rightarrow \xi \in \mathfrak{p} \quad \text{or} \quad \alpha_i \in \mathfrak{p}$$

which contradicts the assumption that $[\xi] \neq [0]$ and $[\alpha_i] \neq [0]$. Next, we have $[\xi][\alpha_i] = [\xi][\alpha_j] \Rightarrow \xi(\alpha_i - \alpha_j) \in \mathfrak{p} \Rightarrow [\alpha_i] = [\alpha_j]$ because $\xi \notin \mathfrak{p}$. Therefore, we see that to multiply $[\xi]$ merely permutes nonzero elements of $\mathfrak{o}_k/\mathfrak{p}$. In particular, for any α_i, one finds ξ such that $[\xi][\alpha_i] = [1]$, i.e., $\xi\alpha_i \equiv 1(\mathfrak{p})$, which shows that $\mathfrak{o}_k/\mathfrak{p}$ is a field. Q.E.D.

Definition 2.10. For ideals $\mathfrak{a}, \mathfrak{b}$ in $\mathfrak{o}_k$, we define

$$\mathfrak{a} \sim \mathfrak{b} \overset{\text{def}}{\Leftrightarrow} (\alpha)\mathfrak{a} = (\beta)\mathfrak{b} \qquad \text{for some } \alpha, \beta \in \mathfrak{o}_k, \qquad \alpha \neq 0, \qquad \beta \neq 0.$$

This relation in the set of all nonzero ideals in $\mathfrak{o}_k$ is an equivalence relation. A class defined by this equivalence relation is called an *ideal class.*

Proposition 2.23 [*Lemma of Hurwitz* (1859–1919)]. *For a number field k, there is a natural number M depending only on basis of $\mathfrak{o}_k$ with the following property. Given α , $\beta \in \mathfrak{o}_k$, $\beta \neq 0$, there is an integer t, $1 \leq t \leq M$, and a number $\theta \in \mathfrak{o}_k$ such that $|N_{k/\mathbb{Q}}(t\alpha - \theta\beta)| < |N_{k/\mathbb{Q}}\beta|$.*

Proof. Let $\gamma = \alpha/\beta \in k$. It is enough to show that for all $\gamma \in k$ there is an M such that $|N_{k/\mathbb{Q}}(t\gamma - \theta)| < 1$ for some t, $1 \leq t \leq M$, and $\theta \in \mathfrak{o}_k$. Let $\omega_1, \ldots, \omega_n$ be a basis of $\mathfrak{o}_k$. Define a mapping $\varphi: k \to \mathbb{R}^n$ by $\varphi(\gamma) = (c_1, \ldots, c_n)$ with $\gamma = \sum_{i=1}^n c_i\omega_i$, $c_i \in \mathbb{Q}$. Put $\|\gamma\| \overset{\text{def}}{=} \max_i |c_i|$. Then we have

$$|N_{k/\mathbb{Q}}\gamma| \leq \prod_j \Big(\sum_i |c_i|\, |\omega_i^{(j)}|\Big) \leq \|\gamma\|^n C, \qquad \text{where } C = \prod_j \Big(\sum_i |\omega_i^{(j)}|\Big).$$

Choose m so that $m > \sqrt[n]{C}$ and put $M = m^n$. We claim this is an M we are looking for. For $\gamma = \sum_i c_i\omega_i$, $c_i \in \mathbb{Q}$, write $c_i = a_i + b_i$, $a_i \in \mathbb{Z}$, $0 \leq b_i < 1$, and put $[\gamma] = \sum_i a_i\omega_i$ and $\{\gamma\} = \sum_i b_i\omega_i$. Then we have $\gamma = [\gamma] + \{\gamma\}$, with $[\gamma] \in \mathfrak{o}_k$ and

$$\varphi(\{\gamma\}) \in U = \{x = (x_1, \ldots, x_n) \in \mathbb{R}^n; 0 \leq x_i < 1\} = \text{the unit cube.}$$

Partition U into m^n subcubes of side $1/m$. For each $\gamma \in k$, consider the points $\varphi(\{\nu\gamma\})$, $1 \leq \nu \leq m^n + 1$; at least two of these points, say $\varphi(\{\nu_1\gamma\})$ and $\varphi(\{\nu_2\gamma\})$, $\nu_1 > \nu_2$, must lie in the same subcube. Then, taking the difference of $\nu_1\gamma = [\nu_1\gamma] + \{\nu_1\gamma\}$ and $\nu_2\gamma = [\nu_2\gamma] + \{\nu_2\gamma\}$, we get $t\gamma = \theta + \delta$, $t = \nu_1 - \nu_2 \leq m^n = M$, $\theta \in \mathfrak{o}_k$, and

$$|N_{k/\mathbb{Q}}\delta| \leq \|\delta\|^n C \leq (1/m)^n C < 1. \qquad \text{Q.E.D.}$$

THEOREM 2.13. *The number of ideal classes of $\mathfrak{o}_k$ is finite.*

PROOF. Let $\mathfrak{a}$ be any ideal. Choose $\beta \in \mathfrak{a}$ so that $|N_{k/\mathbb{Q}}\beta|$ is minimal. For any $\alpha \in \mathfrak{a}$, there is a $t \in \mathbb{Z}$ and $\theta \in \mathfrak{o}_k$ such that $1 \leq t \leq M$ and $|N_{k/\mathbb{Q}}(t\alpha - \theta\beta)| < |N_{k/\mathbb{Q}}\beta|$ (Proposition 2.23). Since $t\alpha - \theta\beta \in \mathfrak{a}$ we must have $t\alpha - \theta\beta = 0$ by the minimality of $|N_{k/\mathbb{Q}}\beta|$. From $1 \leq t \leq M$, it follows that $t \mid M!$ and so $M!\,\mathfrak{a} \subset (\beta)$. Hence we have $\mathfrak{b} = (1/\beta)M!\,\mathfrak{a} \subset \mathfrak{o}_k$, i.e., $(M!)\mathfrak{a} = (\beta)\mathfrak{b}$ which implies $\mathfrak{a} \sim \mathfrak{b}$. As for $\mathfrak{b}$, we have $M!\,\beta \in (M!)\mathfrak{a} = (\beta)\mathfrak{b}$ and hence $(M!) \subset \mathfrak{b}$. Since the ring $\mathfrak{o}_k/(M!)$ is finite (Proposition 2.20), there are only finitely many $\mathfrak{b}$'s. We have proved that for any ideal $\mathfrak{a}$ we have $\mathfrak{a} \sim \mathfrak{b}$ where $\mathfrak{b}$ is one of finitely many ideals. Thus the class number of k is finite. Q.E.D.

DEFINITION 2.11. The number of ideal classes of $\mathfrak{o}_k$ is called the *class number* of k and is written h_k.

REMARK 2.11. we often write $\mathfrak{a} \sim 1$ when $\mathfrak{a} \sim \mathfrak{o}_k = (1)$. Therefore, $\mathfrak{a} \sim 1 \Leftrightarrow \mathfrak{a} = (\alpha)$ and $h_k = 1 \Leftrightarrow$ every ideal is principal.

PROPOSITION 2.24. *For ideals $\mathfrak{a}$ and $\mathfrak{b}$ in $\mathfrak{o}_k$, $\mathfrak{a} = \mathfrak{a}\mathfrak{b} \Rightarrow \mathfrak{b} = \mathfrak{o}_k$.*

PROOF. Let $\mathfrak{a} = [\alpha_1, \ldots, \alpha_n]$. From $\mathfrak{a} = \mathfrak{a}\mathfrak{b}$ it follows that $\alpha_i = \sum_j b_{ij}\alpha_j$, $b_{ij} \in \mathfrak{b}$ and $\det(b_{ij} - \delta_{ij}) = 0$. By developing the determinant, we see that $1 \in \mathfrak{b}$, i.e., $\mathfrak{b} = \mathfrak{o}_k$. Q.E.D.

PROPOSITION 2.25. *Let $\mathfrak{a}$ and $\mathfrak{b}$ be ideals of $\mathfrak{o}_k$. If a number $\beta \in \mathfrak{o}_k$ satisfies $(\beta)\mathfrak{a} = \mathfrak{a}\mathfrak{b}$, then $\mathfrak{b} = (\beta)$.*

PROOF. Take any $\gamma \in \mathfrak{b}$. Since $\gamma\mathfrak{a} \in (\beta)\mathfrak{a}$, we have $(\gamma/\beta)\mathfrak{a} \subset \mathfrak{a}$ which implies that $\gamma/\beta \in \mathfrak{o}_k$ (Proposition 1.20). Therefore $\mathfrak{b} \subset (\beta)$ and so $\beta^{-1}\mathfrak{b} \subset \mathfrak{o}_k$. Applying Proposition 2.24 to the equality $\mathfrak{a} = (\beta^{-1}\mathfrak{b})\mathfrak{a}$, we get $\beta^{-1}\mathfrak{b} = \mathfrak{o}_k$, i.e., $\mathfrak{b} = (\beta)$. Q.E.D.

PROPOSITION 2.26. *For an ideal $\mathfrak{a}$ in $\mathfrak{o}_k$ there is a natural number ν, $1 \leq \nu \leq h_k$, such that $\mathfrak{a}^\nu \sim 1$.*

PROOF. Since the class number is finite, there must be two natural numbers i, j, $i < j$, such that $\mathfrak{a}^i \sim \mathfrak{a}^j$, i.e., $(\alpha)\mathfrak{a}^i =$

$(\beta)\mathfrak{a}^j$, $\alpha, \beta \in \mathfrak{o}_k$. Thus, putting $\nu = j - i$, we have $(\alpha)\mathfrak{a}^i = (\beta)a^\nu\mathfrak{a}^i$ and hence $(\alpha) = (\beta)\mathfrak{a}^\nu$, i.e., $\mathfrak{a}^\nu \sim 1$ (Proposition 2.25). Q.E.D.

THEOREM 2.14. *The set H_k of all ideal classes of $\mathfrak{o}_k$ makes up naturally a finite abelian group of order h_k. The group H_k is called the ideal class group of k.*

PROOF. Denote by $[\mathfrak{a}]$ the class of an ideal $\mathfrak{a}$. If $\mathfrak{a} \sim \mathfrak{a}_1$ and $\mathfrak{b} \sim \mathfrak{b}_1$, then $(\alpha)\mathfrak{a} = (\beta)\mathfrak{a}_1$ and $(\gamma)\mathfrak{b} = (\delta)\mathfrak{b}_1$, and hence $(\alpha\gamma)\mathfrak{a}\mathfrak{b} = (\beta\delta)\mathfrak{a}_1\mathfrak{b}_1$, i.e., $\mathfrak{a}\mathfrak{b} \sim \mathfrak{a}_1\mathfrak{b}_1$. Thus the product of ideal classes is well defined by $[\mathfrak{a}][\mathfrak{b}] = [\mathfrak{a}\mathfrak{b}]$. The associativity of the multiplication of classes follows directly from that of the product of ideals: $(\mathfrak{a}\mathfrak{b})\mathfrak{c} = \mathfrak{a}(\mathfrak{b}\mathfrak{c})$. The associativity of the multiplication of classes follows directly from that of the product of ideals: $(\mathfrak{a}\mathfrak{b})\mathfrak{c} = \mathfrak{a}(\mathfrak{b}\mathfrak{c})$. The class $[\mathfrak{o}_k]$ serves as the identity. Lastly, since $\mathfrak{a}^\nu \sim 1$ for some natural number ν (Proposition 2.26) we have $\mathfrak{a} \cdot \mathfrak{a}^{\nu-1} \sim 1$, which means that the class $[\mathfrak{a}]^{\nu-1}$ serves as the inverse $[\mathfrak{a}]^{-1}$. Q.E.D.

REMARK 2.12. By Proposition 1.7, we have $\mathfrak{a}^{h_k} \sim 1$ for all ideals $\mathfrak{a}$ in $\mathfrak{o}_k$.

PROPOSITION 2.27. *For ideals $\mathfrak{a}$, $\mathfrak{b}$, and $\mathfrak{c}$ in $\mathfrak{o}_k$, we have $\mathfrak{a}\mathfrak{b} = \mathfrak{a}\mathfrak{c} \Rightarrow \mathfrak{b} = \mathfrak{c}$.*

PROOF. There is a natural number ν such that $\mathfrak{a}^\nu = (\alpha)$. Thus, we get $(\alpha)\mathfrak{b} = (\alpha)\mathfrak{c}$ which implies that $\mathfrak{b} = \mathfrak{c}$. Q.E.D.

PROPOSITION 2.28. *For ideals $\mathfrak{a}$ and $\mathfrak{b}$ in $\mathfrak{o}_k$ such that $\mathfrak{a} \subset \mathfrak{b}$, there exists an ideal $\mathfrak{c}$ such that $\mathfrak{a} = \mathfrak{b}\mathfrak{c}$. When that is so, we write $\mathfrak{b} \mid \mathfrak{a}$.*

PROOF. Suppose that $\mathfrak{b}^\nu = (\beta)$. Since $\mathfrak{a} \subset \mathfrak{b}$, we have $\mathfrak{b}^{\nu-1}\mathfrak{a} \subset (\beta)$ and so $\mathfrak{c} = (1/\beta)\mathfrak{b}^{\nu-1}\mathfrak{a} \subset \mathfrak{o}_k$, i.e., $\mathfrak{b}\mathfrak{c} = (1/\beta)\mathfrak{b}^\nu\mathfrak{a}$. Q.E.D.

THEOREM 2.15 (*Fundamental theorem of ideal theory*). *Let k be an algebraic number field of finite degree over $\mathbb{Q}$. Every ideal in $\mathfrak{o}_k$ can be written as a product of prime ideals; it can be so written in only one way except for the order of factors.*

PROOF. Let $\mathfrak{a}$ be an ideal such that $\mathfrak{a} \subsetneq \mathfrak{o}_k$. Since the ring $\mathfrak{o}_k/\mathfrak{a}$ is finite, $\mathfrak{a}$ is contained in a maximal ideal $\mathfrak{p}_1 \subsetneq \mathfrak{o}_k$. We claim that $\mathfrak{p}_1$ is a prime ideal. In fact, suppose that $ab \in \mathfrak{p}_1$ but $a \notin \mathfrak{p}_1$, $a, b \in \mathfrak{o}_k$. Since $\mathfrak{p}_1$ is maximal, a and $\mathfrak{p}_1$ generate the ring $\mathfrak{o}_k$ and hence $1 = \lambda a + \pi$, with $\lambda \in \mathfrak{o}_k$, $\pi \in \mathfrak{p}_1$. Then $b = \lambda ab + b\pi \in \mathfrak{p}_1$ which shows that $\mathfrak{p}_1$ is prime. Since $\mathfrak{a} \subset \mathfrak{p}_1$, we have $\mathfrak{a} = \mathfrak{p}_1\mathfrak{a}_1$ (Proposition

2.28). If $\mathfrak{a}_1 \subsetneqq \mathfrak{o}_k$, one repeats the same argument to get $\mathfrak{a}_1 = \mathfrak{p}_2\mathfrak{a}_2$, etc. and obtains an ascending chain of ideals: $\mathfrak{a} \subset \mathfrak{a}_1 \subset \mathfrak{a}_2 \subset \cdots$. Since $\mathfrak{o}_k$ is noetherian (Proposition 2.21), $\mathfrak{a}_m = \mathfrak{o}_k$ for some m and one gets a decomposition $\mathfrak{a} = \mathfrak{p}_1\mathfrak{p}_2 \cdots \mathfrak{p}_m$. If $\mathfrak{a} = \mathfrak{q}_1\mathfrak{q}_2 \cdots \mathfrak{q}_t$ is a second decomposition of $\mathfrak{a}$ into prime ideals, then as the definition of the prime ideal implies that $\mathfrak{p} \mid \mathfrak{a}\mathfrak{b} \Rightarrow \mathfrak{p} \mid \mathfrak{a}$ or $\mathfrak{p} \mid \mathfrak{b}$ (Proposition 2.28) we can argue exactly in the same way as in the proof of the fundamental theorem of number theory (Theorem 1.6) to prove the uniqueness of the decomposition. Q.E.D.

Remark 2.13. When a prime ideal $\mathfrak{p}$ divides an ideal $\mathfrak{a}$, i.e., when $\mathfrak{p} \mid \mathfrak{a}$, $\mathfrak{p}$ is often called a prime divisor of $\mathfrak{a}$.

2.11. Residue Class Rings

Let k be an algebraic number field of finite degree over $\mathbb{Q}$ and let $\mathfrak{o}_k$ be the ring of integers of k. For an ideal $\mathfrak{a}$ ($\neq \{0\}$) of $\mathfrak{o}_k$, we proved that the residue class ring $\mathfrak{o}_k/\mathfrak{a}$ is finite (Proposition 2.20). In this section, we shall generalize results of Section 1.7 and Section 1.8 where $k = \mathbb{Q}$, $\mathfrak{o}_k = \mathbb{Z}$, and $\mathfrak{a} = m\mathbb{Z}$.

Definition 2.12. The number of elements in the residue class ring $\mathfrak{o}_k/\mathfrak{a}$ is called the *norm of* $\mathfrak{a}$, written $N_k\mathfrak{a}$.

Proposition 2.29. *Let $\mathfrak{a}$ be an ideal in $\mathfrak{o}_k$ and $\alpha \in \mathfrak{o}_k$. Then,*

$$(\alpha) + \mathfrak{a} = \mathfrak{o}_k \Leftrightarrow \alpha x \equiv 1 (\mathfrak{a}) \qquad \textit{for some } x \in \mathfrak{o}_k.\dagger$$

Proof.

($\Rightarrow$) If $(\alpha) + \mathfrak{a} = \mathfrak{o}_k$, then $1 = \lambda\alpha + \mu$, $\lambda \in \mathfrak{o}_k$, $\mu \in \mathfrak{a}$, i.e., $\alpha\lambda \equiv 1\ (\mathfrak{a})$.

($\Leftarrow$) If $\alpha x \equiv 1\ (\mathfrak{a})$, then $\alpha x - 1 \in \mathfrak{a}$ and so $1 \in (\alpha) + \mathfrak{a}$, which implies $(\alpha) + \mathfrak{a} = \mathfrak{o}_k$. Q.E.D.

By Proposition 2.29, we see that the group $(\mathfrak{o}_k/\mathfrak{a})^\times$ of invertible elements of the ring $\mathfrak{o}_k/\mathfrak{a}$ consists of residue classes

† For ideals $\mathfrak{a}$ and $\mathfrak{b}$, we put $\mathfrak{a} + \mathfrak{b} = \{a + b;\ a \in \mathfrak{a},\ b \in \mathfrak{b}\}$. This is also an ideal which is frequently written $(\mathfrak{a}, \mathfrak{b})$: $\mathfrak{a} + \mathfrak{b} = (\mathfrak{a}, \mathfrak{b})$. This is the smallest ideal containing $\mathfrak{a}$ and $\mathfrak{b}$. As $\mathfrak{a} \mid \mathfrak{b} \Leftrightarrow \mathfrak{a} \supset \mathfrak{b}$ (Proposition 2.28), $(\mathfrak{a}, \mathfrak{b})$ is the greatest common divisor of $\mathfrak{a}$, and $\mathfrak{b}$. $(\mathfrak{a}, \mathfrak{b}) = \mathfrak{o}_k$ means that $\mathfrak{a}$ and $\mathfrak{b}$ are relatively prime. We write simply $(\mathfrak{a}, \mathfrak{b}) = 1$ if that is so. When $\mathfrak{a}$ is principal, we write $((\alpha), \mathfrak{b}) = (\alpha, \mathfrak{b})$.

$\alpha \bmod \mathfrak{a}$ such that $(\alpha, \mathfrak{a}) = 1$. The order of the group is the Euler function $\varphi_k(\mathfrak{a})$. By Proposition 1.7, we obtain

$$x^{\varphi_k(\mathfrak{a})} \equiv 1 \quad (\mathfrak{a}),$$

where $x \in \mathfrak{o}_k$ and $(x, \mathfrak{a}) = 1$ which is a generalization of the Fermat–Euler theorem (Theorem 1.10). If $\mathfrak{a} = \mathfrak{p}$, a prime ideal, then $\mathfrak{o}_k/\mathfrak{p}$ is a field (Proposition 2.22). Since $\mathbb{Z} \subset \mathfrak{o}_k$, $\mathbb{Z} \cap \mathfrak{p}$ is an ideal of $\mathbb{Z}$, i.e., a module in $\mathbb{Z}$; we have $\mathbb{Z} \cap \mathfrak{p} = p\mathbb{Z}$ with some $p \in \mathbb{Z}, p \geq 0$ (Theorem 1.1). we claim that p is a prime number. In fact, take an α $(\neq 0)$ in $\mathfrak{p}$. So $N_{k/\mathbb{Q}}\alpha$ $(\neq 0) \in \mathbb{Z}$. Since all conjugates $\alpha^{(i)}$ are in $\bar{\mathbb{Z}}$, we have $\alpha^{(2)} \cdots \alpha^{(n)} = \alpha^{-1} N_{k/\mathbb{Q}}\alpha \in \bar{\mathbb{Z}} \cap k = \mathfrak{o}_k$ and $N_{k/\mathbb{Q}}\alpha = \alpha(\alpha^{(2)} \cdots \alpha^{(n)})$ is a nonzero number of $\mathbb{Z} \cap \mathfrak{p}$, i.e., $p > 0$. Now, for $a, b \in \mathbb{N}$, we have

$$p \mid ab \Rightarrow ab \in \mathfrak{p} \Rightarrow a \in \mathfrak{p} \quad \text{or} \quad b \in \mathfrak{p} \Rightarrow a \in p\mathbb{Z}$$
$$\text{or} \quad b \in p\mathbb{Z} \Rightarrow p \mid a \quad \text{or} \quad p \mid b,$$

which proves that p is a prime number.

The field $\mathbb{F}_p = \mathbb{Z}/p\mathbb{Z}$ is naturally mapped into the field $\mathfrak{o}_k/\mathfrak{p}$ by the homomorphism $a \bmod p\mathbb{Z} \mapsto a \bmod \mathfrak{p}$. As the kernel of the homomorphism is $\mathbb{Z} \cap \mathfrak{p} = p\mathbb{Z}$, it is an embedding; in other words, $\mathfrak{o}_k/\mathfrak{p}$ becomes an extension of $\mathbb{F}_p$. The field degree $f = [\mathfrak{o}_k/\mathfrak{p} : \mathbb{F}_p]$ is called the *degree* of $\mathfrak{p}$, written $\deg \mathfrak{p}$. Hence $N_k\mathfrak{p} = p^f$ and $\varphi_k(\mathfrak{p}) = N_k\mathfrak{p} - 1$. The Fermat theorem (Theorem 1.11) for k is

$$\alpha^{N_k\mathfrak{p} - 1} \equiv 1 \quad (\mathfrak{p}), \qquad \alpha \in \mathfrak{o}_k, \qquad (\alpha, \mathfrak{p}) = 1.$$

$(\mathfrak{o}_k/\mathfrak{p})^\times$ is a cyclic group of order $N_k\mathfrak{p} - 1$ (Theorem 1.18) and a number $\alpha \in \mathfrak{o}_k$ such that $\alpha \bmod \mathfrak{p}$ generates $(\mathfrak{o}_k/\mathfrak{p})^\times$ is a primitive root mod $\mathfrak{p}$ (Definition 1.17). As a generalization of Exercise 1.36, we have the following proposition.

PROPOSITION 2.30. *Let* $\mathfrak{a}_1, \ldots, \mathfrak{a}_l$ *be ideals of* $\mathfrak{o}_k$ *such that* $(\mathfrak{a}_i, \mathfrak{a}_j) = 1$, $i \neq j$. *Then, for any* $\gamma_1, \ldots, \gamma_l$ *in* $\mathfrak{o}_k$, *the system*

$$\begin{cases} x \equiv \gamma_1 \quad (\mathfrak{a}_1) \\ \cdots\cdots \\ x \equiv \gamma_l \quad (\mathfrak{a}_l) \end{cases}$$

has a solution in $\mathfrak{o}_k$ *uniquely determined mod* $\mathfrak{a}_1 \cdots \mathfrak{a}_l$.

PROOF. Put $\mathfrak{a}_i^* = \mathfrak{a}_1 \cdots \mathfrak{a}_{i-1}\mathfrak{a}_{i+1} \cdots \mathfrak{a}_l$. Since $(\mathfrak{a}_i, \mathfrak{a}_i^*) = 1$, there are $\lambda_i \in \mathfrak{a}_i$ and $\mu_i \in \mathfrak{a}_i^*$ such that $\lambda_i + \mu_i = 1$. If $j \neq i$, then $\mu_i \in \mathfrak{a}_i^* \subset \mathfrak{a}_j$. We find that $\xi = \gamma_1\mu_1 + \cdots + \gamma_l\mu_l$ is a solution to the system; in fact,

$$\xi = \gamma_1\mu_1 + \cdots + \gamma_i\mu_i + \cdots + \gamma_l\mu_l \equiv \gamma_i\mu_i = \gamma_i(1 - \lambda_i) \equiv \gamma_i(\mathfrak{a}_i).$$

If η is a second solution to the system, then we have $\eta \equiv \xi(\mathfrak{a}_i)$ and so $\eta \equiv \xi(\mathfrak{a}_1 \cdots \mathfrak{a}_l)$, which proves the uniqueness. Q.E.D.

Using Proposition 2.30 one proves immediately the following theorem which is a generalization of Theorem 1.15.

THEOREM 2.16. *For* $\mathfrak{a} = \mathfrak{a}_1 \cdots \mathfrak{a}_l$, $\mathfrak{a}_i \neq \mathfrak{o}_k$, $(\mathfrak{a}_i, \mathfrak{a}_j) = 1$, $i \neq j$, *put* $R = \mathfrak{o}_k/\mathfrak{a}$ *and* $R_i = \mathfrak{o}_k/\mathfrak{a}_i$. *Then we obtain naturally the isomorphisms:*

$$R \approx R_1 \times \cdots \times R_l \ (\text{as rings}), \qquad R^\times \approx R_1^\times \times \cdots \times R_l^\times \ (\text{as groups}).$$

PROPOSITION 2.31. *Let* $\mathfrak{p}$ *be a prime ideal of* $\mathfrak{o}_k$, Γ *be a complete set of representatives of* $\mathfrak{o}_k/\mathfrak{p}$, *and* π *be a number in* $\mathfrak{o}_k$ *such that* $\mathfrak{p} \mid \pi$ *but* $\mathfrak{p}^2 \nmid \pi$. *Then, for a natural number a, the set* $\{\gamma_0 + \gamma_1\pi + \cdots + \gamma_{a-1}\pi^{a-1};\ \gamma_i \in \Gamma\}$ *makes up a complete set of representatives of* $\mathfrak{o}_k/\mathfrak{p}^a$ *and the set*

$$\{\gamma_0 + \gamma_1\pi + \cdots + \gamma_{a-1}\pi^{a-1};\ \gamma_i \in \Gamma,\ \gamma_0 \not\equiv 0(\mathfrak{p})\}$$

makes up a complete set of representatives of $(\mathfrak{o}_k/\mathfrak{p}^a)^\times$. *Therefore we have* $N_k\mathfrak{p}^a = (N_k\mathfrak{p})^a$ *and* $\varphi_k(\mathfrak{p}^a) = (N_k\mathfrak{p} - 1)(N_k\mathfrak{p})^{a-1}$.

PROOF. For an $\alpha \in \mathfrak{o}_k$, there is a unique $\gamma_0 \in \Gamma$ so that $\alpha \equiv \gamma_0 \ (\mathfrak{p})$. Consider next the equation $\pi x \equiv \alpha - \gamma_0 \ (\mathfrak{p}^2)$; since $\alpha - \gamma_0 \in \mathfrak{p} = (\pi, \mathfrak{p}^2)$ there is a $\gamma_1 \in \Gamma$ such that $\pi\gamma_1 \equiv \alpha - \gamma_0 \ (\mathfrak{p}^2)$ *and* since π is not divisible by $\mathfrak{p}^2$, γ_1 is uniquely determined (Exercise 2.20, below). Consider next the equation $\pi^2 x \equiv \alpha - \gamma_0 - \gamma_1\pi \ (\mathfrak{p}^3)$; since $\alpha - \gamma_0 - \gamma_1\pi \in \mathfrak{p}^2 = (\pi^2, \mathfrak{p}^3)$ there is a unique $\gamma_2 \in \Gamma$ such that $\pi^2\gamma_2 \equiv \alpha - \gamma_0 - \gamma_1\pi \ (\mathfrak{p}^3)$ (Exercise 2.20), etc. The assertion on $(\mathfrak{o}_k/\mathfrak{p}^a)^\times$ is obvious. Q.E.D.

EXERCISE 2.20. For $\alpha, \beta \in \mathfrak{o}_k$, we have

$$\alpha x \equiv \beta \ (\mathfrak{a}) \ \text{ has a solution} \Leftrightarrow \beta \in (\alpha, \mathfrak{a}).$$

When that is so, the solution is unique mod $\mathfrak{a}/(\alpha, \mathfrak{a})$.

PROPOSITION 2.32. *For ideals* $\mathfrak{a}$ *and* $\mathfrak{b}$ *in* $\mathfrak{o}_k$, *we have* $N_k(\mathfrak{a}\mathfrak{b}) = N_k\mathfrak{a}N_k\mathfrak{b}$. *If* $\mathfrak{a} = \prod_{i=1}^{l} \mathfrak{p}_i^{e_i}$ $(e_i > 0)$ (*Theorem* 2.15), *then* $\varphi_k(\mathfrak{a}) = N_k\mathfrak{a} \prod_{i=1}^{l} (1 - 1/N_k\mathfrak{p}_i)$.

PROOF. The assertion follows immediately from Theorem 2.16 and Proposition 2.31. Q.E.D.

PROPOSITION 2.33. *For* $\alpha \in \mathfrak{o}_k$, *we have* $N_k(\alpha) = |N_{k/\mathbb{Q}}\alpha|$.

PROOF. Apply Proposition 2.18 to $\mathfrak{o}_k = [\omega_1, \ldots, \omega_n]$ and $(\alpha) = [\alpha\omega_1, \ldots, \alpha\omega_n]$; we get

$$(\alpha\omega_1, \ldots, \alpha\omega_n) = (\omega_1, \ldots, \omega_n)A, \qquad A \in M_n(\mathbb{Z}) \tag{2.9}$$

and $N_k(\alpha) = [\mathfrak{o}_k : (\alpha)] = |\det A|$. Taking conjugates of both sides of (2.9), we have

$$(\alpha^{(i)}\omega_1^{(i)}, \ldots, \alpha^{(i)}\omega_n^{(i)}) = (\omega_1^{(i)}, \ldots, \omega_n^{(i)})A, \qquad 1 \le i \le n,$$

or, in terms of matrices, we have

$$\begin{pmatrix} \alpha^{(1)} & & \\ & \ddots & \\ & & \alpha^{(n)} \end{pmatrix} \begin{pmatrix} \omega_1^{(1)} & \cdots & \omega_n^{(1)} \\ \vdots & & \vdots \\ \omega_1^{(n)} & \cdots & \omega_n^{(n)} \end{pmatrix} = \begin{pmatrix} \omega_1^{(1)} & \cdots & \omega_n^{(1)} \\ \vdots & & \vdots \\ \omega_1^{(n)} & \cdots & \omega_n^{(n)} \end{pmatrix} A. \tag{2.10}$$

Since $\det(\omega_j^{(i)}) \ne 0$, by taking the determinants of both sides of (2.10), we get $\alpha^{(1)} \cdots \alpha^{(n)} = N_{k/\mathbb{Q}}\alpha = \det A$, and hence $N_k(\alpha) = |N_{k/\mathbb{Q}}\alpha|$. Q.E.D.

2.12. Decomposition of Primes in Number Fields

Let k be an algebraic number field, $n = [k : \mathbb{Q}]$, and $\mathfrak{o}_k$ be the ring of integers of k. Fix a prime p in $\mathbb{Q}$. $\mathfrak{o}_k p$ is an ideal in $\mathfrak{o}_k$; We denote it by $(p)_k$ or (p), or simply by p. By the fundamental theorem of ideal theory (Theorem 2.15), we have the decomposition into prime ideals

$$(p) = \mathfrak{p}_1^{e_1} \cdots \mathfrak{p}_g^{e_g} \qquad (e_i > 0, \ \mathfrak{p}_i \ne \mathfrak{p}_j, \ i \ne j). \tag{2.11}$$

Call f_i the degree of $\mathfrak{p}_i$: $f_i \overset{\text{def}}{=} [\mathfrak{o}_k/\mathfrak{p}_i : \mathbb{F}_p]$. By definition, we have $N_k\mathfrak{p}_i \overset{\text{def}}{=} [\mathfrak{o}_k/\mathfrak{p}_i] = p^{f_i}$. Taking the norm of both sides of (2.11), we have $N_k(p) = |N_{k/\mathbb{Q}}p| = p^n$ (Proposition 2.33) and

$$N_k(\mathfrak{p}_1^{e_1}\cdots\mathfrak{p}_g^{e_g}) = (N_k\mathfrak{p}_1)^{e_1}\cdots(N_k\mathfrak{p}_g)^{e_g} = p^{f_1e_1}\cdots p^{f_ge_g} = p^{e_1f_1+\cdots+e_gf_g}.$$

Therefore we get the following important relation:

$$n = [k:\mathbb{Q}] = e_1f_1 + \cdots + e_gf_g. \tag{2.12}$$

Definition 2.13. e_i is the *ramification index* of $\mathfrak{p}_i$. When $e_i > 1$, $\mathfrak{p}_i$ is *ramified* and when $e_i = 1$, $\mathfrak{p}_i$ is *unramified.* For a prime p, p is *ramified* in k if some $e_i > 1$, i.e., if some prime factor $\mathfrak{p}_i$ is ramified; otherwise, i.e., if all $e_i = 1$, p is *unramified* in k.

Let us pass to the relative extension: K/k. Let $[k:\mathbb{Q}] = n$, $[K, \mathbb{Q}] = m$, and $[K:k] = t$; $m = nt$. Let $\mathfrak{o}_k$ and $\mathfrak{o}_K$ be the ring of integers in k and K, respectively. The lift $\mathfrak{o}_K\mathfrak{a}$ of an ideal $\mathfrak{a}$ of $\mathfrak{o}_k$ *is defined by*

$$\mathfrak{o}_K\mathfrak{a} = \Big(\sum_i \lambda_i\alpha_i;\ \lambda_i \in \mathfrak{o}_K,\ \alpha_i \in \mathfrak{a}\Big).$$

We denote $\mathfrak{o}_K\mathfrak{a}$ by $(\mathfrak{a})_K$ or simply by $\mathfrak{a}$. One verifies easily that $(\mathfrak{a}\mathfrak{b})_K = (\mathfrak{a})_K(\mathfrak{b})_K$.

Proposition 2.34. $N_K(\mathfrak{a})_K = (N_k\mathfrak{a})^{[K:k]}$.

Proof. By Proposition 2.32 and by the multiplicativity of $\mathfrak{a} \mapsto (\mathfrak{a})_K$ it is enough to consider the case $\mathfrak{a} = \mathfrak{p}$, a prime ideal of k. By Exercise 2.21 below, there is an ideal $\mathfrak{c}$ in $\mathfrak{o}_k$ such that $(\mathfrak{c}, (p)_k) = 1$ and $\mathfrak{p}\mathfrak{c} = (\lambda)$ where p is the prime in $\mathbb{Q}$ such that $\mathfrak{p} \mid p$. Then, Proposition 2.33 and Exercise 2.6 imply that

$$\begin{aligned} N_K(\mathfrak{p})_K N_K(\mathfrak{c})_K &= N_K(\lambda)_K = |N_{K/\mathbb{Q}}\lambda| = |N_{k/\mathbb{Q}}(N_{K/k}\lambda)| = |N_{k/\mathbb{Q}}\lambda^t| \\ &= |N_{k/\mathbb{Q}}\lambda|^t = N_k(\lambda)^t = (N_k\mathfrak{p})^t(N_k\mathfrak{c})^t. \end{aligned}$$

Since $N_K(\mathfrak{p})_K$ and $N_k\mathfrak{p}$ are powers of p,† $((\mathfrak{c})_K, (p)_K) = 1$ and $N_K(\mathfrak{c})_K$, $N_k\mathfrak{c}$ are prime to p, we get $N_K(\mathfrak{p})_K = (N_k\mathfrak{p})^t$ by comparing the p-power part of both sides of the equality $N_K(\mathfrak{p})_K N_K(\mathfrak{c})_K = (N_K\mathfrak{p})^t(N_k\mathfrak{c})^t$. Q.E.D.

† Verify that $\mathfrak{o}_K/(\mathfrak{p})_K$ is a vector space over the field $\mathfrak{o}_k/\mathfrak{p}$.

EXERCISE 2.21. Given ideals $\mathfrak{a}$ and $\mathfrak{b}$ in $\mathfrak{o}_k$, there exists an ideal $\mathfrak{c}$ in $\mathfrak{o}_k$ such that $(\mathfrak{b}, \mathfrak{c}) = 1$ and $\mathfrak{a}\mathfrak{c} \sim 1$.

Let us generalize formula (2.12) for $k/\mathbb{Q}$ to the case of any extension K/k. For a prime p in $\mathbb{Q}$, denote by $\mathfrak{p}$ and $\mathfrak{P}$ prime ideals in $\mathfrak{o}_k$ and $\mathfrak{o}_K$, respectively, such that $\mathfrak{p} \mid p$ and $\mathfrak{P} \mid \mathfrak{p}$. Call e the ramification index of $\mathfrak{p}$ for $k/\mathbb{Q}$, e^* the ramification index of $\mathfrak{P}$ for $K/\mathbb{Q}$. Furthermore, let e' be the ramification index of $\mathfrak{P}$ for K/k, i.e., the largest nonnegative exponent e' such that $\mathfrak{P}^{e'} \mid (\mathfrak{p})_K$.

$$\begin{cases} (p)_k = \cdots \mathfrak{p}^{e} \cdots \\ (p)_K = \cdots \mathfrak{P}^{e^*} \cdots \\ (\mathfrak{p})_K = \cdots \mathfrak{P}^{e'} \cdots \end{cases} \tag{2.13}$$

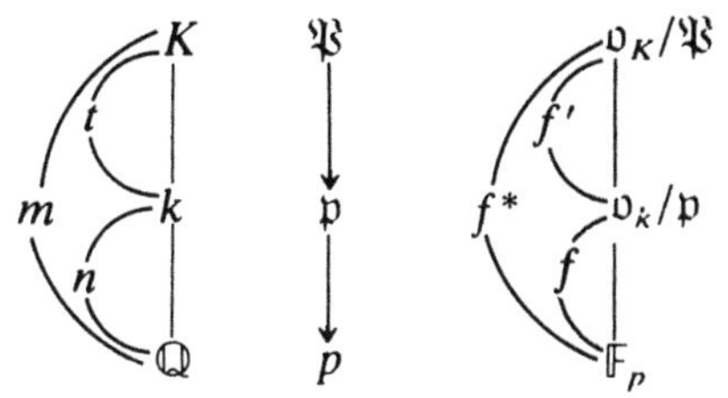

From (2.13), we obtain the relation

$$e^* = ee'.\dagger \tag{2.14}$$

As for degrees, we have $f = [\mathfrak{o}_k/\mathfrak{p} \colon \mathbb{F}_p]$ and $f^* = [\mathfrak{o}_K/\mathfrak{P} \colon \mathbb{F}_p] = [\mathfrak{o}_K/\mathfrak{P} \colon \mathfrak{o}_k/\mathfrak{p}][\mathfrak{o}_k/\mathfrak{p} \colon \mathbb{F}_p] = f'f$, where $f' = [\mathfrak{o}_K/\mathfrak{P} \colon \mathfrak{o}_k/\mathfrak{p}]$. Therefore, we obtain the relation

$$f^* = ff'. \tag{2.15}$$

Decompose $\mathfrak{p}$ in K:

$$(\mathfrak{p})_K = \mathfrak{P}_1^{e'_1} \cdots \mathfrak{P}_{g'}^{e'_{g'}}. \tag{2.16}$$

Taking the norm on both sides of (2.16), we have $N_K(\mathfrak{p})_K = (N_k\mathfrak{p})^t = p^{ft}$ and $N_K(\mathfrak{p})_K = (N_K\mathfrak{P}_1)^{e'_1} \cdots (N_K\mathfrak{P}_{g'})^{e'_{g'}} = p^{e'_1 f_1^* + \cdots + e'_{g'} f_{g'}^*}$. Therefore, using (2.15), we get

$$t = [K \colon k] = e'_1 f'_1 + \cdots e'_{g'} f'_{g'}. \tag{2.17}$$

We shall close this section with a theorem due to Kummer (1810–1893) which is not only important theoretically but also useful practically.

† e' is called the relative ramification index of $\mathfrak{P}$ for K/k.

Theorem 2.17. *Let $k=\mathbb{Q}(\theta)$, $\theta\in\mathfrak{o}_k$ and $f_\theta(X)\in\mathbb{Z}[X]$ be the minimal polynomial of θ. Assume that $\mathfrak{o}_k=\mathbb{Z}[\theta]$. Let*

$$f_\theta(X)\equiv\varphi_1(X)^{e_1}\cdots\varphi_g(X)^{e_g}\ (\operatorname{mod} p) \tag{2.18}$$

be the decomposition of $f_\theta(X)\bmod p$ where each $\varphi_i(X)\in\mathbb{Z}[X]$ is normalized (i.e., the highest coefficient is one) and irreducible $\operatorname{mod} p$. *Then, $\mathfrak{p}_i=(p,\varphi_i(\theta))$ is a prime ideal and we have*

$$(p)_k=\mathfrak{p}_1^{e_1}\cdots\mathfrak{p}_g^{e_g},\qquad \deg\mathfrak{p}_i=\deg\varphi_i,\qquad 1\le i\le g.\dagger$$

Proof. In the proof we put $f=f_\theta$. We denote by $\bar{h}(X)$ the polynomial in $\mathbb{F}_p[X]$ obtained from a polynomial $h(X)$ in $\mathbb{Z}[X]$ by reducing its coefficients $\operatorname{mod} p$. For $\varphi_i(X)$ in (2.18), $\bar{\varphi}_i\in\mathbb{F}_p[X]$ is irreducible by the assumption; if we call η a root of $\bar{\varphi}_i$, then $\bar{\varphi}_i$ becomes the minimal polynomial of η.‡ We shall consider a mapping

$$\rho_i\colon \mathfrak{o}_k=\mathbb{Z}[\theta]\to\mathbb{F}_p[\eta]=\mathbb{F}_p(\eta)$$

defined by $\rho_i(P(\theta))=\bar{P}(\eta)$. To see that ρ_i is well defined, we must verify that $P(\theta)=0$ implies $\bar{P}(\eta)=0$. So suppose $P(\theta)=0$. Since $f(X)$ is the minimal polynomial of θ over $\mathbb{Q}$, we have $P(X)=f(X)h(X)$, $h\in\mathbb{Z}[X]$. (Since $f(X)$ is normalized the division is performed in $\mathbb{Z}[X]$.) Therefore $\bar{P}(X)=\bar{f}(X)\bar{h}(X)$ and hence $\bar{P}(\eta)=\bar{f}(\eta)\bar{h}(\eta)$. Since $\bar{\varphi}_i\mid\bar{f}$, we have $\bar{f}(\eta)=0$ and $\bar{P}(\eta)=0$ which proves our assertion. Obviously, ρ_i is a surjective homomorphism of the ring $\mathfrak{o}_k$ to the field $\mathbb{F}_p(\eta)$. If we put $\mathfrak{p}_i=\operatorname{Ker}\rho_i$, then ρ_i induces the isomorphism

$$\rho_i^*\colon \mathfrak{o}_k/\mathfrak{p}_i\xrightarrow{\sim}\mathbb{F}_p(\eta)\approx\mathbb{F}_p[X]/(\bar{\varphi}_i). \tag{2.19}$$

Since $\mathfrak{o}_k/\mathfrak{p}_i$ is a field by (2.19), $\mathfrak{p}_i$ is prime; we also find that $\deg\mathfrak{p}_i=\deg\varphi_i$ by (2.19). Next, we shall prove that

$$\mathfrak{p}_i=(p,\varphi_i(\theta)). \tag{2.20}$$

Since $\rho_i(p)=\bar{p}=0$, $\rho_i(\varphi_i(\theta))=\bar{\varphi}_i(\eta)=0$, we have $(p,\varphi_i(\theta))\subset$

† We say that $\varphi_i\in\mathbb{Z}[X]$ is irreducible $\operatorname{mod} p$ if $\bar{\varphi}_i\in\mathbb{F}_p[X]$ obtained from φ_i by reducing the coefficients $\operatorname{mod} p$ is irreducible.

‡ Since $\mathbb{F}_p[X]/(\varphi_i)\approx\mathbb{F}_p[X\bmod\varphi_i]=\mathbb{F}_p(X\bmod\varphi_i)$ one can put $\eta=X\bmod\varphi_i$.

$\mathfrak{p}_i$. Conversely, take any $g(\theta) \in \mathfrak{p}_i$; $g(X) \in \mathbb{Z}[X]$. As $\bar{g}(\eta) = 0$ and $\bar{\varphi}_i$ is the minimal polynomial of η over $\mathbb{F}_p$, we have $\bar{\varphi}_i \mid \bar{g}$, i.e., $\bar{g} = \bar{\varphi}_i \bar{h}$ with $h \in \mathbb{Z}[X]$; therefore all coefficients of $g - \varphi_i h \in \mathbb{Z}[X]$ are divisible by p and hence $g(X) = pq(X) + \varphi_i(X)h(X)$, $q(X) \in \mathbb{Z}[X]$. Substituting θ in X, we have $g(\theta) = pq(\theta) + \varphi_i(\theta)h(\theta) \in (p, \varphi_i(\theta))$, which proves that $\mathfrak{p}_i \subset (p, \varphi_i(\theta))$ and (2.20) is proved. It remains to prove the decomposition $(p)_k = \mathfrak{p}_1^{e_1} \cdots \mathfrak{p}_g^{e_g}$. First of all, for $\mathfrak{p}_i$ in (2.20), we have $\mathfrak{p}_i \neq \mathfrak{p}_j$, if $i \neq j$. In fact, if $\mathfrak{p}_i = \mathfrak{p}_j$, then $\bar{\varphi}_j(\eta) = 0$ because $\varphi_j(\theta) \in \mathfrak{p}_i$. But, since $\bar{\varphi}_i(\eta) = 0$ and $\bar{\varphi}_i$, $\bar{\varphi}_j$ are both irreducible, we must have $\bar{\varphi}_i = \bar{\varphi}_j$ which is impossible because $\bar{\varphi}_i$ and $\bar{\varphi}_j$ are distinct prime factors of $\bar{f}$. Next, we shall prove that $(p)_k$ is not divisible by prime ideals other than $\mathfrak{p}_i$. In fact, if $\mathfrak{p} \mid (p)_k$, then, since $\mathbb{Z}[\theta]/\mathfrak{p}$ is a finite field, we can write $\mathbb{Z}[\theta]/\mathfrak{p} = \mathbb{F}_p[\bar{\theta}]$ with $\bar{\theta} = \theta \bmod \mathfrak{p}$. As $f(\theta) = 0$, we have $\bar{f}(\bar{\theta}) = 0$ and so $\bar{\varphi}_i(\bar{\theta}) = 0$ for some i: this means that $\varphi_i(\theta) \in \mathfrak{p}$ and hence $\mathfrak{p}_i = (p, \varphi_i(\theta)) \subset \mathfrak{p}$ which implies that $\mathfrak{p} = \mathfrak{p}_i$. Thus we have

$$(p)_k = \mathfrak{p}_1^{\epsilon_1} \cdots \mathfrak{p}_g^{\epsilon_g},\dagger \tag{2.21}$$

and we are reduced to prove that $\epsilon_i = e_i$ for all i. Now, we have $f(X) - \varphi_1(X)^{e_1} \cdots \varphi_g(X)^{e_g} = ph(X)$ by (2.18), and since $f(\theta) = 0$ it follows that

$$\varphi_1(\theta)^{e_1} \cdots \varphi_g(\theta)^{e_g} \in (p)_k. \tag{2.22}$$

Since (2.20) implies that $\mathfrak{p}_i^{e_i} \subset (p, \varphi_i(\theta)^{e_i})$, it follows from (2.21) and (2.22) that

$$\mathfrak{p}_1^{e_1} \cdots \mathfrak{p}_g^{e_g} \subset (p, \varphi_1(\theta)^{e_1} \cdots \varphi_g(\theta)^{e_g}) \subset (p)_k = \mathfrak{p}_1^{\epsilon_1} \cdots \mathfrak{p}_g^{\epsilon_g}$$

and that $e_i \geq \epsilon_i$, $1 \leq i \leq g$. On the other hand, taking the norm on both sides of (2.21) and comparing the exponents of p, we obtain, with $N_k \mathfrak{p}_i = p^{f_i}$,

$$\epsilon_1 f_1 + \cdots \epsilon_g f_g = n = [k : \mathbb{Q}] = e_1 f_1 + \cdots + e_g f_g$$

which implies $e_i = \epsilon_i$ as we wanted. Q.E.D.

Exercise 2.22. Let $l \neq 2$ be a prime, $\zeta = e^{2\pi i/l}$ and $k = \mathbb{Q}(\zeta)$, the lth cyclotomic field. Prove that any prime $p \neq l$ is unramified for $k/\mathbb{Q}$. When $p = l$, prove that $(l)_k = \mathfrak{l}^{l-1}$, where $\mathfrak{l} = (\zeta - 1)$, a prime ideal in k. (l is *totally ramified* in k.)

† Since $\mathfrak{p}_i \ni p$ we have $\epsilon_i > 0$.

Exercise 2.23. Let $k=\mathbb{Q}(\sqrt{m})$ where m is a square free integer. Let Δ_k be the discriminant of k (Exercise 2.17):

$$\Delta_k=\begin{cases}4m, & m\equiv 2,3 \quad (4),\\ m, & m\equiv 1 \quad (4),\end{cases}$$

Then a prime p in $\mathbb{Q}$ decomposes in k as follows:

(1) $m\equiv 2,3\,(4)$, $\Delta_k=4m$.

$$p\mid\Delta_k\Rightarrow(p)=\mathfrak{p}^2,\ N\mathfrak{p}=p,$$

$$p\nmid\Delta_k\begin{cases}\lambda_p(m)=1\Rightarrow(p)=\mathfrak{p}\mathfrak{p}',\ (\mathfrak{p}\neq\mathfrak{p}'),\ N\mathfrak{p}=N\mathfrak{p}'=p,\\ \lambda_p(m)=-1\Rightarrow(p)=\mathfrak{p},\ N\mathfrak{p}=p^2,\end{cases}$$

(2) $m\equiv 1\quad(4)$, $\Delta_k=m$.

$$p\mid\Delta_k\Rightarrow(p)=\mathfrak{p}^2,\ N\mathfrak{p}=p,$$

$$p\nmid\Delta_k\begin{cases}p\neq 2\begin{cases}\lambda_p(m)=1\Rightarrow(p)=\mathfrak{p}\mathfrak{p}',\ (\mathfrak{p}\neq\mathfrak{p}'),\ N\mathfrak{p}=N\mathfrak{p}'=p,\\ \lambda_p(m)=-1\Rightarrow(p)=\mathfrak{p},\ N\mathfrak{p}=p^2,\end{cases}\\ p=2\begin{cases}m\equiv 1\quad(8)\Rightarrow(2)=\mathfrak{p}\mathfrak{p}',\ \mathfrak{p}\neq\mathfrak{p}',\ N\mathfrak{p}=N\mathfrak{p}'=2,\\ m\equiv 5\quad(8)\Rightarrow(2)=\mathfrak{p},\ N\mathfrak{p}=2^2.\end{cases}\end{cases}$$

Remark 2.14. From Exercise 2.23 we find

$$p \text{ is ramified for } k/\mathbb{Q}\Leftrightarrow p\mid\Delta_k.$$

Generators of $\mathfrak{p}$ and $\mathfrak{p}'$ in Exercise 2.23 are given by Theorem 2.17. This will be used in Section 4.5 to determine the structure of the ideal class group H_k of quadratic field k.

2.13. Discriminant and Ramification

When a vector space A over a field F has a structure of a ring and satisfies the conditions

$$a(xy)=(ax)y=x(ay),\qquad a\in F,\qquad x,y\in A,$$

A is called an *algebra* over F.† The dimension of the vector space A over F is called the *degree* of the algebra A over F. For example,

† In this section, we only need commutative ring A.

an extension K of F is an algebra over F and its degree is the degree $[K: F]$. Suppose that an algebra A has degree n over F and let $\{u_1, \ldots, u_n\}$ be basis of the vector space A over F. For any element $\alpha \in A$ a matrix $M(\alpha) \in M_n(F)$ is determined by

$$(\alpha u_1, \ldots, \alpha u_n) = (u_1, \ldots, u_n)M(\alpha).$$

We put

$$t_A(\alpha) \overset{\text{def}}{=} \operatorname{tr} M(\alpha), \qquad d_A(u_1, \ldots, u_n) \overset{\text{def}}{=} \det(t_A(u_i u_j)).$$

EXERCISE 2.24. If $\{u_1', \ldots, u_n'\}$ is another basis of A, there is a matrix $T \in \mathrm{GL}_n(F)$ such that $(u_1', \ldots, u_n') = (u_1, \ldots, u_n)T$. Prove that $d_A(u_1', \ldots, u_n') = (u_1, \ldots, u_n)(\det T)^2$.

EXERCISE 2.25. Let $F = \mathbb{Q}$ and $A = k$, a number field. Prove that $d_k(u_1, \ldots, u_n) = d_{k/\mathbb{Q}}(u_1, \ldots, u_n)$ where the latter was defined in Section 2.3 (after Exercise 2.6).

PROPOSITION 2.35. *Let A and B be algebras over F of degree n and m, respectively. Then $C = A \times B$ becomes naturally an algebra over F of degree $n + m$. For simplicity, we identify an element $a \in A$ with the element $(a, 0) \in C$ and an element $b \in B$ with the element $(0, b) \in C$. If $\{u_1, \ldots, u_n\}$ is a basis of A and $\{v_1, \ldots, v_m\}$ is a basis of B, then $\{u_1, \ldots, u_n, v_1, \ldots, v_m\}$ is a basis of C. In this situation, we have*

$$d_C(u_1, \ldots, u_n, v_1, \ldots, v_m) = d_A(u_1, \ldots, u_n)d_B(v_1, \ldots, v_m).$$

PROOF. In C, the product of $a \in A$ and $b \in B$ is 0 because $ab = (a, 0)(0, b) = (0, 0) = 0$. From this, we see that $t_A(a) = t_C(a)$, $a \in A \subset C$ and $t_B(b) = t_C(b)$, $b \in B \subset C$. Therefore, we have

$$d_C(u_1, \cdots, u_n, v_1, \ldots, v_m) = \begin{vmatrix} (t_C(u_i u_j)) & 0 \\ 0 & (t_C(v_k v_l)) \end{vmatrix}$$

$$= \begin{vmatrix} (t_A(u_i u_j)) & 0 \\ 0 & (t_B(v_k v_l)) \end{vmatrix} = \det(t_A(u_i u_j)) \det(t_B(v_k v_l))$$

$$= d_A(u_1, \ldots, u_n)\, d_B(v_1, \ldots, v_m).$$ Q.E.D.

EXERCISE 2.26. Let k be a number field, $n = [k: \mathbb{Q}]$, $\mathfrak{o}_k = [\omega_1, \ldots, \omega_n]$, and Δ_k the discriminant of k. For a prime p in $\mathbb{Q}$ and $\alpha \in \mathfrak{o}_k$, put $\bar{\alpha} = \alpha \bmod (p) \in \mathfrak{o}_k/(p)$. Let $A = \mathfrak{o}_k/(p)$, which is an algebra over $\mathbb{F}_p = \mathbb{Z}/p\mathbb{Z}$. In this situation, prove that $d_A(\bar{\omega}_1, \ldots, \bar{\omega}_n) = \bar{\Delta}_k$.

Proposition 2.36. *Let $k/\mathbb{Q}$ be a finite extension, p a prime in $\mathbb{Q}$, $\mathfrak{p}$ a prime ideal in k such that $\mathfrak{p} \mid (p)$. Call e the highest exponent such that $\mathfrak{p}^e \mid (p)$ and f the degree of $\mathfrak{p}$: $N\mathfrak{p} = p^f$. Then the algebra $A = \mathfrak{o}_k/\mathfrak{p}^e$ over $\mathbb{F}_p$ has degree ef. If $e \geq 2$, then, for any basis $\{\eta_1, \ldots, \eta_{ef}\}$ of A over $\mathbb{F}_p$, we have $d_A(\eta_1, \ldots, \eta_{ef}) = 0 \in \mathbb{F}_p$.*

Proof. For $\alpha \in \mathfrak{o}_k$, put $\bar{\alpha} = \alpha \bmod \mathfrak{p}^e$. Since $e \geq 2$, there is a number $\pi \in \mathfrak{p}$ such that $\pi \notin \mathfrak{p}^e$, i.e., $\bar{\pi} \neq 0$ and $\bar{\pi}^e = 0$. Since the property $d_A(\eta_1, \ldots, \eta_{ef}) = 0$ does not depend on the choice of basis η_i's, we can choose a basis with $\eta_1 = \bar{\pi}$. Then we have $(\bar{\pi}\eta_j) = 0$, $1 \leq j \leq ef$, and, if we define matrices $M(\bar{\pi}\eta_j)$ by the relation

$$(\bar{\pi}\eta_j)(\eta_1, \ldots, \eta_{ef}) = (\eta_1, \ldots, \eta_{ef})M(\bar{\pi}\eta_j),$$

we get $M(\bar{\pi}\eta_j)^e = 0$; it then follows that all eigenvalues of matrices $M(\bar{\pi}\eta_j) = M(\eta_1\eta_j)$ are 0 and hence $t_A(\eta_1\eta_j) = 0$, or $d_A(\eta_1, \ldots, \eta_{ef}) = 0$. Q.E.D.

Theorem 2.18. *Let $k/\mathbb{Q}$ be a finite algebraic extension. If a prime p does not divide Δ_k, then p is unramified in k.*

Proof. Let $(p) = \mathfrak{p}_1^{e_1} \cdots \mathfrak{p}_g^{e_g}$, $N\mathfrak{p}_i = p^{f_i}$, $1 \leq i \leq g$, $A = \mathfrak{o}_k/(p)$ and $A_i = \mathfrak{o}_k/\mathfrak{p}_i^{e_i}$. One sees easily that $A \approx A_1 \times \cdots \times A_g$ (Theorem 2.16). Let $\{\eta_{i1}, \ldots, \eta_{ie_if_i}\}$ be a basis of A_i and let

$$\{\cdots; \eta_{i1}, \ldots, \eta_{ie_if_i}; \cdots\}, \qquad 1 \leq i \leq g,$$

be the basis for A obtained by taking those bases of A_i jointly. By Proposition 2.35, we have

$$d_A(\cdots; \eta_{i1}, \ldots, \eta_{ie_if_i}; \cdots) = \prod_{i=1}^{g} d_{A_i}(\eta_{i1}, \ldots, \eta_{ie_if_i}).$$

Now, if p is ramified in k, then there is at least one e_i such that $e_i \geq 2$ and so $d_{A_i}(\eta_{i1}, \ldots, \eta_{ie_if_i}) = 0$ (Proposition 2.36), hence we have $d_A(\cdots; \eta_{i1}, \ldots, \eta_{ie_if_i}; \cdots) = 0 = d_A(\bar{\omega}_1, \ldots, \bar{\omega}_n) = \bar{\Delta}_k$ (Exercise 2.24 and Exercise 2.26) which means that $p \mid \Delta_k$. Q.E.D.

Remark 2.15. In fact, the converse of Theorem 2.18 is true. Namely, we have the equivalence (due to Dedekind)

$$p \mid \Delta_k \Leftrightarrow p \text{ is ramified in } k. \qquad (\#)$$

For two reasons, we omit the proof of the converse. First of all, in order to define the Artin map (Section 2.15), Theorem 2.18 is good enough. Second, in this book, it is only the cases of quadratic fields and cyclotomic fields where we go into depth beyond the general theory and in these cases the statement (#) is already verified (Exercise 2.22 and Exercise 2.23) without even using Theorem 2.18.

2.14. Hilbert Theory

Throughout this section K/k will be a finite Galois extension of algebraic number fields; we denote by G the Galois group: $G = G(K/k)$. Fix a prime ideal $\mathfrak{p}$ in k and consider the finite set of prime ideals in $\mathfrak{o}_K$:

$$P_\mathfrak{p} = \{\mathfrak{P}; \mathfrak{P} \mid \mathfrak{p}\} = \{\mathfrak{P}; \mathfrak{p} = \mathfrak{P} \cap \mathfrak{o}_k\}.\dagger$$

Since $\mathfrak{P}^\sigma$, $\sigma \in G$, also divides $\mathfrak{p}$, G acts on the set $P_\mathfrak{p}$. In view of the exponential notation, we use our convention: $\mathfrak{P}^{\sigma\tau} = (\mathfrak{P}^\tau)^\sigma$, $\sigma, \tau \in G$.

PROPOSITION 2.37. *The group G acts transitively on $P_\mathfrak{p}$.*

PROOF. We must show that for any $\mathfrak{P}_0, \mathfrak{P} \in P_\mathfrak{p}$ there is a $\sigma \in G$ such that $\mathfrak{P}_0^\sigma = \mathfrak{P}$. Suppose, on the contrary, that $\mathfrak{P}_0^\sigma \neq \mathfrak{P}_1$ for any $\sigma \in G$. Then, by Proposition 2.30, there is an $\alpha \in \mathfrak{o}_K$ such that $\alpha \equiv 0 \quad (\mathfrak{P}_1)$ and $\alpha \equiv 1 \quad (\mathfrak{P}_0^\sigma)$, $\sigma \in G$. Therefore, we have $\alpha \in \mathfrak{P}_1$ but $\alpha \notin \mathfrak{P}_0^\sigma$ for all $\sigma \in G$. However, since

$$N_{K/k}\alpha = \prod \alpha^\sigma \in \mathfrak{P}_1 \cap \mathfrak{o}_k = \mathfrak{p} \subset \mathfrak{P}_0,$$

we have $\mathfrak{P}_0 \mid \prod_\sigma \alpha^\sigma$ and so, for some σ_0, $\mathfrak{P}_0 \mid \alpha^{\sigma_0}$, i.e., $\alpha \in \mathfrak{P}_0^{\sigma\sigma_0^{-1}}$, a contradiction, Q.E.D.

Fixing a prime ideal $\mathfrak{P}$ in K such that $\mathfrak{P} \mid \mathfrak{p}$, define a subgroup $G(\mathfrak{P})$ of G by

$$G(\mathfrak{P}) \overset{\text{def}}{=} \{\sigma \in G; \mathfrak{P}^\sigma = \mathfrak{P}\}.$$

$G(\mathfrak{P})$ is the *decomposition group* of $\mathfrak{P}$ for K/k. As G acts

† $\mathfrak{P} \mid p \Leftrightarrow \mathfrak{P} \supset \mathfrak{p} \Leftrightarrow \mathfrak{P} \cap \mathfrak{o}_k = \mathfrak{p}$ (note that $\mathfrak{p}$ is a maximal ideal in $\mathfrak{o}_k$).

transitively on $P_{\mathfrak{p}}$, the correspondence $\sigma G(\mathfrak{P}) \leftrightarrow \mathfrak{P}^\sigma$ gives the bijection

$$G/G(\mathfrak{P}) \approx P_{\mathfrak{p}}.$$

Hence, we have $g = [P_{\mathfrak{p}}] = [G: G(\mathfrak{P})]$. Let $\sigma_1, \ldots, \sigma_g$ be a complete set of representatives of left cosets in $G/G(\mathfrak{P})$ with $\sigma_1 = 1 = 1_G$. Then we can write the decomposition of $\mathfrak{p}$ as $\mathfrak{p} = (\mathfrak{P}^{\sigma_1})^{e_1} \cdots (\mathfrak{P}^{\sigma_g})^{e_g}$. For each $\sigma \in G$, there is $\tau \in G(\mathfrak{P})$ such that $\sigma\sigma_1 = \sigma_i \tau$ and we get

$$\mathfrak{p}^\sigma = (\mathfrak{P}^{\sigma\sigma_1})^{e_1} \cdots (\mathfrak{P}^{\sigma\sigma_g})^{e_g} = (\mathfrak{P}^{\sigma_i})^{e_1} \cdots.$$

As $\mathfrak{p}^\sigma = \mathfrak{p}$, it follows from Theorem 2.15 that $e_i = e_1$. Conversely, since, for any i, $1 \le i \le g$, there is a σ such that $\sigma\sigma_1 \in \sigma_i G(\mathfrak{P})$, we have $e_1 = \cdots = e_g \overset{\text{def}}{=} e$. Therefore, the decomposition takes the form of

$$\mathfrak{p} = (\mathfrak{P}^{\sigma_1} \cdots \mathfrak{P}^{\sigma_g})^e.$$

As for the residue fields, there is a natural bijection

$$\mathfrak{o}_K/\mathfrak{P} \underset{\mathfrak{o}_k/\mathfrak{p}}{\approx} \mathfrak{o}_K/\mathfrak{P}^\sigma, \qquad \sigma \in G,$$

and so the relative degree $[\mathfrak{o}_K/\mathfrak{P}^\sigma : \mathfrak{o}_k/\mathfrak{p}]$ of $\mathfrak{P}^\sigma$ does not depend on $\sigma \in G$. If we denote this number by f, i.e., $f = [\mathfrak{o}_K/\mathfrak{P} : \mathfrak{o}_k/\mathfrak{p}]$, then formula (2.17) in Section 2.12 becomes, for a Galois extension K/k,

$$n = [K: k] = efg.$$

If $\sigma \in G(\mathfrak{P})$, then the above bijection $\mathfrak{o}_K/\mathfrak{P} \approx \mathfrak{o}_K/\mathfrak{P}^\sigma$ over $\mathfrak{o}_k/\mathfrak{p}$ becomes an automorphism of the field extension $(\mathfrak{o}_K/\mathfrak{P})/(\mathfrak{o}_k/\mathfrak{p})$ which we denote by $\bar{\sigma}$. In other words, the automorphism is described as

$$(\alpha \bmod \mathfrak{P})^{\bar{\sigma}} = \alpha^\sigma \bmod \mathfrak{P}, \qquad \bar{\sigma} \in \bar{G} = G((\mathfrak{o}_K/\mathfrak{P})/(\mathfrak{o}_k/\mathfrak{p})).$$

If we put $N_k\mathfrak{p} = q$ and $N_K\mathfrak{P} = q^f$, then $\mathfrak{o}_k/\mathfrak{p} = \mathbb{F}_q$, $\mathfrak{o}_K/\mathfrak{P} = \mathbb{F}_{q^f}$, and $\bar{G}$ is a cyclic group of order f generated by the Frobenius automorphism σ_F defined by

$$\sigma_F x = x^q, \qquad x \in \mathbb{F}_{q^f} \qquad \text{(see Theorem 2.10)}.$$

It is clear that the mapping $\sigma \mapsto \bar{\sigma}$ is a homomorphism $G(\mathfrak{P}) \to \bar{G}$.

PROPOSITION 2.38. *The group homomorphism* $G(\mathfrak{P}) \to \bar{G}$ *is surjective.*

PROOF. Since $\bar{G}$ is a cyclic group generated by σ_F, it is enough to find an element $\tau \in G(\mathfrak{P})$ such that $\sigma_F = \bar{\tau}$. Let now ρ be a number in $\mathfrak{o}_K$ which is a primitive root modulo $\mathfrak{P}$, i.e., $\bar{\rho} = \rho \bmod \mathfrak{P}$ generates the cyclic group $(\mathfrak{o}_K/\mathfrak{P})^\times$ (Theorem 1.18). Since $\mathfrak{o}_K/\mathfrak{P}$ consists of 0 and powers of $\bar{\rho}$, $\bar{\rho}$ is a primitive element for the field extension $(\mathfrak{o}_K/\mathfrak{P})/(\mathfrak{o}_k/\mathfrak{p})$, i.e., $\mathfrak{o}_K/\mathfrak{P} = (\mathfrak{o}_k/\mathfrak{p})[\bar{\rho}]$. Therefore our problem is reduced to finding a $\tau \in G(\mathfrak{P})$ such that $\bar{\rho}^{\sigma_F} = \bar{\rho}^{\bar{\tau}}$. Now, since $\mathfrak{P}^\sigma \neq \mathfrak{P}$ if $\sigma \notin G(\mathfrak{P})$, there is an $\alpha \in \mathfrak{o}_K$ such that $\alpha \equiv \rho \ (\mathfrak{P})$, $\alpha \equiv 0 \ (\mathfrak{P}^\sigma)$, $\forall \sigma \notin G(\mathfrak{P})$ (Proposition 2.30). Consider the polynomial

$$f(X) = \prod_{\sigma \in G} (X - \alpha^\sigma) \in \mathfrak{o}_k[X].$$

Since $\alpha \in \mathfrak{P}^\sigma$, i.e., $\alpha^{\sigma^{-1}} \in \mathfrak{P}$, for $\sigma \notin G(\mathfrak{P})$, by reducing modulo $\mathfrak{P}$ of both sides of the equality, we get

$$f(X) \equiv X^m \prod_{\tau \in G(\mathfrak{P})} (X - \alpha^\tau) \quad (\mathfrak{P}),$$

or

$$\bar{f}(X) = X^m \bar{g}(X), \qquad g(X) = \prod_{\tau \in G(\mathfrak{P})} (X - \alpha^\tau),$$

where $\bar{f}$ and $\bar{g}$ are polynomials in $(\mathfrak{o}_k/\mathfrak{p})[X]$. Since $\alpha \equiv \rho \quad (\mathfrak{P})$, it follows that $0 = \bar{f}(\bar{\alpha}) = \bar{f}(\bar{\rho}) = \bar{\rho}^m \bar{g}(\bar{\rho})$; this implies $\bar{g}(\bar{\rho}) = 0$ because $\bar{\rho} \neq 0$. On the other hand, call $\bar{h}(X)$ the minimal polynomial of the primitive element $\bar{\rho}$ over $\mathfrak{o}_k/\mathfrak{p}$. Since $\bar{g}(\bar{\rho}) = 0$ we have $\bar{h} \mid \bar{g}$. As $\bar{\rho}^{\sigma_F}$ is a root of $\bar{h}(X) = 0$, it is a root of $\bar{g}(X) = 0$, i.e., there must be a $\tau \in G(\mathfrak{P})$ such that $\bar{\rho}^{\sigma_F} = \bar{\alpha}^{\bar{\tau}} = \bar{\rho}^{\bar{\tau}}$. Q.E.D.

Denote by $T(\mathfrak{P})$ the kernel of the homomorphism $G(\mathfrak{P}) \to \bar{G}$:

$$T(\mathfrak{P}) \stackrel{\text{def}}{=} \{\sigma \in G(\mathfrak{P});\, \bar{\sigma} = 1\} = \{\sigma \in G(\mathfrak{P});\, \alpha^\sigma \equiv \alpha \quad (\mathfrak{P}),\ \alpha \in \mathfrak{o}_K\}.$$

This subgroup of $G(\mathfrak{P})$ is the *inertia group* of $\mathfrak{P}$ for K/k. By Proposition 2.38, we get the exact sequence

$$1 \to T(\mathfrak{P}) \to G(\mathfrak{P}) \to \bar{G} \to 1$$

from which we see easily that $[T(\mathfrak{P})] = e$.

EXERCISE 2.27. For any $\tau \in G$, prove that $G(\mathfrak{P}^\tau) = \tau G(\mathfrak{P})\tau^{-1}$.

$$\begin{array}{clcl}
 & K & 1 & \\
e & \Big| & \Big| & \\
 & T(\mathfrak{P})^\flat & T(\mathfrak{P}) & \text{interial group} \\
f & \Big| & \Big| & \\
 & G(\mathfrak{P})^\flat & G(\mathfrak{P}) & \text{decomposition group} \\
g & \Big| & \Big| & \\
 & k & G = G(K/k) &
\end{array}$$

If, in particular, $\mathfrak{P}$ is unramified in K, then we have $e = 1$ and so $T(\mathfrak{P}) = 1$. In this case, the homomorphism $\sigma \mapsto \bar{\sigma}$ becomes an isomorphism $G(\mathfrak{P}) \approx \bar{G}$. The unique automorphism in $G(\mathfrak{P})$ which corresponds to the Frobenius automorphism σ_F in $\bar{G}$ is the *Frobenius automorphism* of $\mathfrak{P}$ for K/k:

$$\left[\frac{K/k}{\mathfrak{P}}\right] \qquad \text{(Frobenius automorphism).}$$

In other words, the Frobenius automorphism of $\mathfrak{P}$ is an element of the Galois group $G(K/k)$ described by the property

$$\alpha^{\left[\frac{K/k}{\mathfrak{P}}\right]} \equiv \alpha^{N_k \mathfrak{p}} \ (\mathfrak{P}), \qquad \alpha \in \mathfrak{O}_K.$$

If $\mathfrak{p} \nmid \Delta_K$, then any prime ideal $\mathfrak{P}$ in K such that $\mathfrak{P} \mid \mathfrak{p}$ is unramified in K (Theorem 2.18) and, *a fortiori*, for K/k [Section 2.12, formula (2.14)]. Therefore $[(K/k)/\mathfrak{P}]$ makes sense for all $\mathfrak{P}$ but a finite number of exceptions.

EXERCISE 2.28. For any $\tau \in G$, prove that

$$\left[\frac{K/k}{\mathfrak{P}^\tau}\right] = \tau\left[\frac{K/k}{\mathfrak{P}}\right]\tau^{-1}.$$

Assume now that K/k is *abelian*, i.e., the Galois group $G(K/k)$ is abelian. In this case, for a prime ideal $\mathfrak{P}$ in K which is unramified for K/k, we have

$$\left[\frac{K/k}{\mathfrak{P}^\tau}\right] = \left[\frac{K/k}{\mathfrak{P}}\right]$$

(Exercise 2.28). Therefore the Frobenius automorphism of $\mathfrak{P}$ for K/k depends only on $\mathfrak{p}$ in k such that $\mathfrak{P} \mid \mathfrak{p}$. So we can write

$$\left[\frac{K/k}{\mathfrak{P}}\right] = \left[\frac{K/k}{\mathfrak{p}}\right],$$

where the latter notation is the Artin (1898–1962) symbol.

EXERCISE 2.29. Let K/k be abelian and K_0 be an intermediate field: $K \supset K_0 \supset k$. Clearly, K_0/k is abelian. If $\mathfrak{p}$ is a prime ideal in k unramified for K/k, then it is unramified for K_0/k. In this situation, prove that $((K_0/k)/\mathfrak{p})$ is the restriction of $((K/k)/\mathfrak{p})$ on K_0.

2.15. Artin Map

In order to introduce the Artin map, it is convenient to extend slightly the concept of ideals.

DEFINITION 2.14. Let k be an algebraic number field of finite degree over $\mathbb{Q}$ and $\mathfrak{o}_k$ be the ring of integers of k. An $\mathfrak{o}_k$-module† $\mathfrak{a} \neq 0$ in k is a *fractional ideal* if there is a $\mu \in \mathfrak{o}_k$, $\mu \neq 0$, such that $\mu\mathfrak{a} \subset \mathfrak{o}_k$.

REMARK 2.16.

1. If $\mathfrak{a}$ is an ideal in $\mathfrak{o}_k$, then it is a fractional ideal because one can take $\mu = 1$ for $\mathfrak{a}$.
2. We can assume that $\mu \in \mathbb{N}$ as $|N_{k/\mathbb{Q}}\mu|\, \mathfrak{a} \subset \mathfrak{o}_k$ if $\mu\mathfrak{a} \subset \mathfrak{o}_k$.
3. If $\mathfrak{a}$ is a fractional ideal, then $\mu\mathfrak{a}$ is an ideal in $\mathfrak{o}_k$; as for bases, $\mu\mathfrak{a} = [u_1, \ldots, u_n]$ implies $\mathfrak{a} = [u_1/\mu, \ldots, u_n/\mu]$.
4. For fractional ideals $\mathfrak{a}$ and $\mathfrak{b}$, the sum $\mathfrak{a} + \mathfrak{b}$ and the product $\mathfrak{a}\mathfrak{b}$ are defined naturally (Section 2.10).

DEFINITION 2.15. For an $\alpha \in k$, $\alpha \neq 0$, the set $(\alpha) = \mathfrak{o}_k\alpha$ makes up a fractional ideal which is called the *principal fractional ideal* generated by α.

PROPOSITION 2.39. *Let P_k be the set of all principal fractional ideals of k. P_k makes up naturally a group where* $(1) = \mathfrak{o}_k$ *serves as*

† For a module M and a commutative ring R with 1, M is called an R module if M is an R^+ module for the additive group R^+ (Definition 2.6) and $(ab)x = a(bx)$, $1x = x$ for $a, b \in R$, $x \in M$.

the unit element. The mapping $\alpha \mapsto (\alpha)$ *is a homomorphism* $k^\times \to P_k$ *and we obtain the exact sequence*

$$1 \to \mathfrak{o}_k^\times \to k^\times \to P_k \to 1$$

where $\mathfrak{o}_k^\times$ *is the group of units of k.*

PROOF. P_k is a group by $(\alpha\beta) = (\alpha)(\beta)$, $(\alpha)^{-1} = (\alpha^{-1})$. By definition, the mapping $\alpha \mapsto (\alpha)$ is surjective. The kernel consists of $\alpha \in k^\times$ such that $(\alpha) = (1)$, i.e., $\alpha \in \mathfrak{o}_k^\times$. Q.E.D.

We shall denote by I_k the set of all fractional ideals of k.

THEOREM 2.19. *I_k makes up naturally an abelian group and P_k is a subgroup of I_k.*

PROOF. The only point is to show that for $\mathfrak{a} \in I_k$ there is a unique $\mathfrak{b} \in I_k$ such that $\mathfrak{a}\mathfrak{b} = (1)$. First, let $\mu \in \mathfrak{o}_k$ be a number $\neq 0$ such that $\mu\mathfrak{a} \subset \mathfrak{o}_k$. By Exercise 2.21 [with $\mathfrak{b} = (1)$] there is an ideal $\mathfrak{c} \subset \mathfrak{o}_k$ such that $(\mu\mathfrak{a})\mathfrak{c} = (\gamma)$, $\gamma \in \mathfrak{o}_k$. Then we have $\mathfrak{a}\mathfrak{b} = (1)$ with $\mathfrak{b} = (\mu\gamma^{-1})\mathfrak{c}$. Suppose now that $\mathfrak{a}\mathfrak{b}' = (1)$ with $\mathfrak{b}' \in I_k$. We can find $\nu \in \mathfrak{o}_k$ such that $\nu\mathfrak{b} \subset \mathfrak{o}_k$, $\nu\mathfrak{b}' \subset \mathfrak{o}_k$ and so $(\mu\mathfrak{a})(\nu\mathfrak{b}) = (\mu\mathfrak{a})(\nu\mathfrak{b}')$. Since the cancellation works for ideals in $\mathfrak{o}_k$ (Proposition 2.27), we get $\nu\mathfrak{b} = \nu\mathfrak{b}'$ and hence $\mathfrak{b} = \mathfrak{b}' = \mathfrak{a}^{-1}$. Q.E.D.

DEFINITION 2.16. The group $H_k = I_k/P_k$ is called the *ideal class group* of K.

REMARK 2.17. We have already introduced the ideal class group H_k in Theorem 2.14. In order to verify the consistency of two definitions of the ideal class groups, let us write temporarily $H_k^* = I_k/P_k$ and check that $H_k \approx H_k^*$. An element of H_k was a class $[\mathfrak{a}]$, $\mathfrak{a} \subset \mathfrak{o}_k$, defined by the equivalence relation

$$\mathfrak{a} \sim \mathfrak{b} \Leftrightarrow (\alpha)\mathfrak{a} = (\beta)\mathfrak{b}, \qquad \alpha, \beta \in \mathfrak{o}_k.$$

Therefore a mapping $f: H_k \to H_k^*$ can be defined by $f([\mathfrak{a}]) = \mathfrak{a} \bmod P_k$. Since $f([\mathfrak{a}][\mathfrak{b}]) = f([\mathfrak{a}\mathfrak{b}]) = (\mathfrak{a}\mathfrak{b}) \bmod P_k = (\mathfrak{a} \bmod P_k)(\mathfrak{b} \bmod P_k) = f([\mathfrak{a}])f([\mathfrak{b}])$, f is a homomorphism. Next, let $\mathfrak{a}$ be any fractional ideal in I_k. Since $\mu\mathfrak{a} \subset \mathfrak{o}_k$ for some $\mu \in \mathfrak{o}_k$, any class in H_k^* can be represented by $\mathfrak{a} \subset \mathfrak{o}_k$; then, with this $\mathfrak{a}$, $f([\mathfrak{a}]) = \mathfrak{a} \bmod P_k$, i.e., f is surjective. Finally suppose that $f([\mathfrak{a}]) = \mathfrak{a} \bmod P_k = 1$. Then $\mathfrak{a} = (\alpha) \subset \mathfrak{o}_k$ which implies that $\alpha \in \mathfrak{o}_k$, hence

$(1)\mathfrak{a}=(\alpha)\mathfrak{o}_k$ or $\mathfrak{a}\sim\mathfrak{o}_k$, i.e., $[\mathfrak{a}]=[\mathfrak{o}_k]$, the unit element of the group H_k; hence f is injective. Therefore we have $H_k\approx H_k^*$. From now on, we shall, of course, identify two groups H_k and H_k^* and call the order h_k of the group the *class number* of k (Definition 2.11).

THEOREM 2.20 (*Fundamental theorem of fractional ideal theory*). *Every fractional ideal* $\mathfrak{a}$ $(\neq\mathfrak{o}_k)$ *can be written uniquely as a product of prime ideals except for the order of factors*:

$$\mathfrak{a}=\mathfrak{p}_1^{e_1}\cdots\mathfrak{p}_l^{e_l},\qquad e_i\in\mathbb{Z},\qquad e_i\neq 0,\qquad 1\leq i\leq l.$$

PROOF. For $\mathfrak{a}$, there is a $\mu\in\mathfrak{o}_k$ such that $\mu\mathfrak{a}\subset\mathfrak{o}_k$. Our assertion follows by applying Theorem 2.15 to ideals $(\mu)\mathfrak{a}$ and (μ).

Q.E.D.

REMARK 2.18. Theorem 2.20 means that the group I_k is a *free abelian group* generated by all prime ideals $\mathfrak{p}$ of k. When a fractional ideal $\mathfrak{a}$ is written as in Theorem 2.20, by grouping prime ideals $\mathfrak{p}_i$ with positive (negative) e_i, we get a unique expression $\mathfrak{a}=\mathfrak{b}\mathfrak{c}^{-1}$ such that $\mathfrak{b}$, $\mathfrak{c}\subset\mathfrak{o}_k$. When $\mathfrak{m}$ is an ideal of $\mathfrak{o}_k$, we say that $\mathfrak{a}$ is prime to $\mathfrak{m}$, written $(\mathfrak{a},\mathfrak{m})=1$, when $(\mathfrak{b},\mathfrak{m})=1$ and $(\mathfrak{c},\mathfrak{m})=1$. We can define the norm $N_k\mathfrak{a}$ for a fractional ideal $\mathfrak{a}$ by $N_k\mathfrak{a}=N_k\mathfrak{b}/N_k\mathfrak{c}$. Clearly, the mapping $\mathfrak{a}\mapsto N_k\mathfrak{a}$ is a homomorphism of I_k to the group of positive rational numbers. More importantly, let K/k be an extension. Then we obtain a natural homomorphism $f: I_k\to I_K$ defined by $f(\mathfrak{a})=(\mathfrak{a})_K$. We claim that f is injective. In fact, suppose that $f(\mathfrak{a})=\mathfrak{o}_K$. Then we have $\mathfrak{o}_k=k\cap\mathfrak{o}_K=k\cap(\mathfrak{a})_K\supset\mathfrak{a}$. Since $f(\mathfrak{a}^{-1})=f(\mathfrak{a})^{-1}=\mathfrak{o}_K$, by the same argument, we have $\mathfrak{o}_k\supset\mathfrak{a}^{-1}$ and hence $\mathfrak{a}=\mathfrak{o}_k$. An extension K/k induces two exact sequences:

$$\begin{array}{ccccccccc}
1 & \longrightarrow & \mathfrak{o}_k^\times & \longrightarrow & k^\times & \xrightarrow{(\ )_k} & I_k & \longrightarrow & H_k & \longrightarrow & 1\\
 & & \downarrow & & \downarrow & & \downarrow & & \downarrow{\scriptstyle\delta} & & \\
1 & \longrightarrow & \mathfrak{o}_K^\times & \longrightarrow & K^\times & \xrightarrow{(\ )_K} & I_K & \longrightarrow & H_K & \longrightarrow & 1
\end{array}$$

By what we have just proved, the first three vertical mappings are all injective. As for the fourth mapping δ, it can be trivial which means that every ideal of k becomes principal in K. Such a proposition is known to be true when K is the Hilbert class field over k, i.e., when K is the maximal unramified abelian extension over k.

Consider now an abelian extension K/k. Let S be a finite set of prime ideals of k such that all prime ideals outside S are unramified for K/k. For example, we can take $S=\{\mathfrak{p};\mathfrak{p}\mid\Delta_K\}$ (Theorem 2.18). Obviously, any finite set of prime ideals in k containing S satisfies the required property. Now put

$$I_k(S)=\{\mathfrak{a}\in I_k;\,(\mathfrak{a},\mathfrak{p})=1,\ \forall\mathfrak{p}\in S\}.$$

The set $I_k(S)$ clearly forms a subgroup of I_k. Write $\mathfrak{a}\in I_k(S)$ as

$$\mathfrak{a}=\mathfrak{p}_1^{e_1}\cdots\mathfrak{p}_l^{e_l},\qquad e_i\in\mathbb{Z}.$$

By the choice of S, each $\mathfrak{p}_i$ is unramified for the abelian extension K/k and so the Artin symbol makes sense (Section 2.14, after Exercise 2.28). Now we extend the symbol for ideal in $I_k(S)$:

$$\left(\frac{K/k}{\mathfrak{a}}\right)\overset{\text{def}}{=}\left(\frac{K/k}{\mathfrak{p}_1}\right)^{e_1}\cdots\left(\frac{K/k}{\mathfrak{p}_l}\right)^{e_l}.$$

Since the Galois group $G(K/k)$ is abelian, we have

$$\left(\frac{K/k}{\mathfrak{a}\mathfrak{b}}\right)=\left(\frac{K/k}{\mathfrak{a}}\right)\left(\frac{K/k}{\mathfrak{b}}\right),\qquad \mathfrak{a},\mathfrak{b}\in I_k(S)$$

which means that the mapping

$$\mathfrak{a}\mapsto\left(\frac{K/k}{\mathfrak{a}}\right)$$

is a homomorphism of $I_k(S)$ to $G(K/k)$. We shall denote this homomorphism by $\alpha_{K/k}$ and call it the *Artin map*:

$$\alpha_{K/k}(\mathfrak{a})\overset{\text{def}}{=}\left(\frac{K/k}{\mathfrak{a}}\right),\qquad \mathfrak{a}\in I_k(S).\dagger$$

REMARK 2.19. The homomorphism $\alpha_{K/k}$ seems to be important! In fact, to determine Ker $\alpha_{K/k}$ and Im $\alpha_{K/k}$ is the content of the Artin reciprocity in class field theory. Although we shall not prove the Artin reciprocity for an arbitrary abelian extension K/k, we shall determine Ker $\alpha_{K/k}$ and Im $\alpha_{K/k}$ for subfields of the lth cyclotomic field and quadratic fields.

† Strictly speaking, $\alpha_{K/k}$ depends on the choice of S.

2.16. Artin Maps of Subfields of the lth Cyclotomic Field

Let l be an odd prime, $\zeta = e^{2\pi i/l}$ and $k = \mathbb{Q}(\zeta)$. Since $\Delta_k = (-1)^{(l-1)/2} l^{l-2}$ (Exercise 2.18), any prime number $p \neq l$ is unramified for $k/\mathbb{Q}$, and, *a fortiori*, unramified for any $k'/\mathbb{Q}$, $k \supset k' \supset \mathbb{Q}$. The group $I_{\mathbb{Q}}$ may be identified with the group of positive rational numbers by the correspondence $a > 0 \Leftrightarrow (a)$ and so we have

$$I_{\mathbb{Q}}(l) \stackrel{\text{def}}{=} \{\mathfrak{a} \in I_{\mathbb{Q}}; (\mathfrak{a}, l) = 1\} = \{a \in \mathbb{Q}^{\times}; a > 0, (a, l) = 1\}.$$

Now, for $a, a' \in I_{\mathbb{Q}}(l)$, we define the following relation:

$$a \equiv a' \quad (l) \stackrel{\text{def}}{\Leftrightarrow} bc' \equiv b'c \quad (l)$$

where $a = b/c$, b, $c \in \mathbb{N}$, $(l, b) = (l, c) = 1$ and $a' = b'/c'$, b', $c' \in \mathbb{N}$, $(l, b') = (l, c') = 1$.

Exercise 2.30. Verify that $a \equiv a' \quad (l)$ is an equivalence relation in $I_{\mathbb{Q}}(l)$. Verify also that for $a, a' \in \mathbb{N}$, we have

$$a \equiv a' \quad (l) \Leftrightarrow a \equiv a' \quad (l).$$

For a natural number n, consider the following set:

$$H_n(l) \stackrel{\text{def}}{=} \{a \in I_{\mathbb{Q}}(l); a \equiv x^n \quad (l), \exists x \in \mathbb{N}, (l, x) = 1\}.$$

If, in particular, $n = l - 1$, then we have

$$H_{l-1}(l) = \{a \in I_{\mathbb{Q}}(l); a \equiv 1 \quad (l)\}$$

because $x^{l-1} \equiv 1 \quad (l)$ (Theorem 1.11). On the other hand, it is clear that $H_1(l) = I_{\mathbb{Q}}(l)$.

Exercise 2.31. For any $n \in \mathbb{N}$, the set $H_n(l)$ forms a subgroup of $I_{\mathbb{Q}}(l)$. There is an isomorphism $I_{\mathbb{Q}}(l)/H_n(l) \approx \mathbb{F}_l^{\times}/(\mathbb{F}_l^{\times})^n$; this group is cyclic of order $(l-1, n)$.

From now on, we assume that $n \mid l - 1$ and denote by k_n the subfield of $k = \mathbb{Q}(\zeta)$ such that $[k_n : \mathbb{Q}] = n$ (the situation in Section 2.6). Since p $(\neq l)$ is unramified for $k_n/\mathbb{Q}$, the Artin map $\alpha_{k_n/\mathbb{Q}}$ makes sense:

$$\alpha_{k_n/\mathbb{Q}}(p) = \left(\frac{k_n/\mathbb{Q}}{p}\right) \in G(k_n/\mathbb{Q}) = G(k/\mathbb{Q})/k_n^{\#}.$$

By Exercise 2.5, there is an isomorphism $\psi: \mathbb{F}_l^\times \xrightarrow{\sim} G(k/\mathbb{Q})$ given by $\psi([a]) = \sigma_a$ ($\zeta^{\sigma_a} = \zeta^a$) and it induces, for each n, the following isomorphism:

$$\psi_n: \mathbb{F}_l^\times/(\mathbb{F}_l^\times)^n \xrightarrow{\sim} G(k_n/\mathbb{Q}), \qquad \psi_n([a] \bmod (\mathbb{F}_l^\times)^n) = \sigma_a \bmod k_n^\#.$$

Furthermore, by Exercise 2.31, we have a homomorphism φ_n:

$$\varphi_n: I_\mathbb{Q}(l) \to \mathbb{F}_l^\times/(\mathbb{F}_l^\times)^n, \qquad \varphi_n(p) = [p] \bmod (\mathbb{F}_l^\times)^n.$$

We now clarify, for a fixed n, relations among three homomorphisms $\alpha_{k_n/\mathbb{Q}}$, ψ_n, and φ_n.

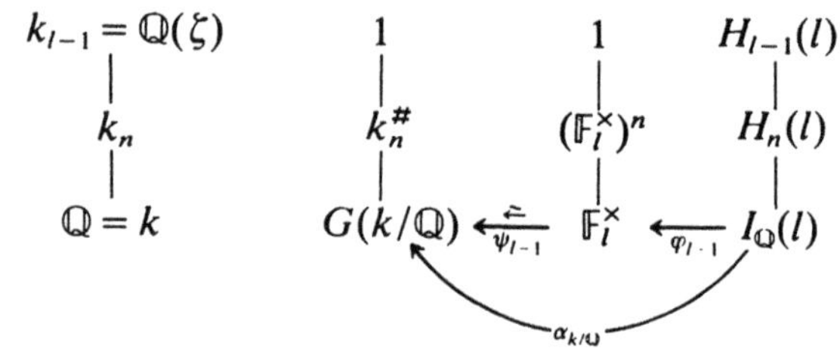

Proposition 2.40. *For* $n \mid l-1$, *we have* $\psi_n\varphi_n = \alpha_{k_n/\mathbb{Q}}$, *i.e., the following diagram is "commutative":*

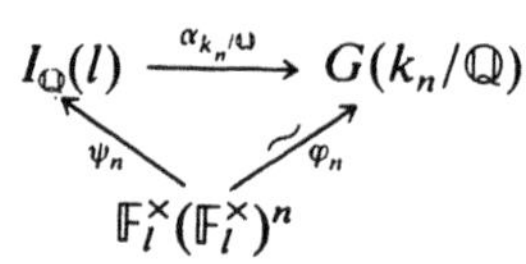

Proof. Since φ_n, ψ_n, and $\alpha_{k_n/\mathbb{Q}}$ are obtained naturally from $\varphi = \varphi_{l-1}$, $\psi = \psi_{l-1}$, and $\alpha_{k/\mathbb{Q}}$, respectively (Exercise 2.29), it is enough to verify that $\psi\varphi(p) = \alpha_{k/\mathbb{Q}}(p)$, $p \neq l$. However, the left-hand side $= \psi\varphi(p) = \psi([p]) = \sigma_p$, the right-hand side $= \alpha_{k/\mathbb{Q}}(p) = ((k/\mathbb{Q})/p)$, and so we are reduced to proving that

$$\zeta^{\left(\frac{k/\mathbb{Q}}{p}\right)} = \zeta^p. \tag{2.23}$$

In fact, put

$$\zeta^{\left(\frac{k/\mathbb{Q}}{p}\right)} = \zeta^a, \qquad (l, a) = 1.$$

By definition of the Frobenius automorphism, we have $\zeta^a \equiv \zeta^p \ (\mathfrak{p})$, $\mathfrak{p} \mid p$, or $\zeta^a - \zeta^p = \zeta^p(\zeta^{a-p} - 1) \in \mathfrak{p}$. But, since $\zeta^p \in$

$\mathfrak{o}_k^\times$, we have $\zeta^{a-p}-1 \in \mathfrak{p}$, i.e., $\zeta^b - 1 \in \mathfrak{p}$ with $b = a - p$. Let

$$X^{l-1} + \cdots + X + 1 = \prod_{i=1}^{l-1} (X - \zeta^i)$$

be the decomposition of the minimal polynomial of ζ over $\mathbb{Q}$. If we put $X = 1$, the equality implies $\prod_{i=1}^{l-1} (1-\zeta^i) = l$. If we have here $l \nmid b$, then we must have $\mathfrak{p} \mid \zeta^b - 1 \mid l$ and hence $p = l$, a contradiction. Therefore, we have $l \mid b$, i.e., $a \equiv p \quad (l)$ which settles (2.23). Q.E.D.

As the mapping ψ_n in Proposition 2.40 is an isomorphism, to determine $\operatorname{Ker} \alpha_{k_n/\mathbb{Q}}$ and $\operatorname{Im} \alpha_{k_n/\mathbb{Q}}$ is the same problem as to determine $\operatorname{Ker} \varphi_n$ and $\operatorname{Im} \varphi_n$; and the latter problem has been solved in Exercise 2.31. Namely, $\operatorname{Im} \varphi_n = \mathbb{F}_l^\times / (\mathbb{F}_l^\times)^n$, i.e., φ_n is surjective and

$$\operatorname{Ker} \varphi_n = H_n(l) = \{a \in I_{\mathbb{Q}}(l);\, a \equiv x^n \quad (l),\, x \in \mathbb{N},\, (l, x) = 1\}.$$

In other words, we obtain the exact sequence

$$1 \to H_n(l) \to I_{\mathbb{Q}}(l) \xrightarrow{\alpha_{k_n/\mathbb{Q}}} G(k_n/\mathbb{Q}) \to 1. \tag{2.24}$$

Remark 2.20. When a in $H_n(l)$ is in $\mathbb{N}$, one can replace the unsightly notation $\equiv$ by ordinary $\equiv$. In particular, for a prime $p \neq l$, the exactness of (2.24) implies that

$$p \in H_n(l) \Leftrightarrow \alpha_{k_n/\mathbb{Q}}(p) = \left(\frac{k_n/\mathbb{Q}}{p}\right) = 1$$

where the right-hand side means that p *splits completely* for $k_n/\mathbb{Q}$, i.e., $p = \mathfrak{p}_1 \cdots \mathfrak{p}_n$, $n = [k_n \colon \mathbb{Q}]$. In terms of Remark 1.22, we have, for $n \mid l - 1$,

$$p \text{ is an } n\text{th power residue mod } l \Leftrightarrow p \text{ splits completely for } k_n/\mathbb{Q}. \tag{2.25}$$

When $n = 2$ (note that $2 \mid l - 1$), we have

$$p \text{ is a quadratic residue mod } l \Leftrightarrow p \text{ splits completely for } \mathbb{Q}(\sqrt{l^*})/\mathbb{Q}. \tag{2.26}$$

We used here the fact that $k_2 = \mathbb{Q}(\sqrt{l^*})$ with $l^* = (-1)^{(l-1)/2}l$ [Exercise 2.5(4)]. In terms of Legendre character $\lambda_l(x) = (x/l)$, the left-hand side of (2.26) means that $(p/l) = 1$. Since, for $k_2 = \mathbb{Q}(\sqrt{l^*})$, we find $l^* \equiv 1 \quad (4)$ and $\Delta_{k_2} = l^*$, we have, by Exercise 2.23,

$$\begin{cases} p \neq 2\colon \lambda_p(l^*) = (l^*/p) = 1 \Leftrightarrow p \text{ splits completely for } k_2/\mathbb{Q}, \\ p = 2\colon l^* \equiv 1 \quad (8) \Leftrightarrow l \equiv \pm 1 \quad (8) \Leftrightarrow 2 \text{ splits completely for } k_2/\mathbb{Q}. \end{cases}$$

Thus, (2.26) is equivalent to the statement:

$$\begin{cases} p \neq 2\colon (p/l) = 1 \Leftrightarrow (l^*/p) = 1, \\ p = 2\colon (2/l) = 1 \Leftrightarrow l \equiv \pm 1 \quad (8). \end{cases} \tag{2.27}$$

Therefore, if we take Theorem 1.25 (the first supplementary law: $(-1/p) = (-1)^{(p-1)/2}$) for granted, we see that (2.27) is nothing but a combination of Theorem 1.26 (the second supplementary law) and Theorem 1.27 (the Gauss reciprocity law):

$$\left(\frac{2}{l}\right) = (-1)^{(l^2-1)/8} \quad \text{and} \quad \left(\frac{p}{l}\right)\left(\frac{l}{p}\right) = (-1)^{(l-1)/2\cdot(p-1)/2}.$$

In other words, we have shown that the Gauss reciprocity follows from the exact sequence (2.24) and the fact that the quadratic field $k_2 = \mathbb{Q}(\sqrt{l^*})$ is a subfield of the lth cyclotomic field $k_{l-1} = \mathbb{Q}(\zeta)$, $\zeta^l = 1$.

2.17. The Artin Map in Quadratic Fields

Let $k = \mathbb{Q}(\sqrt{m})$ be a quadratic field where m is a square free integer. We know that $\Delta_k = 4m$ if $m \equiv 2, 3 \quad (4)$, $= m$ if $m \equiv 1 \quad (4)$. Therefore, if we put $S = \{p; p \mid \Delta_k\}$, then p is unramified in k outside S. In what follows we put $I_{\mathbb{Q}}(\Delta_k) \overset{\text{def}}{=} I_{\mathbb{Q}}(S) = \{a \in \mathbb{Q}^\times; a > 0, (a, \Delta_k) = 1\}$. The Artin map is given by

$$\alpha_{k/\mathbb{Q}}(p) = \left(\frac{k/\mathbb{Q}}{p}\right).$$

Since the Galois group $G(k/\mathbb{Q})$ is of order 2, there is a unique isomorphism $\psi\colon G(k/\mathbb{Q}) \xrightarrow{\sim} \{\pm 1\}$. Therefore, there is a homomorphism $\chi_k\colon I_{\mathbb{Q}}(\Delta_k) \to \{\pm 1\}$ so that the following diagram is commutative: $\alpha_{k/\mathbb{Q}} = \psi\chi_k$.

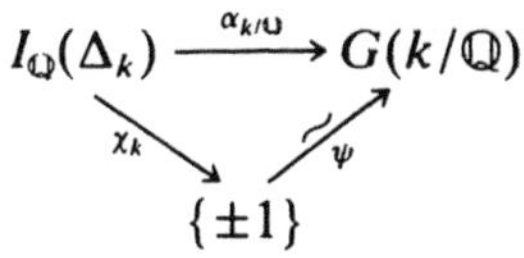

By Exercise 2.23, we can determine the values $\chi_k(p)$. First of all, for $p \nmid \Delta_k$,

$$\chi_k(p) = 1 \Leftrightarrow \alpha_{k/\mathbb{Q}}(p) = 1 \Leftrightarrow p \text{ splits completely for } k/\mathbb{Q}. \qquad (2.28)$$

Next, we have

$$\begin{cases} p \neq 2\colon (\Delta_k/p) = 1 \Leftrightarrow p \text{ splits completely for } k/\mathbb{Q}, \\ p = 2\colon \Delta_k \equiv 1 \pmod 8 \Leftrightarrow 2 \text{ splits completely for } k/\mathbb{Q}. \end{cases} \qquad (2.29)$$

[Note that $2 \nmid \Delta_k$ if and only if $m \equiv 1 \pmod 4$ and in this case we have $\Delta_k = m \equiv 1$ or $5 \pmod 8$.] Finally, we get

$$\chi_k(p) = \begin{cases} \left(\dfrac{\Delta_k}{p}\right), & p \neq 2, \\ (-1)^{(\Delta_k^2 - 1)/8}, & p = 2. \end{cases} \qquad (2.30)$$

The homomorphism χ_k is the *Kronecker* (1823–1891) *character* of k. As the above diagram shows, it is the same thing as the Artin map in view of the canonical identification of the Galois group $G(k/\mathbb{Q})$ with $\{\pm 1\}$. For $p \mid \Delta_k$, we put $\chi_k(p) = 0$.

In order to study χ_k in more detail, we need the Jacobi characters η_N. Let N be an odd integer >0, $x \in \mathbb{Z}$ such that $(x, N) = 1$, and let $N = p_1^{e_1} \cdots p_t^{e_t}$ be the canonical decomposition of N. Since the Legendre character $\lambda_{p_i}(x)$ makes sense for each i, $1 \leq i \leq t$, we can introduce a function η_N by

$$\eta_N(x) \overset{\text{def}}{=} \lambda_{p_1}(x)^{e_1} \cdots \lambda_{p_t}(x)^{e_t}, \qquad (x, N) = 1, \qquad x \in \mathbb{Z}.\dagger$$

† Usually Jacobi character $\eta_N(x)$ is written (x/N).

EXERCISE 2.32.

1. η_N is a homomorphism: $(\mathbb{Z}/N\mathbb{Z})^\times \to \{\pm 1\}$. ($\eta_N$ is called the *Jacobi character*. When $N = p \neq 2$, we have $\eta_p = \lambda_p$, the Legendre character).
2. Assume that $M, N \in \mathbb{N}$ are both odd and that $(N, M) = 1$. Then we have $\eta_M(N)\eta_N(M) = (-1)^{(M-1)/2(N-1)/2}$.
3. $\eta_M(-1) = (-1)^{(M-1)/2}$.
4. $\eta_M(2) = (-1)^{(M^2-1)/8}$.

Back to χ_k, since $\Delta_k = m \equiv 1 \pmod 4$ when $2 \nmid \Delta_k$, the Jacobi character $\eta_{|\Delta_k|}$ makes sense and, by (2.30) and Exercise 2.32 (1), (4), we get

$$\chi_k(p) = \begin{cases} \eta_p(\Delta_k) = \lambda_p(\Delta_k), & p \neq 2, \\ \eta_{|\Delta_k|}(2), & p = 2. \end{cases}$$

EXERCISE 2.33.

1. Assume that $\Delta_k \equiv 1 \quad (4)$. Then $\chi_k(x) = \eta_{|\Delta_k|}(x)$, $x \in \mathbb{N}$, $(x, \Delta_k) = 1$.
2. Assume that $\Delta_k \equiv 0 \quad (4)$ and write $\Delta_k = 2^\beta \Delta_k'$, $2 \nmid \Delta_k'$ with

$$\beta = \begin{cases} 3, & m \equiv 2 \quad (4), \\ 2, & m \equiv 3 \quad (4). \end{cases}$$

Then $\chi_k(x) = \eta_x(2)^\beta \eta_{|\Delta_k'|}(x)(-1)^{(x-1)/2 \cdot (\Delta_k'-1)/2}$, $x \in \mathbb{N}$, $(x, \Delta_k) = 1$.

3. There exists $x \in \mathbb{N}$, $(x, \Delta_k) = 1$ such that $\chi_k(x) = -1$. [This shows that the Artin map $\alpha_{k/\mathbb{Q}}: I_\mathbb{Q}(\Delta_k) \to G(k/\mathbb{Q})$ is surjective.]
4. If $x, y \in \mathbb{N}$, $(x, \Delta_k) = (y, \Delta_k) = 1$, and $x \equiv y \quad (\Delta_k)$, then $\chi_k(x) = \chi_k(y)$.

Since the equality $\operatorname{Im} \alpha_{k/\mathbb{Q}} = G(k/\mathbb{Q})$ is established [Exercise 2.33(3)], it remains to determine $\operatorname{Ker} \alpha_{k/\mathbb{Q}} = \operatorname{Ker} \chi_k$. As in Section 2.16, for $a, a' \in I_\mathbb{Q}(\Delta_k)$ we define

$$a \equiv a' \quad (\Delta_k) \overset{\text{def}}{\Leftrightarrow} bc' \equiv b'c \quad (\Delta_k)$$

where $a = b/c$, $b, c \in \mathbb{N}$, $(\Delta_k, c) = 1$, and similarly $a' = b'/c'$. If we put

$$S(\Delta_k) \overset{\text{def}}{=} \{a \in I_{\mathbb{Q}}(\Delta_k); a \equiv 1 \quad (\Delta_k)\},$$

then, as in Exercise 2.31, the set $S(\Delta_k)$ forms a subgroup of $I_{\mathbb{Q}}(\Delta_k)$.

PROPOSITION 2.41. $S(\Delta_k) \subset \operatorname{Ker} \chi_k$.

PROOF. If $a \equiv 1 \quad (\Delta_k)$, then $a = b/c$, $b, c \in \mathbb{N}$, $(\Delta_k, b) = (\Delta_k, c) = 1$, and $b \equiv c \quad (\Delta_k)$. By Exercise 2.33 (4), we have $\chi_k(b) = \chi_k(c)$ which implies that $\chi_k(a) = \chi_k(b)/\chi_k(c) = 1$ Q.E.D.

Let I_k be the group of fractional ideals of k and let

$$I_k(\Delta_k) = \{\mathfrak{a} \in I_k; (\mathfrak{a}, \Delta_k) = 1\}.$$

Then we see that the norm $N = N_k$ (Remark 2.18) induces a homomorphism of $I_k(\Delta_k)$ to $I_{\mathbb{Q}}(\Delta_k)$.†

PROPOSITION 2.42. $N_k(I_k(\Delta_k)) \subset \operatorname{Ker} \chi_k$.

PROOF. Take a fractional ideal: $\mathfrak{a} = \prod_{\mathfrak{p}} \mathfrak{p}^{e_{\mathfrak{p}}}$. Since $(\mathfrak{p}, \Delta_k) = 1$, $\mathfrak{p}$ is unramified for $k/\mathbb{Q}$ and, by (2.28)–(2.30),

$$N\mathfrak{p} = p \Leftrightarrow p \text{ splits completely for } k/\mathbb{Q} \Leftrightarrow \chi_k(p) = 1.$$

On the other hand, if $\chi_k(p) = -1$, then $N\mathfrak{p} = p^2$; so $\chi_k(N\mathfrak{p}) = 1$ in all cases, i.e., $\chi_k(N\mathfrak{a}) = 1$, $\mathfrak{a} \in I_k(\Delta_k)$. Q.E.D.

Combining Proposition 2.41 and Proposition 2.42, we get

PROPOSITION 2.43. $S(\Delta_k)N_k(I_k(\Delta_k)) \subset \operatorname{Ker} \chi_k$..

Actually, we have the equality here; this is Theorem 2.21 below. In its proof, we need the *theorem on arithmetical progressions* due to Dirichlet (1805–1859): for any $N \in \mathbb{N}$, each coset of the group $(\mathbb{Z}/N\mathbb{Z})^\times$ contains infinitely many primes. A proof will be given in Chapter 3 (Theorem 3.12).

THEOREM 2.21. *For any quadratic field $k/\mathbb{Q}$, we have* $\operatorname{Ker} \chi_k = S(\Delta_k)N_k(I_k(\Delta_k))$.

† Let $\mathfrak{a} = \mathfrak{b}\mathfrak{c}^{-1}$ and $(\mathfrak{b}, \Delta_k) = (\mathfrak{c}, \Delta_k) = 1$, $\mathfrak{b}, \mathfrak{c} \subset \mathfrak{o}_k$. If a prime p such that $p \mid \Delta_k$ divides $N\mathfrak{b}$ or $N\mathfrak{c}$, say, $N\mathfrak{b}$, then $\mathfrak{p} \,|p|\, \Delta_k$ as $N\mathfrak{p} = p^f$, contradicting $(\mathfrak{b}, \Delta_k) = 1$.

Proof. Take any $a \in \operatorname{Ker} \chi_k$ and write $a = b/c$, b, $c \in \mathbb{N}$, $(b, \Delta_k) = (c, \Delta_k) = 1$. Choose $g \in \mathbb{N}$ such that $cg \equiv b \pmod{\Delta_k}$; hence $(g, \Delta_k) = 1$. As $a^{-1}g = cg/b \equiv 1 \pmod{\Delta_k}$, we have $a^{-1}g \in S(\Delta_k)$. Therefore, it is enough to show that $g \in S(\Delta_k)N(I_k(\Delta_k))$. Since $a \in \operatorname{Ker} \chi_k$ and $a^{-1}g \in S(\Delta_k) \subset \operatorname{Ker} \chi_k$ (Proposition 2.41), we have $g \in \operatorname{Ker} \chi_k$. Now, by the theorem of arithmetical progressions, one can choose an odd prime p such that $p \equiv g \pmod{\Delta_k}$. Then, by (2.30) and Exercise 2.33(4), we get

$$\left(\frac{\Delta_k}{p}\right) = \chi_k(p) = \chi_k(g) = 1;$$

hence p splits completely for $k/\mathbb{Q}$ and so $p = N\mathfrak{p}$, i.e., $p \in N(I_k(\Delta_k))$. On the other hand, we have $g/p \equiv 1 \pmod{\Delta_k}$, or $g/p \in S(\Delta_k)$, and so $g \in S(\Delta_k)N(I_k(\Delta_k))$. Q.E.D.

3

Analytic Methods

In Chapter 2, we used Dirichlet's theorem on arithmetical progressions in order to determine the kernel $\mathrm{Ker}_{k/\mathbb{Q}}\,\alpha$ of the Artin map of a quadratic field $k/\mathbb{Q}$ (Theorem 2.21) and promised the reader its proof in Chapter 3. Therefore this is at least one raison d'être of this chapter in the book. To prove Dirichlet's theorem he invented the L-functions. The use of Dirichlet L-functions, however, goes beyond the proof of the theorem on arithmetical progressions. It turns out that the L-functions are closely related to the law of decomposition of rational primes in algebraic number fields. The reader will learn in Chapter 4 how to use L-functions to handle class numbers of some cyclotomic and quadratic fields. For this reason, we will not rush to a proof of the theorem of arithmetical progressions. We will rather start with basic topics such as geometry of numbers due to Minkowski (1864–1909) and the famous Dirichlet's unit theorem.

3.1. Lattices in $\mathbb{R}^n$

Let $\mathbb{R}^n$ denote the n-dimensional euclidean space. An element x in $\mathbb{R}^n$ is therefore a vector $x = (x_1, \ldots, x_n)$. For $x, y \in \mathbb{R}^n$, the inner product is $\langle x, y\rangle \stackrel{\text{def}}{=} \sum_{i=1}^{n} x_i y_i$ and the length of x is $|x| \stackrel{\text{def}}{=} \langle x, x\rangle^{1/2}$. We denote by $U_r(x)$ the open ball of radius $r > 0$ centered at x:

$$U_r(x) \stackrel{\text{def}}{=} \{y \in \mathbb{R}^n;\, |x - y| < r\} = x + U_r(0).$$

A nonempty set $M \subset \mathbb{R}^n$ is said to be *discrete* if, for any $x \in M$, there exists $r_x > 0$ such that $U_{r_x}(x) \cap M = \{x\}$.

Exercise 3.1. For a set $M \subset \mathbb{R}^n$, if $U_r(0) \cap M$ is finite for all $r > 0$, then M is discrete. However, the converse of the statement is not necessarily true. (As will be proved in Proposition 3.1, if M is a module in $\mathbb{R}^n$, the converse is also true.)

Proposition 3.1. *For a module Γ in $\mathbb{R}^n$, we have*

$$\Gamma \text{ is discrete} \Leftrightarrow U_r(0) \cap \Gamma \text{ is finite for all } r > 0.$$

Proof. In view of Exercise 3.1, it is enough to prove ($\Rightarrow$). So assume that Γ is discrete. Note first that Γ is closed. In fact, if Γ is not closed, there is an $x \in \bar{\Gamma} - \Gamma$ where $\bar{\Gamma}$ denotes the closure of Γ. As we can find a sequence $\{\gamma_\nu\} \subset \Gamma$ such that $\lim_{\nu\to\infty} \gamma_\nu = x$, for any $\epsilon > 0$, there is a positive integer N such that $|\gamma_\nu - \gamma_\mu| < \epsilon$ whenever $\nu, \mu > N$; in other words, we have $U_\epsilon(0) \cap \Gamma \supsetneq \{0\}$ for any $\epsilon > 0$ which contradicts that Γ is discrete. Therefore the set $\overline{U_r(0)} \cap \Gamma$ is discrete and compact, i.e., finite, and so is its subset $U_r(0) \cap \Gamma$. Q.E.D.

Definition 3.1. Let Γ be a module in $\mathbb{R}^n$. When Γ can be expressed as $\Gamma = [v_1, \ldots, v_m] = \mathbb{Z}v_1 + \cdots + \mathbb{Z}v_m$ with independent vectors v_i, $1 \le i \le m \le n$, Γ is called a *sublattice* of rank m in $\mathbb{R}^n$. If, in particular, Γ is a sublattice or rank n in $\mathbb{R}^n$, then Γ is called a *lattice* in $\mathbb{R}^n$.

Definition 3.2. For a discrete module Γ in $\mathbb{R}^n$, when there is a bounded subset B of $\mathbb{R}^n$ such that $\mathbb{R}^n = B + \Gamma$ the quotient group $\mathbb{R}^n/\Gamma$ is said to be *compact.*†

Proposition 3.2. *For a module Γ in $\mathbb{R}^n$, we have*

1. *Γ is a sublattice $\Leftrightarrow \Gamma$ is discrete,*
2. *Γ is a lattice $\Leftrightarrow \Gamma$ is discrete and $\mathbb{R}^n/\Gamma$ is compact.*

Proof.

1. ($\Rightarrow$) Let $m \le n$ be the rank of Γ and let $\{v_i\}$, $1 \le i \le n$, be a basis of $\mathbb{R}^n$ such that $\Gamma = [v_1, \ldots, v_m]$. As Γ is a subset of the lattice $\Gamma' = [v_1, \ldots, v_n]$ and a subset of a discrete set is discrete, it is enough to show that Γ' is discrete. So, without loss of generality,

† This is equivalent to the compactness of $\mathbb{R}^n/\Gamma$ with respect to the quotient topology.

we may assume that $m=n$. Now, let $\{v_i^*\}$ be the basis for $\mathbb{R}^n$ which is dual to $\{v_i\}$; by definition, $\{v_i^*\}$ is determined by the relation $\langle v_i^*, v_j\rangle=\delta_{ij}$. If we write $x\in\Gamma$ as $x=\sum a_jv_j$, $a_j\in\mathbb{Z}$, we have $\langle v_i^*, x\rangle=a_i$ and so $|a_i|=|\langle v_i^*, x\rangle|\le|v_i^*|\,|x|$. If, furthermore, $x\in\Gamma\cap U_r(0)$ for an $r>0$, then we have $|a_i|\le r\,|v_i^*|$ and since the number of such a_i's are finite, the set $\Gamma\cap U_r(0)$ is also finite. Our assertion then follows from Proposition 3.1.

($\Leftarrow$) Let Γ be a discrete module in $\mathbb{R}^n$ and let L be the subspace of the vector space $\mathbb{R}^n$ consisting of all linear combinations of elements of Γ. Call m the dimension of L and write $L=\mathbb{R}v_1+\cdots+\mathbb{R}v_m$, $v_i\in\Gamma$. Put $\Gamma_0=[v_1,\ldots,v_m]\subset\Gamma$. We claim that $[\Gamma:\Gamma_0]<\infty$. First of all, for the independent vectors $v_1,\ldots,v_m$, we introduce the *parallelotope* spanned by v_i's:

$$\Pi=\Pi(v_1,\ldots,v_m)\overset{\text{def}}{=}\left\{v=\sum_{i=1}^{m}\alpha_iv_i\in\mathbb{R}^n; 0\le\alpha_i<1\right\}.$$

Since Π is a bounded set in $\mathbb{R}^n$, we have $\Pi\subset U_r(0)$ for some $r>0$. Next, for $x\in\Gamma\subset L$, write $x=\sum\beta_iv_i$, $\beta_i\in\mathbb{R}$. If we write $\beta_i=[\beta_i]+\{\beta_i\}$ with $[\beta_i]\in\mathbb{Z}$, $0\le\{\beta_i\}<1$, then we get, for $x\in\Gamma$,

$$x=\sum[\beta_i]v_i+\sum\{\beta_i\}v_i\in\Gamma_0+\Pi.$$

Since $\Gamma_0\subset\Gamma$, we have $\sum\{\beta_i\}v_i\in\Gamma\cap\Pi\subset\Gamma\cap U_r(0)$, where the latter set is finite because Γ is discrete (Proposition 3.1). This shows that any $x\in\Gamma$ is equivalent to a finite set of vectors in Γ module Γ_0 and so $[\Gamma:\Gamma_0]<\infty$. Now, for any $x\in\Gamma$, we have $Nx\in\Gamma_0$, $N=[\Gamma:\Gamma_0]$, hence

$$\Gamma\subset N^{-1}\Gamma_0=[v_1/N,\ldots,v_m/N].$$

Since $(1/N)\Gamma_0$ is a free abelian group of rank m, i.e., $N^{-1}\Gamma_0\approx\mathbb{Z}^m$, Γ is a free abelian group of rank $l\le m$: $\Gamma=[u_1,\ldots,u_l]$ (Exercise 3.2). However, as $\Gamma\supset\Gamma_0=[v_1,\ldots,v_m]$, we must have $l=m$, i.e., Γ is a sublattice of rank m.

2. ($\Rightarrow$) Let $\Gamma=[v_1,\ldots,v_n]$ be a lattice. Write $x\in\mathbb{R}^n$ as $x=\sum\lambda_iv_i$, $\lambda_i\in\mathbb{R}$. Then, since we have

$$x=\sum[\lambda_i]v_i+\sum\{\lambda_i\}v_i\in\Gamma+\Pi,\qquad \Pi=\Pi(v_1,\ldots,v_n),$$

with a parallelotope Π which is clearly a bounded set in $\mathbb{R}^n$, we see that the quotient $\mathbb{R}^n/\Gamma$ is compact.

($\Leftarrow$) Suppose that a module Γ is discrete and $\mathbb{R}^n/\Gamma$ is compact. From (1), Γ is a sublattice; $\Gamma = [v_1, \ldots, v_m]$. We must show that $m = n$. So suppose, on the contrary, that $m < n$. We shall prove that $\mathbb{R}^n \supsetneqq B + \Gamma$ for any bounded set $B \subset \mathbb{R}^n$. For such a B, there is $r > 0$ such that $B \subset U_r(0)$. Put $L = \mathbb{R}v_1 + \cdots + \mathbb{R}v_m$ and let $L^\perp$ be the subspace in $\mathbb{R}^n$ which is perpendicular to L:

$$L^\perp \stackrel{\text{def}}{=} \{v \in \mathbb{R}^n;\ \langle v, x\rangle = 0,\ \forall x \in L\},$$

this being a subspace of $\mathbb{R}^n$ of dimension $n - m > 0$. Then a vector $v \in L^\perp$ such that $|v| \geq r$ cannot belong to the set $B + \Gamma$; in fact, if $v \in B + \Gamma$ then $v = b + \gamma$, $b \in B$, $\gamma \in \Gamma$, and we have

$$|v|^2 = \langle v, v\rangle = \langle v, b + \gamma\rangle = \langle v, b\rangle \leq |v|\,|b| < r\,|v|$$

or $|v| < r$, a contradiction, so we have proved that $\mathbb{R}^n/\Gamma$ is not compact when $m < n$. Q.E.D.

Exercise 3.2. Let M be a free module of rank n and N be any submodule of M. Prove that N is a free module of rank $m \leq n$.

In the sequel, let $dx = dx_1 \cdots dx_n$ be the standard volume element in the euclidean space $\mathbb{R}^n$ and $\mathrm{vol}(B)$ denote the volume of a bounded set B in $\mathbb{R}^n$: $\mathrm{vol}(B) = \int_B dx$.

Exercise 3.3. Let $\Gamma = [v_1, \ldots, v_n]$ be a lattice in $\mathbb{R}^n$. Regarding v_i as a column vector, put $A = (v_1, \ldots, v_n)$. Prove that $\mathrm{vol}(\Pi(v_1, \ldots, v_n)) = |\det A|$.

Exercise 3.4. Let $s, t \in \mathbb{Z}$, $s \geq 0$, $t \geq 0$, and put $E^{s,t} = \mathbb{R}^s \times \mathbb{C}^t = \mathbb{R}^n$, $n = s + 2t$. Write $x = (x_j)$, $x_j \in \mathbb{R}$, $1 \leq j \leq s$, $x_{s+j} = u_j + iv_j$, $u_j, v_j \in \mathbb{R}$, $1 \leq j \leq t$. Given $c_j > 0$, $1 \leq j \leq s + t$, put

$$\begin{aligned} B = \{x = (x_j) \in E^{s,t},\ |x_1| \leq c_1, \ldots, |x_s| \leq c_s,\ &|x_{s+1}|^2 \\ \leq c_{s+1}, \ldots, &|x_{s+t}|^2 \leq c_{s+t}\}. \end{aligned}$$

Prove that $\mathrm{vol}(B) = 2^s \pi^t \prod_{i=1}^{s+t} c_i$.

Exercise 3.5. With the notation as above (except for B), for $\rho > 0$, put

$$B = \{x = (x_j) \in E^{s,t};\ |x_1| + \cdots + |x_s| + 2\,|x_{s+1}| + \cdots + 2\,|x_{s+t}| \leq \rho\}.$$

Prove that $\mathrm{vol}(B) = 2^s(\pi/2)^t(1/n!)\rho^n$.

EXERCISE 3.6. For a lattice $\Gamma = [v_1, \ldots, v_n] = [v_1', \ldots, v_n']$ verify that $\mathrm{vol}(\Pi(v_1, \ldots, v_n)) = \mathrm{vol}(\Pi(v_1', \ldots, v_n'))$.

3.2. Minkowski's Theorem

A bounded set $B \subset \mathbb{R}^n$ is *central symmetric* if $-B = B$, i.e., $x \in B \Rightarrow -x \in B$, and *convex* if $x, y \in B \Rightarrow \lambda x + (1-\lambda)y \in B$, $0 \leq \lambda \leq 1$. We shall denote by $v(\Gamma)$ the volume of the parallelotope $\Pi = \Pi(v_1, \ldots, v_n)$ of a lattice $\Gamma = [v_1, \ldots, v_n]$ in $\mathbb{R}^n$. (This definition makes sense by Exercise 3.6.)

THEOREM 3.1 (Minkowski). *Let B be a bounded set in $\mathbb{R}^n$. Assume that B is central symmetric and convex.*† *Let Γ be a lattice in $\mathbb{R}^n$. If $\mathrm{vol}(B) > 2^n v(\Gamma)$, then we have $B \cap (\Gamma - \{0\}) \neq \varnothing$.*

PROOF. By the proof of Proposition 3.2 [($\Rightarrow$) of (2)], we have $\mathbb{R}^n = \Pi + \Gamma$. Let us first prove that, for $z, z' \in \Gamma$, $\Pi + z = \Pi + z' \Rightarrow z = z'$. It is enough to show that $\Pi = \Pi + z \Rightarrow z = 0$. So take any element $x = \Sigma(\alpha_i + z_i)v_i \in \Pi + z$, $z_i \in \mathbb{Z}$, $0 \leq \alpha_i < 1$, $z = \Sigma z_i v_i$. Since x also belongs to Π, we must have $0 \leq \alpha_i + z_i$; but this is possible only if $z = 0$. Next, for any bounded set B_0 in $\mathbb{R}^n$ we shall prove that

$$\text{the set } \{z \in \Gamma;\ B_0 \cap (\Pi + z) \neq \varnothing\} \text{ is finite.} \tag{3.1}$$

As both of B_0 and Π are bounded, there is $r > 0$ such that $B_0 \subset U_r(0)$ and $\Pi \subset U_r(0)$. If we take a $z \in \Gamma$ from the set in (3.1), then there is a $v \in \Pi$ such that $r > |v + z| \geq |z| - |v|$; hence $|z| < r + |v| < 2r$ which implies (3.1). We have therefore proved that

$$B_0 \bigcup_{z \in \Gamma} B_0 \cap (\Pi + z) \text{ is a finite disjoint union.} \tag{3.2}$$

We then find that

$$\begin{aligned} \mathrm{vol}(B_0) = \sum_{z \in \Gamma} \mathrm{vol}(B_0 \cap (\Pi + z)) &= \sum_{z \in \Gamma} \mathrm{vol}((B_0 - z) \cap \Pi) \\ &= \sum_{z \in \Gamma} \mathrm{vol}((B_0 + z) \cap \Pi). \end{aligned}$$

† B ($\neq \varnothing$) always contains 0. In fact, $u \in B \Rightarrow -u \in B \Rightarrow 0 = \frac{1}{2}u + (1 - \frac{1}{2})(-u) \in B$.

Now, we return to the bounded set $B \subset \mathbb{R}^n$ which is central symmetric and convex and put $B_0 = \frac{1}{2}B$; hence we have $\operatorname{vol}(B_0) = 2^{-n} \operatorname{vol}(B) > v(\Gamma)$. We claim that

$$(B_0 + z) \cap (B_0 + z') \neq \varnothing \qquad \text{for some } z, z' \in \Gamma. \tag{3.3}$$

In fact, if $(B_0 + z) \cap (B_0 + z') = \varnothing$ for all $z, z' \in \Gamma$, then we would have

$$\operatorname{vol}(B_0) = \sum_{z \in \Gamma} \operatorname{vol}((B_0 + z) \cap \Pi) \leq \operatorname{vol}(\Pi) = v(\Gamma)$$

which contradicts the assumption $\operatorname{vol}(B_0) > v(\Gamma)$. Thus from (3.3) it follows that there are $b, b' \in B$ such that $\frac{1}{2}b + z = \frac{1}{2}b' + z'$ and, since $z - z' = \frac{1}{2}b' - \frac{1}{2}b = \frac{1}{2}b' + (1 - \frac{1}{2})(-b) \in B$, we get $z - z' \in B \cap (\Gamma - \{0\})$. Q.E.D.

From now on, we shall apply our geometric results to some problems of algebraic number field $k/\mathbb{Q}$, $n = [k\colon \mathbb{Q}]$. Let $k^{(j)} \subset \mathbb{C}$ denote the jth conjugate of k, $1 \leq j \leq n$. Some of them may already be contained in $\mathbb{R}$. When $k^{(j)} \not\subset \mathbb{R}$, then $\alpha \mapsto \alpha^{(j)}$ and $\alpha \mapsto \bar{\alpha}^{(j)}$ are distinct embeddings of the field k. We shall assume that $k^{(j)} \subset \mathbb{R}$, $1 \leq j \leq r_1$ and $k^{(r_1+j)} \subset \mathbb{C}$ ($\not\subset \mathbb{R}$), $1 \leq j \leq r_2$; we shall arrange all conjugates of a number $\alpha \in k$ as follows:

$$\alpha^{(j)} \in k^{(j)}, \qquad 1 \leq j \leq r_1, \qquad \alpha^{(r_1+j)} \text{ and } \bar{\alpha}^{(r_1+j)} \in k^{(r_1+j)}, \qquad 1 \leq j \leq r_2.$$

THEOREM 3.2. *Let k be a number field, $n = [k\colon \mathbb{Q}]$, and $\mathfrak{a}$ be an ideal of $\mathfrak{o}_k$. There exists a number $\mu \neq 0$ in $\mathfrak{a}$ such that $|N_{k/\mathbb{Q}}\mu| \leq M_k(N_k\mathfrak{a})$ where $M_k = (4/\pi)^{r_2}(n!/n^n)\,|\Delta_k|^{1/2}$, the Minkowski constant of k.*

PROOF. For $\alpha \in k$, the first r_1 conjugates $\alpha^{(j)}$ are real and the rest are r_2 pairs of complex conjugates, $\alpha^{(j)} = \beta^{(j)} + i\gamma^{(j)}$ and $\bar{\alpha}^{(j)} = \beta^{(j)} - i\gamma^{(j)}$. Define a mapping $x\colon k \to E^{r_1,r_2} = \mathbb{R}^{r_1} \times \mathbb{C}^{r_2}$ by

$$x(\alpha) = (\alpha^{(1)}, \ldots, \alpha^{(r_1)}, \alpha^{(r_1+1)}, \ldots, \alpha^{(r_1+r_2)}). \tag{3.4}$$

Clearly x is injective. The natural identification $\mathbb{C} = \mathbb{R}^2$ identifies E^{r_1,r_2} with $\mathbb{R}^n$. Let $\mathfrak{a} = [\alpha_1, \ldots, \alpha_n] \subset \mathfrak{o}_k$ be an ideal and let v_j be the vector in $\mathbb{R}^n$ corresponding to $x(\alpha_j) \in E^{r_1,r_2}$ defined by (3.4):

$$v_j = (\alpha_j^{(1)}, \ldots, \alpha_j^{(r_1)}, \beta_j^{(r_1+1)}, \gamma_j^{(r_1+1)}, \ldots, \beta_j^{(r_1+r_2)}, \gamma_j^{(r_1+r_2)}).$$

Then

$$\Gamma = [v_1, \ldots, v_n] \subset \mathbb{R}^n$$

becomes a lattice in $\mathbb{R}^n$ because $\det(\alpha_j^{(i)}) \neq 0$. By Exercise 3.3 and Proposition 2.18, we get

$$\begin{aligned} v(\Gamma) = 2^{-r_2} \det(\alpha_j^{(i)}) &= 2^{-r_2} |d_{k/\mathbb{Q}}(\alpha_1, \ldots, \alpha_n)|^{1/2} \\ &= 2^{-r_2} |\Delta_k|^{1/2} N_k \mathfrak{a}.\dagger \end{aligned} \tag{3.5}$$

Now, determine $\rho > 0$ so that

$$2^{r_1}(\pi/2)^{r_2}(1/n!)\rho^n = 2^n v(\Gamma) \tag{3.6}$$

and, for any $\epsilon > 0$, put

$$B_\epsilon = \{x = (x_j) \in E^{r_1, r_2}; |x_1| + \cdots + |x_{r_1}| + 2|x_{r_1+1}| + \cdots + 2|x_{r_1+r_2}| \leq \rho + \epsilon\}. \tag{3.7}$$

Then, by Exercise 3.5, we have

$$\operatorname{vol}(B_\epsilon) = 2^{r_1}(\pi/2)^{r_2}(1/n!)(\rho + \epsilon)^n. \tag{3.8}$$

Hence, from (3.6) and (3.8), we find that

$$\operatorname{vol}(B_\epsilon) > 2^n v(\Gamma). \tag{3.9}$$

Since B_ϵ is central symmetric and convex as one verifies easily, we get, by (3.9) and Theorem 3.1,

$$B_\epsilon \cap (\Gamma - \{0\}) \neq \emptyset, \qquad \epsilon > 0.$$

† Note that the value of the Jacobian determinant due to the change of variables

$$\begin{pmatrix} \beta^{(j)} \\ \gamma^{(j)} \end{pmatrix} = \begin{pmatrix} \frac{1}{2} & \frac{1}{2} \\ \frac{1}{2i} & -\frac{1}{2i} \end{pmatrix} \begin{pmatrix} \alpha^{(j)} \\ \bar{\alpha}^{(j)} \end{pmatrix}, \qquad r_1 + 1 \leq j \leq r_1 + r_2$$

is

$$\begin{vmatrix} \frac{1}{2} & \frac{1}{2} \\ \frac{1}{2i} & -\frac{1}{2i} \end{vmatrix}^{r_2} = \left(\frac{i}{2}\right)^{r_2}.$$

Now put $B_0 = \bigcap_{\epsilon>0} B_\epsilon$; actually, B_0 is obtained by putting $\epsilon = 0$ in (3.7) and we have $B_0 \cap (\Gamma - \{0\}) = \bigcap_{\epsilon>0} (B_\epsilon \cap (\Gamma - \{0\}))$. Since B_ϵ is compact and $\Gamma - \{0\}$ is discrete, for any $\epsilon > 0$, $B_\epsilon \cap (\Gamma - \{0\})$ is a nonempty finite set and, for any finite number of ϵ_i's, $1 \le i \le l$, we have

$$\bigcap_{i=1}^{l} B_{\epsilon_i} \cap (\Gamma - \{0\}) \neq \varnothing. \tag{3.10}$$

If we take $\epsilon < 1$, since $B_\epsilon \cap (\Gamma - \{0\})$ is a closed subset of the compact set B_1 and satisfies the finite intersection property (3.10), we see that $B_0 \cap (\Gamma - \{0\}) \neq \varnothing$. This means that there is a number $\mu \neq 0$ in $\mathfrak{a}$ such that

$$|\mu^{(1)}| + \cdots + |\mu^{(r_1)}| + 2\,|\mu^{(r_1+1)}| + \cdots + 2\,|\mu^{(r_1+r_2)}| \le \rho.$$

On the other hand, we have

$$\begin{aligned} |N_{k/\mathbf{Q}}\mu| &= |\mu^{(1)}| \cdots |\mu^{(r_1)}|\,|\mu^{(r_1+1)}|^2 \cdots |\mu^{(r_1+r_2)}|^2 \\ &\le [n^{-1}(|\mu^{(1)}| + \cdots + |\mu^{(r_1)}| + 2\,|\mu^{(r_1+1)}| + \cdots + 2\,|\mu^{(r_1+r_2)}|]^n \\ &\le n^{-n}\rho^n, \end{aligned}$$

and, substituting here the values of ρ^n in (3.6) and $v(\Gamma)$ in (3.5), we conclude that

$$|N_{k/\mathbf{Q}}\mu| \le (n!/n^n)(4/\pi)^{r_2}\,|\Delta_k|^{1/2}\,N_k\mathfrak{a} = M_k(N_k\mathfrak{a}). \qquad \text{Q.E.D.}$$

Theorem 3.3 *Any ideal class of the group H_k contains an ideal $\mathfrak{a} \subset \mathfrak{o}_k$ such that $N_k\mathfrak{a} \le M_k$.*

Proof. Let C be any ideal class in H_k and $\mathfrak{b} \subset \mathfrak{o}_k$ be an ideal in the class C^{-1}. By Theorem 3.2, there is a $\mu \neq 0$ in $\mathfrak{b}$ such that $|N_{k/\mathbf{Q}}\mu| \le (N_k\mathfrak{b})M_k$. Since $\mu \in \mathfrak{b}$, we can find an ideal $\mathfrak{a} \subset \mathfrak{o}_k$ such that $(\mu) = \mathfrak{a}\mathfrak{b}$; hence $\mathfrak{a} \sim \mathfrak{b}^{-1}$ and so $\mathfrak{a} \in C$. As $N_k\mathfrak{a}N_k\mathfrak{b} = |N_{k/\mathbf{Q}}\mu| \le (N_k\mathfrak{b})M_k$, we have $N_k\mathfrak{a} \le M_k$. Q.E.D.

Remark 3.1. If $M_k < 2$ for a number field k, any ideal class must contain an ideal $\mathfrak{a}$ in $\mathfrak{o}_k$ such that $N_k\mathfrak{a} = 1$, i.e., $\mathfrak{a} = \mathfrak{o}_k$; therefore we have $h_k = 1$.

Remark 3.2. The finiteness of h_k (Theorem 2.13) also follows from Theorem 3.3. Since the number of ideals $\mathfrak{a}$ in $\mathfrak{o}_k$ such that

$N_k\mathfrak{a} \le M_k$ is finite, we can determine h_k in some cases as in Exercise 3.7 and Exercise 3.8.

EXERCISE 3.7. Prove that $h_k = 1$ for the following 9 imaginary quadratic fields: $k = \mathbb{Q}(\sqrt{m})$, $-m = 1, 2, 3, 7, 11, 19, 43, 67, 163$.†

EXERCISE 3.8. Prove that $h_k = 1$ for the following real quadratic fields: $k = \mathbb{Q}(\sqrt{m})$, $m = 2, 3, 5, 13$.

EXERCISE 3.9. Let us construct an imaginary quadratic field with $h_k \ge 2$. Let a and b be square free positive integers >1 such that $(a, b) = 1$ and $-ab \not\equiv 1 \pmod 4$, e.g., $a = 2$ and $b = 3$. Prove that $h_k \ge 2$ for $k = \mathbb{Q}(\sqrt{-ab})$.

3.3. Dirichlet's Unit Theorem

The set $E^{s,t} = \mathbb{R}^s \times \mathbb{C}^t$ has a natural structure of a commutative algebra of dimension $s + 2t$ over $\mathbb{R}$. Its group of invertible elements is

$$(E^{s,t})^\times = (\mathbb{R}^\times)^s \times (\mathbb{C}^\times)^t.$$

Define a mapping $l\colon (E^{s,t})^\times \to \mathbb{R}^{s+t}$ by

$$l(x) = (l_1(x), \ldots, l_s(x), l_{s+1}(x), \ldots, l_{s+t}(x)) \tag{3.11}$$

where

$$\begin{cases} l_i(x) = \log |x_i|, & 1 \le i \le s, \\ l_i(x) = \log |x_i|^2, & s+1 \le i \le s+t. \end{cases}$$

PROPOSITION 3.3. *The mapping l defined by (3.11) is a surjective homomorphism whose kernel is the group $\{\pm 1\}^s \times T^t$, $T = \{z \in \mathbb{C}; |z| = 1\}$.*

PROOF. It is clear that $l(xy) = l(x) + l(y)$, $x, y \in (E^{s,t})^\times$. For any $a = (a_i) \in \mathbb{R}^{s+t}$, determine $x = (x_i) \in (E^{s,t})^\times$ so that $x_i = e^{a_i}$, $1 \le i \le s$, $x_i = e^{a_i/2}$, $s+1 \le i \le s+t$. Then we have $l(x) = a$, i.e., l is surjective. The statement about Ker l is obvious. Q.E.D.

† Conversely, we know that these nine fields exhaust all imaginary quadratic fields with $h_k = 1$ (Stark, 1939–). By the way, there are exactly 18 imaginary quadratic fields with $h_k = 2$:

$-m = 5, 6, 10, 13, 15, 22, 35, 37, 51, 58, 91, 115, 123, 187, 235, 267, 403, 427.$

Let k be a number field, $n=[k:\mathbb{Q}]$. Combining the mapping $x: k \to E^{r_1,r_2}$ [see formula (3.4)] with the mapping l in (3.11), we obtain the mapping $l^*: k^\times \to \mathbb{R}^{r_1+r_2}$ given by

$$l^*(\alpha)=l(x(\alpha))=(\log|\alpha^{(1)}|,\ldots,\log|\alpha^{(r_1)}|,\log|\alpha^{(r_1+1)}|^2,\ldots,\log|\alpha^{(r_1+r_2)}|^2). \quad (3.12)$$

If we put $l_i^*=l_i\cdot x$, $1\le i\le r_1+r_2$, we have

$$\sum_{i=1}^{r_1+r_2} l_i^*(\alpha)=\log|N_{k/\mathbb{Q}}\alpha|. \quad (3.13)$$

Consider a hyperplane in $\mathbb{R}^{r_1+r_2}$:

$$H^r \overset{\text{def}}{=} \left\{v=(v_1,\ldots,v_{r_1+r_2});\ \sum_{i=1}^{r_1+r_2} v_i=0\right\}, \qquad r=r_1+r_2-1.$$

Since $N_{k/\mathbb{Q}}\alpha=\pm 1$ for $\alpha\in\mathfrak{o}_k^\times$, we see from (3.13) that $l^*(\mathfrak{o}_k^\times)\subset H^r$. Denote by $l^\#$ the restriction of l^* on the subgroup $\mathfrak{o}_k^\times$ of $k^\times$. We have thus a homomorphism

$$l^\#: \mathfrak{o}_k^\times \to H^r, \qquad r=r_1+r_2-1.$$

The study of the structure of the group $\mathfrak{o}_k^\times$ is reduced to that of $\operatorname{Ker} l^\#$ and $\operatorname{Im} l^\#$.

Proposition 3.4. $\operatorname{Ker} l^\# = W_k \overset{\text{def}}{=} \{\alpha\in\mathfrak{o}_k^\times;$ *order of* α *is finite*$\}$.

Proof. Take an $\alpha\in W_k$. As $\alpha^N=1$, for some $N\in\mathbb{N}$ we have $|\alpha^{(i)}|=1$, $1\le i\le r_1+r_2$, hence $l^\#(\alpha)=l^*(\alpha)=0$, i.e., $\alpha\in\operatorname{Ker} l^\#$. Conversely, assume that $l^\#(\alpha)=0$. Then $|\alpha^{(i)}|=1$, $1\le i\le r_1+r_2$, which implies that the set $x(\operatorname{Ker} l^\#)$ is a bounded set in $\mathbb{R}^n=E^{r_1,r_2}$, $n=r_1+2r_2$. On the other hand, as $x(\mathfrak{o}_k)$ is a lattice in $\mathbb{R}^n$, it is discrete in $\mathbb{R}^n$ and so is the subset $x(\operatorname{Ker} l^\#)$. Therefore $x(\operatorname{Ker} l^\#)$ is finite. Since x is injective, $\operatorname{Ker} l^\#$ is a finite group and so $\operatorname{Ker} l^\#\subset W_k$. Q.E.D.

Remark 3.3. By Theorem 1.18, the group W_k is cyclic. Since -1 is contained in W_k, the order of W_k is even and W_k consists of all roots of unity in k. For any k, it is difficult to determine the order of W_k; $W_\mathbb{Q}=\{\pm 1\}$, of course.

Exercise 3.10. Let $k=\mathbb{Q}(\sqrt{m})$ be a quadratic field. If $m>0$, then $W_k=\{\pm1\}$. For $m<0$, $W_k=\{\pm1, \pm i\}$ if $m=-1$, $W_k=\{\pm1, \pm\rho, \pm\rho^2\}$, $\rho=e^{2\pi i/3}$, if $m=-3$, and $W_k=\{\pm1\}$ otherwise.

Exercise 3.11. For an integer $m\geq 2$, let $\zeta=e^{2\pi i/m}$, $k=\mathbb{Q}(\zeta)$, the mth cyclotomic field. Prove that the minimal polynomial of ζ over $\mathbb{Q}$ is $f(X)=\prod_{(a,m)=1}(X-\zeta^a)$, $0\leq a\leq m-1$. Thus we have $\deg f=[\mathbb{Q}(\zeta):\mathbb{Q}]=\varphi(m)$.

Exercise 3.12. Let l be an odd prime, $k=\mathbb{Q}(\zeta)$, $\zeta=e^{2\pi i/l}$. Prove that $W_k=\{\pm\zeta^a; 0\leq a\leq l-1\}$; hence the order of W_k is $2l$.

Proposition 3.5. *$\operatorname{Im} l^{\#}$ is a sublattice of H^r.*

Proof. Since $\operatorname{Im} l^{\#}\subset H^r$, it is enough to prove that $\operatorname{Im} l^{\#}$ is discrete in $\mathbb{R}^{r_1+r_2}$. To do this, we must prove that, for any $r>0$, $\operatorname{Im} l^{\#}\cap U_r(0)$ is finite. Take any vector $l^{\#}(\alpha)\in\operatorname{Im} l^{\#}\cap U_r(0)$. Then (3.12) implies

$$(\log|\alpha^{(1)}|)^2+\cdots+(\log|\alpha^{(r_1+r_2)}|^2)^2<r^2$$

and hence $|\alpha^{(1)}|<e^r,\ldots,|\alpha^{(r_1)}|<e^r$, $|\alpha^{(r_1+1)}|^2<e^r,\ldots,|\alpha^{(r_1+r_2)}|^2<e^r$. Therefore the set

$$S=\{x(\alpha); \alpha\in\mathfrak{o}_k^{\times}, l^{\#}(\alpha)\in\operatorname{Im} l^{\#}\cap U_r(0)\} \tag{3.14}$$

is bounded in $E^{r_1,r_2}=\mathbb{R}^n$, $n=r_1+2r_2$. On the other hand, $x(\mathfrak{o}_k)$ is a lattice in $\mathbb{R}^n$ and so its subset S is discrete; hence S is finite. As x is injective, the above argument shows that $\operatorname{Im} l^{\#}\cap U_r(0)$ is finite. Q.E.D.

Proposition 3.6. *Let ν be a natural number. There are numbers $\alpha_1,\ldots,\alpha_N$ in $\mathfrak{o}_k$ with $|N_{k/\mathbb{Q}}\alpha_i|=\nu$, $1\leq i\leq N$, such that any $\alpha\in\mathfrak{o}_k$ with $|N_{k/\mathbb{Q}}\alpha|=\nu$ can be expressed as $\alpha=\epsilon\alpha_i$ for some i and $\epsilon\in\mathfrak{o}_k^{\times}$.*

Proof. If we write $\mathfrak{o}_k=[\omega_1,\ldots,\omega_n]$, then $\nu\mathfrak{o}_k\subset\mathfrak{o}_k$ and clearly $[\mathfrak{o}_k:\nu\mathfrak{o}_k]=\nu^n$. Call $\alpha_1,\ldots,\alpha_N$ the set of representatives of the set $\{\alpha\in\mathfrak{o}_k; |N_{k/\mathbb{Q}}\alpha|=\nu\}$ considered modulo $\nu\mathfrak{o}_k$. Then for any number α in that set there is an i such that $\alpha\equiv\alpha_i\ (\nu\mathfrak{o}_k)$. We must show that α/α_i is a unit of k. In fact, if $\alpha-\alpha_i=\beta\nu$, $\beta\in\mathfrak{o}_k$, then

$$\alpha/\alpha_i=1+\beta(\nu/\alpha_i)=1\pm\beta(N\alpha_i/\alpha_i), \qquad \text{where } (N\alpha_i)/\alpha_i\in\mathfrak{o}_k.\dagger$$

† For $\alpha\in\mathfrak{o}_k$, $N\alpha=\alpha(\alpha^{(2)}\cdots\alpha^{(n)})\Rightarrow N\alpha/\alpha=\alpha^{(2)}\cdots\alpha^{(n)}\in k\cap\bar{\mathbb{Z}}=\mathfrak{o}_k$.

Therefore we have $\alpha/\alpha_i \in \mathfrak{o}_k$. Interchanging α and α_i, we get $\alpha_i/\alpha \in \mathfrak{o}_k$ and so $\alpha/\alpha_i \in \mathfrak{o}_k^\times$. Q.E.D.

PROPOSITION 3.7. *Im $l^\#$ is a lattice of H^r.*

PROOF. Let $l: (E^{r_1,r_2})^\times \to \mathbb{R}^{r_1+r_2}$ be the surjective homomorphism given by (3.11) (Proposition 3.3). Put

$$T \overset{\text{def}}{=} l^{-1}(H^r) = \{y \in (E^{r_1,r_2})^\times; l(y) \in H^r\}. \tag{3.15}$$

For $y = (y_1, \ldots, y_{r_1}, y_{r_1+1}, \ldots, y_{r_1+r_2}) \in E^{r_1,r_2} = \mathbb{R}^{r_1} \times \mathbb{C}^{r_2}$, put

$$N_{E^{r_1,r_2}/\mathbb{R}}(y) \overset{\text{def}}{=} y_1 \cdots y_{r_1} |y_{r_1+1}|^2 \cdots |y_{r_1+r_2}|^2.$$

Then we have

$$y \in T \Leftrightarrow l(y) \in H^r \Leftrightarrow \sum_{i=1}^{r_1+r_2} l_i(y) = 0 \Leftrightarrow$$
$$\log |N_{E^{r_1,r_2}/\mathbb{R}}(y)| = 0 \Leftrightarrow N_{E^{r_1,r_2}/\mathbb{R}}(y) = \pm 1. \tag{3.16}$$

Let $\Gamma = x(\mathfrak{o}_k)$, a lattice in $E^{r_1,r_2} = \mathbb{R}^n$ [see formula (3.4)] and let $v(\Gamma)$ be, as before, the volume of the parallelotope $\Pi(v_1, \ldots, v_n)$ for $\Gamma = [v_1, \ldots, v_n]$. Take any $y \in T$. For a lattice $y\Gamma = [yv_1, \ldots, yv_n]$, define a matrix $M(y) \in \mathrm{GL}_n(\mathbb{R})$ by

$$(yv_1, \ldots, yv_n) = (v_1, \ldots, v_n)M(y). \tag{3.17}$$

Then we find that $\det M(y) = N_{E^{r_1,r_2}/\mathbb{R}}(y)$ (Exercise 3.13) and since $\det M(y) = \pm 1$ by (3.16) we have $v(\Gamma) = v(y\Gamma)$, $y \in T$.

Now, find positive numbers $c_1, \ldots, c_{r_1+r_2}$ so that

$$\gamma = c_1 \cdots c_{r_1+r_2} > (4/\pi)^{r_2} v(\Gamma) \tag{3.18}$$

and put

$$B = \{x \in E^{r_1,r_2}; |x_1| \le c_1, \ldots, |x_{r_1}| \le c_{r_1}, |x_{r_1+1}|^2$$
$$\le c_{r_1+1}, \ldots, |x_{r_1+r_2}|^2 \le c_{r_1+r_2}\}.$$

Then B is a bounded central symmetric convex subset of $E^{r_1,r_2} = \mathbb{R}^n$ and we have, by Exercise 3.4,

$$\mathrm{vol}(B) = 2^{r_1}\pi^{r_2} \prod_{i=1}^{r_1+r_2} c_i. \tag{3.19}$$

As $v(\Gamma)=v(y\Gamma)$ for $y\in T$, (3.18) and (3.19) imply $\mathrm{vol}(B)>2^n v(y\Gamma)$ and we get $B\cap(y\Gamma-\{0\})\neq\varnothing$ by Theorem 3.1, i.e., we can find z such that

$$z\in B\cap y\Gamma,\qquad z\neq 0. \tag{3.20}$$

If we write $z=yx(\alpha)$, $\alpha\in\mathfrak{o}_k$, then since $\alpha\neq 0$, we have $z\in(E^{r_1,r_2})^\times$. On the other hand, since $z\in B$, we have

$$|N_{E^{r_1,r_2}/\mathbb{R}}(z)|=|z_1\cdots z_{r_1}|z_{r_1+1}|^2\cdots|z_{r_1+r_2}|^2|\leq c_1\cdots c_{r_1+r_2}=\gamma$$

or $|N_{k/\mathbb{Q}}\alpha|\leq\gamma$ because $N_{E^{r_1,r_2}/\mathbb{R}}(z)=N_{E^{r_1,r_2}/\mathbb{R}}(y)N_{k/\mathbb{Q}}\alpha=\pm N_{k/\mathbb{Q}}\alpha$. Applying Proposition 3.6 to integers $\nu=1,2,\ldots[\gamma]$, we find $\alpha_1,\ldots,\alpha_N$, in $\mathfrak{o}_k$ with $|N_{k/\mathbb{Q}}\alpha_i|\leq\gamma$ so that $\epsilon\alpha=\alpha_i$ with some i and $\epsilon\in\mathfrak{o}_k^\times$. Hence the z in (3.20) can be written $z=yx(\alpha_i)x(\epsilon)^{-1}$; in other words, any $y\in T$ can be written $y=zx(\alpha_i^{-1})x(\epsilon)$. Now, put

$$B_0\overset{\text{def}}{=}T\cap\left(\bigcup_{i=1}^{N}Bx(\alpha_i^{-1})\right)\subset E^{r_1,r_2}=\mathbb{R}^n,$$

which is clearly a bounded set. We shall prove the following equality:

$$T=B_0x(\mathfrak{o}_k^\times). \tag{3.21}$$

Since the inclusion $T\supset B_0x(\mathfrak{o}_k^\times)$ is obvious, it is enough to show that $T\subset B_0x(\mathfrak{o}_k^\times)$. In fact, take any $y\in T$ and write, as above, $y=zx(\alpha_i^{-1})x(\epsilon)$. First, we have $zx(\alpha_i^{-1})\in Bx(\alpha_i^{-1})$. Second, the equality $zx(\alpha_i^{-1})=yx(\epsilon^{-1})$ implies that $zx(\alpha_i^{-1})\in B_0$ as $zx(\alpha_i^{-1})\in T$. Therefore we get $y=zx(\alpha_i^{-1})x(\epsilon)\in B_0x(\mathfrak{o}_k^\times)$ which proves (3.21). Taking l of both sides of (3.21) yields

$$l(T)=l(B_0)+l^{\#}(\mathfrak{o}_k^\times)\subset H^r. \tag{3.22}$$

Now, since the mapping $l\colon(E^{r_1,r_2})^\times\to\mathbb{R}^{r_1+r_2}$ is surjective (Proposition 3.3), for any $\lambda\in H^r$, there is a $y\in(E^{r_1+r_2})^\times$ such that $\lambda=l(y)$. Then we have $y\in T$ by (3.15), i.e., $l(T)=H^r$ and so $H^r=l(B_0)+l^{\#}(\mathfrak{o}_k^\times)$ by (3.22). Since B_0 is bounded, $\bar{B}_0$ is compact and so is the image $l(\bar{B}_0)$, which implies that $l(B_0)$ is bounded. Since $l^{\#}(\mathfrak{o}_k^\times)$ is a discrete subgroup of H^r (Proposition 3.5), $l^{\#}(\mathfrak{o}_k^\times)$ is a lattice in H^r (Proposition 3.2). Q.E.D.

EXERCISE 3.13. For a matrix $M(y)$ in (3.17), prove that $\det M(y) = N_{E^{r_1,r_2}/\mathbb{R}}(y)$.

LEMMA 3.4 (*Dirichlet's unit theorem*). *Let k be a number field and $\mathfrak{o}_k^\times$ be the group of units of $\mathfrak{o}_k$. Then we have*

$$\mathfrak{o}_k^\times = W_k \times \mathbb{Z}^r, \qquad r = r_1 + r_2 - 1\dagger$$

where W_k is the group of roots of unity in k, r_1 is the number of real conjugates of k, and r_2 is the number of pairs of complex conjugates of k.

PROOF. From Proposition 3.4, we obtain the following exact sequence:

$$1 \to W_k \to \mathfrak{o}_k^\times \to l^\#(\mathfrak{o}_k^\times) \to 1.$$

By Proposition 3.7, $l^\#(\mathfrak{o}_k^\times)$ is a lattice of H^r and we have $l^\#(\mathfrak{o}_k^\times) = [l^\#(\epsilon_1), \ldots, l^\#(\epsilon_r)]$, $\epsilon_i \in \mathfrak{o}_k^\times$. Hence, for any unit $\epsilon \in \mathfrak{o}_k^\times$, integers a_i are uniquely determined so that

$$l^\#(\epsilon) = \sum_{i=1}^{r} a_i l^\#(\epsilon_i) = l^\#\left(\prod_{i=1}^{r} \epsilon_i^{a_i}\right).$$

Therefore we have $\epsilon = \zeta \Pi \epsilon_i^{a_i}$, $\zeta \in W_k$. Q.E.D.

REMARK 3.4. We see easily that

$$r = r_1 + r_2 - 1 = 0 \Leftrightarrow \begin{matrix} r_1 = 1, & r_2 = 0 \Leftrightarrow k = \mathbb{Q} & \text{or} & \\ r_1 = 0, & r_2 = 1 \Leftrightarrow k = \mathbb{Q}(\sqrt{m}), & & m < 0. \end{matrix}$$

In these cases, we have $\mathfrak{o}_k^\times = W_k$, a finite group. On the other hand, if $k/\mathbb{Q}$ is a Galois extension, we have

$$k \text{ is real} \Leftrightarrow r_1 = n, \qquad r_2 = 0, \qquad r = n - 1,$$

$$k \text{ is imaginary} \Leftrightarrow r_1 = 0, \qquad r_2 = n/2, \qquad r = n/2 - 1.$$

For example, if k is a real quadratic field, then $r = 1$. In this case, $\mathfrak{o}_k^\times$ is closely related to continued fractions (see Chapter 4).

† We wrote $\mathfrak{o}_k^\times = W_k \times \mathbb{Z}^r$ as we identified $\mathbb{Z}^r$ with a subgroup of $\mathfrak{o}_k^\times$ isomorphic with it.

EXERCISE 3.14. For $\gamma \in \mathfrak{o}_k$, prove that $\gamma \in W_k \Leftrightarrow |\gamma^{(i)}| = 1$, $1 \le i \le r_1 + r_2$.

EXERCISE 3.15. Let $K = \mathbb{Q}(\zeta)$, $\zeta = e^{2\pi i/l}$, l being a prime $\neq 2$, and $k = \mathbb{Q}(\zeta + \zeta^{-1})$, the maximal real subfield of K. Prove that $\mathfrak{o}_K^\times = \langle \zeta \rangle \times \mathfrak{o}_k^\times$.

3.4. Pre-Zeta Functions

Let $F(X) \in \mathbb{R}[X]$, and $X = (X_1, \ldots, X_n)$ be a real homogeneous polynomial of degree d with n variables. Put

$$V = \{x \in \mathbb{R}^n; F(x) \neq 0\}.$$

This set V is a nonempty cone.† Next, put

$$V_1 = \{x \in V; |F(x)| \le 1\}.$$

This is not empty, too. Consider a subset C of $\mathbb{R}^n$ with the following properties:

$$\begin{cases} 1. \ C \subset \mathbb{R}^n \text{ is a cone,} \\ 2. \ C \subset V, \text{ i.e., } x \in C \Rightarrow F(x) \neq 0, \\ 3. \ C_1 = C \cap V_1 \text{ is bounded and has the volume } \mathrm{vol}(C_1) > 0. \end{cases} \tag{3.23}$$

Next, consider a lattice Γ in $\mathbb{R}^n$ such that

$$C \cap \Gamma \neq \varnothing.‡ \tag{3.24}$$

For a real variable s, put

$$\zeta(F, C, \Gamma; s) \overset{\text{def}}{=} \sum_{x \in C \cap \Gamma} \frac{1}{|F(x)|^{(n/d)s}}. \tag{3.25}$$

The function of the real variable s defined by (3.25) is called a *pre-zeta function* determined by F, C, and Γ.

† A set $C \subset \mathbb{R}^n$ is a cone $\Leftrightarrow x \in C \Rightarrow \lambda x \in C$, $\forall \lambda > 0$.
‡ Note that $x \in C \cap \Gamma \Rightarrow F(x) \neq 0 \Rightarrow x \neq 0$.

THEOREM 3.5.

1. $\zeta(F, C, \Gamma; s)$ *converges for* $s > 1$.
2. $\lim_{s\to 1+0}(s-1)\zeta(F, C, \Gamma; s) = \mathrm{vol}(C_1)/v(\Gamma)$, *where*

$$v(\Gamma) = \mathrm{vol}(\Pi(v_1, \ldots, v_n)), \qquad \Gamma = [v_1, \ldots, v_n].$$

REMARK 3.5. $s \to 1+0$ means that s approaches in $\mathbb{R}$ to 1 from the right. Although in this book we do not need the complex variable s, we can extend the definition of $\zeta(F, C, \Gamma; s)$ for a complex s and prove the theorem as above. In Theorem 3.5(1) we have only to replace s by $\mathrm{Re}\, s$. The left-hand side of Theorem 3.5(2) is nothing but the *residue* at $s=1$ of the function $\zeta(F, C, \Gamma; s)$ of complex variable s which has a pole of order 1 at $s=1$. Having these in mind, we shall adopt the term "residue" even in the case of real variable s; hence Theorem 3.5(2) says that "$\zeta(F, C, \Gamma; s)$ has the residue $\mathrm{vol}(C_1)/v(\Gamma)$ at $s=1$."

REMARK 3.6. As we shall need the *Riemann* (1826–1866) *zeta function* in the proof of Theorem 3.5, we prepare some basic facts here. Notation being as above, let $n=1$, $F(X)=X$, $d=1$, $V=\mathbb{R}^\times$, $V_1=\{x \in \mathbb{R}^\times; |x| \le 1\} = [-1, 1] - \{0\}$, $C = \mathbb{R}^\times_+ = \{x \in \mathbb{R}^\times; x > 0\}$, $C_1 = C \cap V_1 = (0, 1]$, and $\Gamma = \mathbb{Z} = [1]$. Hence $\Pi(1) = \{\alpha; 0 \le \alpha < 1\} = [0, 1)$ and $v(\Gamma) = \mathrm{vol}\, \Pi(1) = 1$. The Riemann zeta function is

$$\zeta_{\mathbb{Q}}(s) \stackrel{\text{def}}{=} \zeta(F, C, \Gamma; s) = \sum_{x \in C\cap\Gamma} \frac{1}{|x|^s} = \sum_{n=1} \frac{1}{n^s}.$$

The exercise below is a special case of Theorem 3.5.

EXERCISE 3.16. Prove that

1. $\zeta_{\mathbb{Q}}(s)$ converges for $s > 1$, and
2. the residue of $\zeta_{\mathbb{Q}}(s)$ at $s=1$ is 1.

PROOF OF THEOREM 3.5. For simplicity, we put $\zeta(s) = \zeta(F, C, \Gamma; s)$ and $\varphi(x) = |F(x)|^{n/d}$; note that $\varphi(\lambda x) = \lambda^n \varphi(x)$, for $\lambda > 0$, and that

$$\zeta(s) = \sum_{x \in C\cap\Gamma} \frac{1}{\varphi(x)^s}.$$

If we put $\Gamma_r = r^{-1}\Gamma$, for $r > 0$, then we have $v(\Gamma_r) = r^{-n}v(\Gamma)$ by Exercise 3.3. Since Γ_r is discrete and C_1 is bounded, we can define a number $N(r) \stackrel{\text{def}}{=} [\Gamma_r \cap C_1]$ where $[*]$ means the cardinality of a finite set $*$ (Figure 3.1). By the principle of integral calculus, we have

$$\mathrm{vol}(C_1) = \lim_{r \to \infty} N(r)v(\Gamma_r) = v(\Gamma) \lim_{r \to \infty} \frac{N(r)}{r^n}.$$

Since $N(r) = [\Gamma \cap rC_1]$ and $rC_1 = \{x \in C; \varphi(x) \leq r^n\}$, we see that $N(r) = [\{x \in \Gamma \cap C; \varphi(x) \leq r^n\}]$. Taking $r = 1, 2, 3, \ldots,$ we can order the elements of $\Gamma \cap C$ as a sequence $\Gamma \cap C = \{x_k\}$ so that

$$0 < \varphi(x_1) \leq \varphi(x_2) \leq \cdots \leq \varphi(x_k) \leq \cdots \tag{3.26}$$

Put $r_k \stackrel{\text{def}}{=} \varphi(x_k)^{1/n}$; so we have

$$\varphi(x_k) = r_k^n. \tag{3.27}$$

Since $x_1, \ldots, x_k$ belong to $\Gamma \cap r_k C_1$, the equality $N(r_k) = [\Gamma \cap r_k C_1]$ implies that

$$k \leq N(r_k). \tag{3.28}$$

On the other hand, for any $\epsilon > 0$, $k' \geq k$ implies $x_{k'} \notin (r_k - \epsilon)C_1$; in fact, suppose that $x_{k'} = (r_k - \epsilon)y$, $y \in C_1$. Then from (3.27) it follows that

$$r_{k'}^n = \varphi(x_{k'}) = (r_k - \epsilon)^n \varphi(y) \leq (r_k - \epsilon)^n < r_k^n,$$

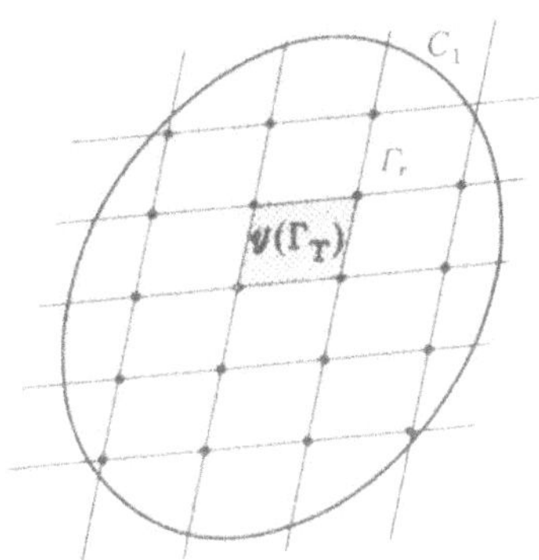

number of black points = N(r)

Figure 3.1

or $r_{k'} < r_k$, contradicting (3.26). Therefore, at most $k-1$ elements $x_1, \ldots, x_{k-1}$ belong to $\Gamma \cap (r_k - \epsilon)C_1$. In other words, we have

$$N(r_k - \epsilon) \leq k - 1 < k, \qquad \forall \epsilon > 0. \tag{3.29}$$

From (3.28) and (3.29) we get

$$\frac{N(r_k - \epsilon)}{(r_k - \epsilon)^n} \left(\frac{r_k - \epsilon}{r_k}\right)^n < \frac{k}{r_k^n} \leq \frac{N(r_k)}{r_k^n}. \tag{3.30}$$

Since $\lim_{k\to\infty} N(r_k)/r_k^n = \text{vol}(C_1)/v(\Gamma)$ and $\epsilon > 0$ is arbitrary, (3.30) implies

$$\lim_{k\to\infty} \frac{k}{\varphi(x_k)} = \frac{\text{vol}(C_1)}{v(\Gamma)}, \tag{3.31}$$

and hence for any $s \in \mathbb{R}$,

$$\lim_{k\to\infty} \frac{k^s}{\varphi(x_k)^s} = \left(\frac{\text{vol}(C_1)}{v(\Gamma)}\right)^s \neq 0.$$

Now, since $\sum_{k=1}^{\infty} 1/k^s$ converges for $s > 1$ [Exercise 3.16(1)], so does $\zeta(s)$. Next, we shall determine the residue of $\zeta(s)$ at $s = 1$. First of all, by (3.31), for any $\epsilon > 0$, we have

$$\left| \frac{k}{\varphi(x_k)} - \frac{\text{vol}(C_1)}{v(\Gamma)} \right| < \epsilon, \qquad k > k_0,$$

and so

$$\left(\frac{\text{vol}(C_1)}{v(\Gamma)} - \epsilon\right) \frac{1}{k} < \frac{1}{\varphi(x_k)} < \left(\frac{\text{vol}(C_1)}{v(\Gamma)} + \epsilon\right) \frac{1}{k}. \tag{3.32}$$

Taking the sth power of both sides of (3.32) and summing from $k = k_0$ to ∞, we get

$$\left(\frac{\text{vol}(C_1)}{v(\Gamma)} - \epsilon\right)^s (s-1) \sum_{k=k_0}^{\infty} \frac{1}{k^s} < (s-1) \sum_{k=k_0}^{\infty} \frac{1}{\varphi(x_k)^s}$$
$$< \left(\frac{\text{vol}(C_1)}{v(\Gamma)} + \epsilon\right)^s (s-1) \sum_{k=k_0}^{\infty} \frac{1}{k^s}. \tag{3.33}$$

Since $\lim_{s\to 1+0}(s-1)\sum_{k=1}^{k_0-1}(*)=0$, we can replace the three summations $\sum_{k=k_0}^{\infty}$ in (3.33) by the full summations $\sum_{k=1}^{\infty}$ and, by Exercise 3.16(2), we get

$$\frac{\mathrm{vol}(C_1)}{v(\Gamma)}-\epsilon\le \lim_{s\to 1+0}(s-1)\sum_{k=1}^{\infty}\frac{1}{\varphi(x_k)^s}\le\frac{\mathrm{vol}(C_1)}{v(\Gamma)}+\epsilon,$$

which implies that the residue at $s=1$ of $\sum_{k=1}^{\infty}\varphi(x_k)^{-s}$ is $\mathrm{vol}(C_1)/v(\Gamma)$. Q.E.D.

3.5. Dedekind Zeta Function

Let k be a number field, $n=[k:\mathbb{Q}]$. For an ideal $\mathfrak{a}$, $\mathfrak{a}\neq 0$, of $\mathfrak{o}_k$, the norm $N_k\mathfrak{a}$ is the cardinality of the quotient ring $\mathfrak{o}_k/\mathfrak{a}$. The *Dedeking zeta function* $\zeta_k(s)$ is defined by

$$\zeta_k(s)=\sum_{\mathfrak{a}}\frac{1}{(N_k\mathfrak{a})^s},\qquad \mathfrak{a}\subset\mathfrak{o}_k.$$

REMARK 3.7. When $k=\mathbb{Q}$, an ideal of $\mathfrak{o}_{\mathbb{Q}}=\mathbb{Z}$ is identified with a natural number $a\in\mathbb{N}$. As $[\mathbb{Z}/a\mathbb{Z}]=a$, we have $\zeta_k(s)=\zeta_{\mathbb{Q}}(s)$, the Riemann zeta function.

Call $C_1,\ldots,C_{h_k}$ all ideal classes of k. For each C_j, $1\le j\le h_k$ put

$$\zeta_j(s)=\sum_{\mathfrak{a}\in C_j}\frac{1}{(N_k\mathfrak{a})^s},\qquad \mathfrak{a}\subset\mathfrak{o}_k.$$

Obviously, we have

$$\zeta_k(s)=\sum_{j=1}^{h_k}\zeta_j(s). \tag{3.34}$$

For a j, $1\le j\le h_k$, take an ideal $\mathfrak{a}_0\subset\mathfrak{o}_k$ such that $C_j^{-1}\ni\mathfrak{a}_0$ and define a mapping $\varphi\colon C_j\cap\mathfrak{o}_k\to P_k$ by $\varphi(\mathfrak{a})=\mathfrak{a}\mathfrak{a}_0=(\alpha)$; obviously, φ is injective. We have

$$\operatorname{Im}\varphi=\{(\alpha)\in P_k\colon \mathfrak{a}_0\mid(\alpha)\}.$$

The left-hand side is clearly contained in the right-hand side. Conversely, if $\mathfrak{a}_0 \mid (\alpha)$, then there is an ideal $\mathfrak{a} \subset \mathfrak{o}_k$ such that $\mathfrak{a}\mathfrak{a}_0 = (\alpha)$ and hence $\mathfrak{a} \in C_j$, which proves the equality. Since $N_k(\mathfrak{a}\mathfrak{a}_0) = N_k\mathfrak{a}N_k\mathfrak{a}_0 = N_k(\alpha) = |N_{k/\mathbb{Q}}\alpha|$ (Proposition 2.33), we can write

$$\zeta_j(s) = (N_k\mathfrak{a}_0)^s \sum_{\mathfrak{a}_0 \mid (\alpha)} \frac{1}{|N_{k/\mathbb{Q}}\alpha|^s}, \qquad (\alpha) \subset \mathfrak{o}_k. \tag{3.35}$$

We shall interpret the sum in (3.35) as a pre-zeta function (Section 3.4), i.e., we shall find a homogeneous polynomial F, a cone C, and a lattice Γ so that we have

$$\sum_{\mathfrak{a}_0 \mid (\alpha)} \frac{1}{|N_{k/\mathbb{Q}}\alpha|^s} = \zeta(F, C, \Gamma; s). \tag{3.36}$$

First of all, for $x = (x_1, \ldots, x_{r_1}, x_{r_1+1}, \ldots, x_{r_1+r_2}) = (x_1, \ldots, x_{r_1}, y_1, z_1, \ldots, y_{r_2}, z_{r_2})$, $x_{r_1+j} = y_j + iz_j$, $1 \leq j \leq r_2$, a vector in $E^{r_1,r_2} = \mathbb{R}^{r_1} \times \mathbb{C}^{r_2} = \mathbb{R}^n$, define a homogeneous polynomial $F(x)$ in $\mathbb{R}[x]$ of degree n with n variables:

$$F(x) \stackrel{\text{def}}{=} N_{E^{r_1,r_2}/\mathbb{R}}(x) = x_1 \cdots x_{r_1}(y_1^2 + z_1^2) \cdots (y_{r_2}^2 + z_{r_2}^2).$$

We put

$$V = \{x \in E^{r_1,r_2}; F(x) \neq 0\} = (E^{r_1,r_2})^\times,$$
$$V_1 = \{x \in (E^{r_1,r_2})^\times; |F(x)| \leq 1\}.$$

Next, we shall define a cone C in V which satisfies the conditions in (3.23). We remind the reader about some notation in Section 3.3:

$$k^\times \xrightarrow{x} (E^{r_1,r_2})^\times \xrightarrow{l} \mathbb{R}^{r_1+r_2} \qquad (l \text{ is surjective}),$$

$$l(x) = (\log|x_1|, \ldots, \log|x_{r_1}|, \log|x_{r_1+1}|^2, \ldots, \log|x_{r_1+r_2}|^2),$$

$$x(\alpha) = (\alpha^{(1)}, \ldots, \alpha^{(r_1)}, \alpha^{(r_1+1)}, \ldots, \alpha^{(r_1+r_2)}),$$

$$\begin{aligned} l^*(\alpha) &= l(x(\alpha)) \\ &= (\log|\alpha^{(1)}|, \ldots, \log|\alpha^{(r_1)}|, \log|\alpha^{(r_1+1)}|^2, \ldots, \log|\alpha^{(r_1+r_2)}|^2). \end{aligned}$$

Write W_k for the group of roots of unity in k, and $l^{\#}$ the restriction of l^* on $\mathfrak{o}_k^\times$; we have the exact sequence

$$1 \to W_k \to \mathfrak{o}_k^\times \xrightarrow{l^{\#}} l^{\#}(\mathfrak{o}_k^\times) \to 1$$

where $l^{\#}(\mathfrak{o}_k^\times)$ is a lattice in the hyperplane

$$H^r = \left\{\lambda;\ \sum_{i=1}^{r_1+r_2} \lambda_i = 0\right\} \subset \mathbb{R}^{r_1+r_2} = \mathbb{R}^{r+1}$$

(Proposition 3.4, Proposition 3.7). Let $l^{\#}(\mathfrak{o}_k^\times) = [l^{\#}(\epsilon_1), \ldots, l^{\#}(\epsilon_r)]$ with $\epsilon_i \in \mathfrak{o}_k^\times$. Since the vector

$$v_0 = (\overbrace{1, \ldots, 1}^{r_1}, \overbrace{2, \ldots, 2}^{r_2})$$

does not belong to the hyperplane H^r, $\mathbb{R}^{r_1+r_2}$ is spanned by $\{v_0, l^{\#}(\epsilon_1), \ldots, l^{\#}(\epsilon_r)\}$. In other words, for any $x \in (E^{r_1,r_2})^\times$, the numbers ξ_i, $0 \le i \le r$, are uniquely determined so that

$$l(x) = \xi_0 v_0 + \xi_1 l^{\#}(\epsilon_1) + \cdots + \xi_r l^{\#}(\epsilon_r), \qquad \xi_i \in \mathbb{R}.$$

Now, we put

$$\begin{aligned} C \stackrel{\text{def}}{=} \{x \in (E^{r_1,r_2})^\times;\ &(1)\ 0 \le \xi_1, \ldots, \xi_r < 1, \\ &(2)\ 0 \le \arg x_1 < 2\pi/w_k\}, \end{aligned} \tag{3.37}$$

where $w_k = [W_k]$. (When $r_1 \ge 1$, we have $x_1 \in \mathbb{R}$, and since $w_k = 2$ in this case, the condition (2) means that $x_1 > 0$.)

EXERCISE 3.17. Verify that the set C in (3.37) is a cone.

Lastly, we define a lattice Γ in $\mathbb{R}^n$. For the ideal $\mathfrak{a}_0$ which we needed to write $\zeta_j(s)$ in (3.35), we see that the set $x(\mathfrak{a}_0)$ makes up a lattice in $E^{r_1,r_2} = \mathbb{R}_n$, $n = r_1 + 2r_2$. We shall denote this lattice by Γ. In order that the function $\zeta(F, C, \Gamma; s)$ makes sense, we need to verify conditions (1), (2), and (3) in formula (3.23) and condition (3.24). Among these, (1) is settled by Exercise 3.17 and (2) is trivial. The following Exercise 3.18 takes care of (3.24) and Exercise 3.19 takes care of the boundedness of C_1 in (3).

EXERCISE 3.18. Verify that $C \cap \Gamma \neq \emptyset$.

EXERCISE 3.19. Prove that the set $C_1 = C \cap V_1$ is bounded.

Now that the pre-zeta function $\zeta(F, C, \Gamma; s)$ makes sense, by Theorem 3.5, it converges for $s > 1$ and its residue at $s = 1$ is given by $\mathrm{vol}(C_1)/v(\Gamma)$; so our next task is to determine $\mathrm{vol}(C_1)$ and $v(\Gamma)$ and to prove equality (3.36).

PROPOSITION 3.8. $V(\Gamma) = 2^{-r_2} N_k \mathfrak{a}_0 |\Delta_k|^{1/2}$.

PROOF. This is obtained by setting $\mathfrak{a} = \mathfrak{a}_0$ in (3.5). Q.E.D.

To prove equality (3.36) we need a definition.

DEFINITION 3.3 Let G be a group and H a subgroup of G. A subset S of G is said to be a *fundamental set* for H in G if:

1. $G = SH$,
2. any $g \in G$ can be written uniquely as $g = sh$, $s \in S$, $h \in H$.

REMARK 3.8. Notation being as above, $s \mapsto sH$ gives a bijection $S \approx G/H$; in other words, S consists of a complete set of representatives of G modulo H.

PROPOSITION 3.9. *The cone C in* (3.37) *is a fundamental set for $x(\mathfrak{o}_k^\times)$ in $(E^{r_1,r_2})^\times$.*

PROOF.

1. We must prove that $(E^{r_1,r_2})^\times = Cx(\mathfrak{o}_k^\times)$. Take any $v \in (E^{r_1,r_2})^\times$ and write

$$l(v) = \gamma_0 v_0 + \gamma_1 l^\#(\epsilon_1) + \cdots + \gamma_r l^\#(\epsilon_r), \qquad \gamma_i \in \mathbb{R}.$$

Let $\gamma_i = z_i + \xi_i$, with $z_i \in \mathbb{Z}$, $0 \le \xi_i < 1$, $1 \le i \le r$, and let $\epsilon = \epsilon_1^{z_1} \cdots \epsilon_r^{z_r} \in \mathfrak{o}_k^\times$. Put $u = vx(\epsilon^{-1})$. Then we have

$$\begin{aligned} l(u) &= l(v) + l^\#(\epsilon^{-1}) = l(v) - z_1 l^\#(\epsilon_1) - \cdots - z_r l^\#(\epsilon_r) \\ &= \gamma_0 v_0 + \xi_1 l^\#(\epsilon_1) + \cdots + \xi_r l^\#(\epsilon_r), \qquad 0 \le \xi_i < 1. \end{aligned}$$

Now let $u = (u_1, \ldots, u_{r_1+r_2})$ and call ν the integral part of the real number $w_k \arg u_1/2\pi$ where $w_k = [W_k]$. Thus, we have

$$0 \le \arg u_1 - 2\pi\nu/w_k < 2\pi/w_k. \tag{3.38}$$

Choose a primitive w_kth root ζ of unity so that the first conjugate $\zeta^{(1)} = e^{2\pi i/w_k}$ and put finally $y = ux(\zeta^{-v})$. As $W_k = \operatorname{Ker} l^{\#}$, we have

$$l(y) = l(u) + l^{\#}(\zeta^{-v}) = l(u) = \gamma_0 v_0 + \xi_1 l^{\#}(\epsilon_1) + \cdots + \xi_r l^{\#}(\epsilon_r).$$

Since $y_1 = u_1(\zeta^{(1)})^{-v}$, (3.38) implies $0 \le \arg y_1 < 2\pi/w_k$. Hence, we see that $y \in C$ and $v = ux(\epsilon) = yx(\epsilon\zeta^v) \in Cx(\mathfrak{o}_k^\times)$, which proves (1).

2. We must show that $yx(\epsilon) = y'x(\epsilon') \Rightarrow y = y'$, $\epsilon = \epsilon'$ where $y, y' \in C$ and $\epsilon, \epsilon' \in \mathfrak{o}_k^\times$. Taking the l of both sides of the first equality, we have $l(y) + l^{\#}(\epsilon) = l(y') + l^{\#}(\epsilon')$. If we write $\epsilon = \zeta\epsilon_1^{z_1} \cdots \epsilon_r^{z_r}$, $\epsilon' = \zeta'\epsilon_1^{z_1'} \cdots \epsilon_r^{z_r'}$, $\zeta, \zeta' \in W_k$, and $z_i, z_i' \in \mathbb{Z}$, we get

$$\begin{aligned} l(y) + l^{\#}(\epsilon) &= \xi_0 v_0 + \xi_1 l^{\#}(\epsilon_1) + \cdots + \xi_r l^{\#}(\epsilon_r) + z_1 l^{\#}(\epsilon_1) + \cdots + z_r l^{\#}(\epsilon_r) \\ &= \xi_0 v_0 + (\xi_1 + z_1) l^{\#}(\epsilon_1) + \cdots + (\xi_r + z_r) l^{\#}(\epsilon_r). \end{aligned}$$

Similarly

$$l(y') + l^{\#}(\epsilon') = \xi_0' v_0 + (\xi_1' + z_1') l^{\#}(\epsilon_1') + \cdots + (\xi_r' + z_r') l^{\#}(\epsilon_r').$$

By the uniqueness of the expression, we have $\xi_i + z_i = \xi_i' + z_i'$ and so $\xi_i = \xi_i'$ and $z_i = z_i'$, because $\xi_i \equiv \xi_i' \bmod \mathbb{Z}$ and $0 \le \xi_i$, $\xi_i' < 1$. Therefore, we have $l^{\#}(\epsilon) = l^{\#}(\epsilon')$, i.e., $\epsilon' = \zeta_0\epsilon$, $\zeta_0 \in W_k$, and hence $yx(\epsilon) = y'x(\epsilon') = y'x(\epsilon)x(\zeta_0)$, or $y = y'x(\zeta_0)$ which implies $y_1 = y_1'\zeta_0^{(1)}$. Since $y, y' \in C$, we have $0 \le \arg y_1 < 2\pi/w_k$, $0 < \arg y_1' < 2\pi/w_k$; from this we see that $\zeta_0^{(1)} = 1 = \zeta_0$ because $\arg \zeta_0^{(1)} = 2\pi v/w_k$ for some $v \in \mathbb{Z}$ and $y_1 = y_1'\zeta_0^{(1)}$. Therefore we conclude that $y = y'$ and $x(\epsilon) = x(\epsilon')$, i.e., $\epsilon = \epsilon'$. Q.E.D.

PROPOSITION 3.10 [Equality (3.36)].

$$\zeta(F, C, \Gamma; s) = \sum_{\mathfrak{a}_0 \mid (\alpha)} \frac{1}{|N_{k/\mathbb{Q}}\alpha|^s}, \qquad (\alpha) \subset \mathfrak{o}_k.$$

PROOF. By Proposition 3.9, for $\alpha \in \mathfrak{a}_0$, $\alpha \ne 0$, $x(\alpha)$ can be expressed uniquely as $x(\alpha) = c(\alpha)x(\epsilon)$, $c(\alpha) \in C$, $x(\epsilon) \in x(\mathfrak{o}_k^\times)$. If $\alpha' = \alpha\eta$, $\eta \in \mathfrak{o}_k^\times$, we have $c(\alpha) = c(\alpha')$ because $x(\alpha') = x(\alpha)x(\eta) = c(\alpha)x(\epsilon\eta)$. Hence, for (α) such that $\mathfrak{a}_0 \mid (\alpha)$, the mapping $(\alpha) \mapsto c(\alpha) \in C \cap \Gamma$, $\Gamma = x(\mathfrak{a}_0)$, is well defined and we have

$$F(c(\alpha)) = |N_{E^{r_1,r_2}/\mathbb{R}} c(\alpha)| = |N_{E^{r_1,r_2}/\mathbb{R}} x(\alpha)| = |N_{k/\mathbb{Q}}\alpha|$$

as $|N_{E^{r_1,r_2}/\mathbb{R}} x(\epsilon)| = |N_{k/\mathbb{Q}}\epsilon| = 1$. The mapping $(\alpha) \mapsto c(\alpha)$ gives a bijection of the set of (α)'s such that $\mathfrak{a}_0 \mid (\alpha)$ and the set $C \cap \Gamma$. In fact, if $c(\alpha') = c(\alpha)$, then we have $x(\alpha') = c(\alpha')x(\epsilon') = c(\alpha)x(\epsilon') = x(\alpha)x(\epsilon^{-1})x(\epsilon') = x(\alpha\epsilon^{-1}\epsilon')$ and so $\alpha' = \alpha\epsilon^{-1}\epsilon'$, or $(\alpha') = (\alpha)$; this shows c is injective. Now, take $x(\alpha) \in C \cap \Gamma = C \cap x(\mathfrak{a}_0)$. As $\alpha \in \mathfrak{a}_0$ and $x(\alpha) \in C$, we have $x(\alpha) = c(\alpha)$, i.e., c is surjective. Therefore we obtain

$$\sum_{\mathfrak{a}_0 \mid (\alpha)} \frac{1}{|N_{k/\mathbb{Q}}\alpha|^s} = \sum_{x \in C \cap \Gamma} \frac{1}{|F(x)|^s}. \qquad \text{Q.E.D.}$$

The last problem is to determine $\mathrm{vol}(C_1)$ where $C_1 = C \cap V_1$ and $V_1 = \{x \in (E^{r_1,r_2})^\times; |F(x)| \leq 1\}$. To do this, we need to introduce the regulator of k.

DEFINITION 3.4. Let $l^{\#}(\mathfrak{o}_k^\times) = [l^{\#}(\epsilon_1), \ldots, l^{\#}(\epsilon_r)]$ with $l^{\#}(\epsilon_i) = (l_1^{\#}(\epsilon_i), \ldots, l_{r+1}^{\#}(\epsilon_i)) \in \mathbb{R}^{r_1+r_2} = \mathbb{R}^{r+1}$, $1 \leq i \leq r$. Consider the following $(r, r+1)$-matrix:

$$\begin{pmatrix} l_1^{\#}(\epsilon_1) & l_2^{\#}(\epsilon_1) & \cdots & l_{r+1}^{\#}(\epsilon_1) \\ l_1^{\#}(\epsilon_2) & l_2^{\#}(\epsilon_2) & \cdots & l_{r+1}^{\#}(\epsilon_2) \\ \vdots & \vdots & & \vdots \\ l_1^{\#}(\epsilon_r) & l_2^{\#}(\epsilon_r) & \cdots & l_{r+1}^{\#}(\epsilon_r) \end{pmatrix} \tag{3.39}$$

One proves that the absolute values of all minors of degree r of matrix (3.39) are equal and are independent of the choice of ϵ_i, $1 \leq i \leq r$ (Exercise 3.20). This common value of the absolute value of minors of degree r of (3.39) is called the *regulator* of k, written R_k. [When $r = 0$, we agree to put $R_k = 1$. When $r \geq 1$, we have $R_k \neq 0$ because $\{l^{\#}(\epsilon_1), \ldots, l^{\#}(\epsilon_r)\}$ is independent.]

EXERCISE 3.20. Prove that the absolute values of all minors of degree r of (3.39) are equal and are independent of the choice of ϵ_i's.

PROPOSITION 3.11. $\mathrm{vol}(C_1) = \dfrac{2^{r_1}\pi^{r_2}R_k}{w_k}$.

PROOF. By volume preserving linear transformations we shall reduce the computation of $\mathrm{vol}(C_1)$ to that of $\mathrm{vol}(D^+)$ where D^+ is a set easier to handle. We need two such linear transformations: each one is of the form $\mu(y)\colon x \mapsto yx$, $x \in E^{r_1,r_2} = \mathbb{R}^n$, $y \in (E^{r_1,r_2})^\times$ with $|F(y)| = 1$. If we call $M(y)$ the matrix corresponding to $\mu(y)$,

we have

$$\det J_{\mu(y)} = \det M(y) = N_{E^{r_1,r_2}/\mathbf{R}}\, y = F(y).\dagger$$

Therefore, for any bounded set B we have $\mathrm{vol}(yB) = \mathrm{vol}(B)$. We consider the following two y's:

$$y = x(\epsilon), \qquad \epsilon \in \mathfrak{o}_k^\times,$$

$$y = y^{(\delta)} = (\delta_1, \ldots, \delta_{r_1}, \overbrace{1, \ldots, 1}^{r_2}) \in (E^{r_1,r_2})^\times, \qquad \delta_i = \pm 1.$$

First of all, for each ν, $0 \le \nu \le w_k - 1$, put

$$C_1^{(\nu)} \stackrel{\text{def}}{=} \{x \in (E^{r_1,r_2})^\times;\ (1)\ |F(x)| \le 1,\ (2)\ 0 \le \xi_i < 1,\ 1 \le i \le r,$$
$$(3)\ 2\pi\nu/w_k \le \arg x_1 < 2\pi(\nu+1)/w_k\},$$

where ξ_i are determined by $l(x) = \xi_0 v_0 + \xi_1 l^\#(\epsilon_1) + \cdots + \xi_r l^\#(\epsilon_r)$. If we choose a primitive w_kth root ζ of unity so that $\zeta^{(1)} = e^{2\pi i/w_k}$, then we find that $C_1^{(\nu)} = x(\zeta^\nu)C_1$; we have $C_1^{(\nu)} \cap C_1^{(\mu)} = \varnothing$, $\nu \ne \mu$, in view of (3). Hence we get

$$D \stackrel{\text{def}}{=} \bigcup_{\nu=0}^{w_k-1} C_1^{(\nu)} = \{x \in (E^{r_1,r_2})^\times;\ (1)\ |F(x)| \le 1,$$
$$(2)\ 0 \le \xi_i < 1,\ 1 \le i \le r\}. \tag{3.40}$$

Formula (3.40) and the equality $C_1^{(\nu)} = x(\zeta^\nu)C_1$ (Exercise 3.21) imply

$$\mathrm{vol}(D) = w_k\, \mathrm{vol}(C_1). \tag{3.41}$$

Exercise 3.21. Verify that $C_1^{(\nu)} = x(\zeta^\nu)C_1$.

Next, put

$$D^+ = \{x \in D;\ x_i > 0,\ 1 \le i \le r_1\}. \tag{3.42}$$

Since

$$y^{(\delta)}D^+ = \{x \in D;\ \mathrm{sgn}\, x_i = \delta_i,\ 1 \le i \le r_1\},$$

† $J_{\mu(y)}$ is the Jacobian of $\mu(y)$. See Exercise 3.13.

we have

$$D = \bigcup_{\delta} y^{(\delta)} D^+$$

and so

$$\mathrm{vol}(D) = 2^{r_1}\,\mathrm{vol}(D^+). \tag{3.43}$$

From (3.41) and (3.43) we have

$$\mathrm{vol}(C_1) = (2^{r_1}/w_k)\,\mathrm{vol}(D^+). \tag{3.44}$$

By (3.40) and (3.42) we can write

$$D^+ = \{x \in (\mathbb{R}_+^\times)^{r_1} \times (\mathbb{C}^\times)^{r_2};\ F(x) = x_1 \cdots x_{r_1} |x_{r_1+1}|^2 \cdots |x_{r_1+r_2}|^2 \le 1,$$
$$0 \le \xi_i < 1,\ 1 \le i \le r\}.$$

In what follows, we put

$$x_{r_1+j} = y_j + iz_j = \rho_j e^{i\varphi_j}, \qquad 1 \le j \le r_2,$$
$$v = F(x) = x_1 \cdots x_{r_1} \rho_1^2 \cdots \rho_{r_2}^2, \qquad x \in (\mathbb{R}_+^\times)^{r_1} \times (\mathbb{C}^\times)^{r_2}.$$

Note that, for $x \in (\mathbb{R}_+^\times)^{r_1} \times (\mathbb{C}^\times)^{r_2}$, we have

$$\begin{aligned} l(x) &= (\log x_1, \ldots, \log x_{r_1}, 2\log \rho_1, \ldots, 2\log \rho_{r_2}) \\ &= (n^{-1}\log v)v_0 + \xi_1 l^{\#}(\epsilon_1) + \cdots + \xi_r l^{\#}(\epsilon_r) \in \mathbb{R}^{r_1+r_2} = \mathbb{R}^{r+1}. \end{aligned} \tag{3.45}$$

If we write (3.45) componentwise, we have

$$\begin{cases} \log x_1 = n^{-1}\log v + \xi_1 l_1^{\#}(\epsilon_1) + \cdots + \xi_r l_1^{\#}(\epsilon_r) \\ \qquad\cdots \qquad \cdots \qquad \cdots \\ \log x_{r_1} = n^{-1}\log v + \xi_1 l_{r_1}^{\#}(\epsilon_1) + \cdots + \xi_r l_{r_1}^{\#}(\epsilon_r) \\ 2\log \rho_1 = 2n^{-1}\log v + \xi_1 l_{r_1+1}^{\#}(\epsilon_1) + \cdots + \xi_r l_{r_1+1}^{\#}(\epsilon_r) \\ \qquad\cdots \qquad \cdots \qquad \cdots \\ 2\log \rho_{r_2} = 2n^{-1}\log v + \xi_1 l_{r_1+r_2}^{\#}(\epsilon_1) + \cdots + \xi_r l_{r_1+r_2}^{\#}(\epsilon_r). \end{cases}$$

Now, by the following mappings

$$\begin{array}{l} x = (x_1, \ldots, x_{r_1}, y_1, z_1, \ldots, y_{r_2}, z_{r_2}) \\ \quad \downarrow \varphi \\ x' = (x_1, \ldots, x_{r_1}, \rho_1, \ldots, \rho_{r_2}, \varphi_1, \ldots, \varphi_{r_2}) \\ \quad \downarrow \psi \\ \xi = (v, \xi_1, \ldots, \xi_r, \varphi_1, \ldots, \varphi_{r_2}) \end{array}$$

the set D^+ is mapped to the set G:

$$G \stackrel{\text{def}}{=} \{\xi = (v, \xi_1, \ldots, \xi_r, \varphi_1, \ldots, \varphi_{r_2}), \quad 0 < v \leq 1, 0 \leq \xi_i < 1, 0 \leq \varphi_i < 2\pi\}.$$

The mapping $\psi \circ \varphi$ is a differentiable bijection of D^+ with G and so

$$\mathrm{vol}(D^+) = \int_{D^+} dx = \int_G |\det J_{(\psi\circ\varphi)^{-1}}|\, d\xi = \int_G |\det J_{\psi^{-1}}|\, |\det J_{\varphi^{-1}}|\, d\xi \tag{3.46}$$

where

$$|\det J_{\varphi^{-1}}| = \rho_1 \cdots \rho_{r_2}, \tag{3.47}$$

and

$$|\det J_{\psi^{-1}}| = R_k/(2^{r_2}\rho_1 \cdots \rho_{r_2}) \tag{3.48}$$

as one verifies them (Exercise 3.22). Hence we have

$$\begin{aligned} \mathrm{vol}(D^+) &= (R_k/2^{r_2}) \int_0^{2\pi} d\varphi_1 \cdots \int_0^{2\pi} d\varphi_{r_2} \int_0^1 dv \int_0^1 d\xi_1 \cdots \int_0^1 d\xi_r \\ &= R_k(2\pi)^{r_2}/2^{r_2} = R_k\pi^{r_2}. \end{aligned}$$

Substituting this in (3.44), we obtain

$$\mathrm{vol}(C_1) = 2^{r_1}\pi^{r_2}R_k/w_k. \qquad \text{Q.E.D.}$$

Exercise 3.22. Verify the equalities (3.47) and (3.48).

Theorem 3.6. *The Dedekind zeta function $\zeta_k(s)$ converges for $s > 1$ and its residue at $s = 1$ is given by*

$$\lim_{s\to 1+0} (s-1)\zeta_k(s) = (2^{r_1+r_2}\pi^{r_2}R_k h_k)/(w_k\, |\Delta_k|^{1/2}).$$

PROOF. By (3.34), we have $\zeta_k(s) = \sum_{j=1}^{h_k} \zeta_j(s)$ and by (3.35)

$$\zeta_j(s) = (N_k \mathfrak{a}_0)^s \sum_{\mathfrak{a}_0 \mid (\alpha)} \frac{1}{|N_{k/\mathbb{Q}}\alpha|^s}, \qquad \mathfrak{a}_0 \in C_j^{-1} \cap \mathfrak{o}_k, \qquad (\alpha) \subset \mathfrak{o}_k.$$

On the other hand, since $\zeta_j(s) = (N_k \mathfrak{a}_0)^s \zeta(F, C, \Gamma; s)$ by Proposition 3.10, $\zeta_k(s)$ converges for $s > 1$. From Proposition 3.8 and Proposition 3.11, the residue at $s = 1$ of $\zeta_j(s)$ is given by

$$\lim_{s \to 1+0} (s-1)\zeta_j(s) = (2^{r_1+r_2}\pi^{r_2} R_k)/(w_k\, |\Delta_k|^{1/2})$$

which is independent of j; hence multiplying it by h_k we get the residue for $\zeta_k(s)$. Q.E.D.

3.6. The *m*th Cyclotomic Field

For the time being, we have been in an analytic mood; here we insert an algebraic section before we introduce Dirichlet L-functions and prove our homework, the theorem on arithmetical progressions. By doing so, we shall have an excellent chance to review some basic algebraic facts in Chapter 2 and to prepare ourselves to understand algebraic structures of L-functions in later sections.

Let $k = \mathbb{Q}(\zeta)$ be the mth cyclotomic field, $m \geq 2$, $\zeta = e^{2\pi i/m}$. As we know, the minimal polynomial of ζ over $\mathbb{Q}$ is $f_\zeta(X) = \prod_{(a,m)=1} (X - \zeta^a)$ (Exercise 3.11) and so $[k:\mathbb{Q}] = \varphi(m)$. Since all conjugates over $\mathbb{Q}$ of ζ belong to k, $k/\mathbb{Q}$ is a Galois extension; we have an isomorphism

$$(\mathbb{Z}/m\mathbb{Z})^\times \approx G(k/\mathbb{Q}) \tag{3.49}$$

given by $a \bmod m \mapsto (\zeta \mapsto \zeta^a)$.

Let p be a prime number such that $p \nmid m$. Denoting by Δ_k the discriminant of k, we have the following equalities:

$$d_{k/\mathbb{Q}}(1, \zeta, \ldots, \zeta^{\varphi(m)-1}) = \prod_{1 \leq i \leq j \leq \varphi(m)} (\zeta^{(i)} - \zeta^{(j)})^2 \tag{3.50}$$

and

$$d_{k/\mathbb{Q}}(1, \zeta, \ldots \zeta^{\varphi(m)-1}) = \Delta_k[\mathfrak{o}_k : M]^2, \qquad M = [1, \zeta, \ldots, \zeta^{\varphi(m)-1}] \tag{3.51}$$

(Section 2.3 and Proposition 2.18). Since $X^m - 1 = f(X)g(X)$, $g(X) \in \mathbb{Z}[X]$, and $X^m - 1 \equiv 0\ (p)$ has no multiple roots because $p \nmid m$, we have $\zeta^{(i)} \not\equiv \zeta^{(j)}\ (\mathfrak{p})$ for any prime ideal $\mathfrak{p}$ in k such that $\mathfrak{p} \mid p$; hence, from (3.50) and (3.51), we see that $p \nmid \Delta_k$. Therefore p is unramified for $k/\mathbb{Q}$ by Theorem 2.18. Since $k/\mathbb{Q}$ is abelian, the Artin symbol $((k/\mathbb{Q})/p)$ makes sense:

$$\alpha^{\left(\frac{k/\mathbb{Q}}{p}\right)} \equiv \alpha^p \quad (\mathfrak{p}), \qquad \alpha \in \mathfrak{O}_k.$$

In particular, for $\alpha = \zeta$, as both sides of the congruence are primitive mth roots of unity, we find by (3.50) and by the assumption $p \nmid m$ that

$$\zeta^{\left(\frac{k/\mathbb{Q}}{p}\right)} = \zeta^p. \tag{3.52}$$

Therefore, from (3.49) and (3.52), it follows that the order of p in $(\mathbb{Z}/m\mathbb{Z})^\times$ coincides with that of $((k/\mathbb{Q})/p)$ in $G(k/\mathbb{Q})$, i.e., the integer f determined by the prime decomposition: $p = \mathfrak{p}_1 \cdots \mathfrak{p}_g$, $N_k\mathfrak{p}_i = p^f$, $fg = \varphi(m)$. In order words, f is described inside $\mathbb{Q}$ as the smallest positive integer such that $p^f \equiv 1\ (m)$.

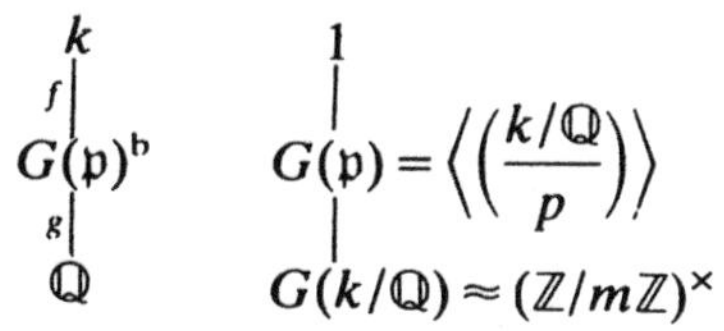

REMARK 3.9. When $m = l$, an odd prime, we have the isomorphism $G(k/\mathbb{Q}) \approx \mathbb{F}_l^\times$, a fact we cherish in this book as a liaison between elementary number theory and algebraic number theory. In view of this isomorphism we could easily study subfields of lth cyclotomic field (Sections 2.6 and 2.16). For a general m, we can also use the isomorphism $G(k/\mathbb{Q}) \approx (\mathbb{Z}/m\mathbb{Z})^\times$ to study subfields of the mth cyclotomic field. In this context, we have the following remarkable theorem (Kronecker–Weber):† *for any abelian extension $k/\mathbb{Q}$, there is an integer m such that $k \subset \mathbb{Q}(\zeta)$, $\zeta = e^{2\pi i/m}$.*

† Weber (1842–1913).

3.7. Dirichlet L-Functions

In Section 1.12, we introduced the character group $\hat{G}$ of a finite abelian group G and proved the orthogonality

$$\sum_{x \in G} \chi(x) = \begin{cases} [G], & \chi = \epsilon \\ 0, & \chi \neq \epsilon \end{cases} \tag{3.53}$$

(Proposition 1.22). We also proved, assuming G cyclic, that

$$G \approx \hat{G} \approx (\mathbb{Z}/n\mathbb{Z})^+, \qquad n = [G]$$

and the following formula dual to (3.53):

$$\sum_{x \in \hat{G}} \chi(x) = \begin{cases} [G], & x = e \\ 0, & x \neq e \end{cases} \tag{3.54}$$

(Exercise 1.44). As the group $(\mathbb{Z}/m\mathbb{Z})^\times$ is not necessarily cyclic for general $m \geq 2$, we need to extend the above results for arbitrary finite abelian groups.

Theorem 3.7 (*Fundamental theorem of finite abelian groups*). *Any finite abelian group G can be written as $G = G_1 \times \cdots \times G_l$ where G_i, $1 \leq i \leq l$, are all cyclic. Furthermore, we can arrange G_i's so that $[G_i] \mid [G_{i+1}]$.*

Proof. Let $g_1, \ldots, g_n$ be a set of generators of G. Writing additively, any element g of the group G may be expressed (not necessarily uniquely) as $g = c_1 g_1 + \cdots + c_n g_n$, $c_i \in \mathbb{Z}$. Consider now a free abelian group G^* of rank n with basis $u_1, \ldots, u_n$. Hence any element g^* of G^* may be expressed uniquely as $g^* = c_1 u_1 + \cdots + c_n u_n$, $c_i \in \mathbb{Z}$. Call f a homomorphism $G^* \mapsto G$ defined by $f(u_i) = g_i$, $1 \leq i \leq n$, and put $H^* = \operatorname{Ker} f$. By Exercise 3.2, H^* is also a free abelian group: $H^* = [v_1, \ldots, v_m]$, $m \leq n$. Let A be a matrix in $M_n(\mathbb{Z})$ determined by the relation

$$(v_1, \ldots, v_m, 0, \ldots, 0) = (u_1, \ldots, u_n)A. \tag{3.55}$$

From Exercise 2.15, we can find matrices $U, V \in \mathrm{GL}_n(\mathbb{Z})$ so that

$$UAV = \begin{bmatrix} e_1 & & & & & \\ & \ddots & & & & \\ & & e_r & & & \\ & & & 0 & & \\ & & & & \ddots & \\ & & & & & 0 \end{bmatrix} = D, \qquad r = \operatorname{rank} A, \qquad e_i \mid e_{i+1}. \tag{3.56}$$

If we choose new generators of G^* and H^* by the relations

$$\begin{aligned}(u_1^*, \ldots, u_n^*) &= (u_1, \ldots, u_n)U^{-1}, \\ (v_1^*, \ldots, v_n^*) &= (v_1, \ldots, v_m, 0, \ldots, 0)V,\end{aligned} \tag{3.57}$$

we have, by (3.55)–(3.57),

$$(v_1^*, \ldots, v_n^*) = (u_1^*, \ldots, u_n^*)\begin{bmatrix} e_1 & & & & & \\ & \ddots & & & & \\ & & e_r & & & \\ & & & 0 & & \\ & & & & \ddots & \\ & & & & & 0 \end{bmatrix},$$

or $v_1^* = e_1u_1^*, \ldots, v_r^* = e_ru_r^*, \; v_{r+1}^* = \cdots = v_n^* = 0$. As the homomorphism f is obviously surjective, we have

$$G^*/G^* \approx (\mathbb{Z}/e_1\mathbb{Z})^+ \times \cdots \times (\mathbb{Z}/e_r\mathbb{Z})^+ \times \mathbb{Z}^+ \times \cdots \times \mathbb{Z}^+ \approx G.$$

However, since G is finite, we must have $r = n$ and

$$G \approx (\mathbb{Z}/e_1\mathbb{Z})^+ \times \cdots \times (\mathbb{Z}/e_n\mathbb{Z})^+. \qquad \text{Q.E.D.}$$

Let G and G' be finite abelian groups. Since $\widehat{G \times G'} \approx \hat{G} \times \hat{G}'$ by a natural mapping and $G \approx \hat{G}$ when G is cyclic, we have, by Theorem 3.7,

$$\hat{G} \approx \hat{G}_1 \times \cdots \times \hat{G}_l \approx G_1 \times \cdots \times G_l = G$$

for any finite abelian group G.

Exercise 3.23. Let G be a finite abelian group and H be a subgroup of G. Then the set $H^\# = \{\chi \in \hat{G}; \chi(h) = 1, \; \forall h \in H\}$ makes up a subgroup of G and

1. $\widehat{G/H} \approx H^\#$,
2. $\hat{G}/H^\# \approx \hat{H}$.

(The latter means that every character of H can be extended to characters of G in $[G: H]$ ways.)

Exercise 3.24. For any finite abelian group G prove that $G \approx \hat{\hat{G}}$. Using this, prove formula (3.54):

$$\sum_{\chi \in \hat{G}} \chi(x) = \begin{cases} [G], & x = e, \\ 0, & x \neq e. \end{cases}$$

Consider the finite abelian group $G=(\mathbb{Z}/m\mathbb{Z})^\times$, $m\geq 2$. A character $\chi\in\hat{G}$ which is a mapping $(\mathbb{Z}/m\mathbb{Z})^\times\to\mathbb{C}$ can be extended to a mapping $\mathbb{Z}/m\mathbb{Z}\to\mathbb{C}$, written still χ, by setting χ (a mod m) $=0$ when $(a, m)>1$. We shall use the same notation χ for the mapping $\mathbb{Z}\to\mathbb{C}$ obtained from χ naturally. We point out the property: $a\equiv b \quad (m)\Rightarrow\chi(a)=\chi(b)$, $\chi(ab)=\chi(a)\chi(b)$, and $\chi(1)=1$.

DEFINITION 3.5. For $\chi\in\widehat{(\mathbb{Z}/m\mathbb{Z})^\times}$, the function of $s\in\mathbb{R}$ given by

$$L(s,\chi)\overset{\text{def}}{=}\sum_{n=1}^{\infty}\frac{\chi(n)}{n^s}$$

is called the *Dirichlet L-function.*

PROPOSITION 3.12.

1. *$L(s,\chi)$ converges for $s>1$.*
2. *$L(s,\chi)=\prod_p(1-\chi(p)/p^s)^{-1}$, $s>1$.*

PROOF. Since $|\chi(a)|\leq 1$, (1) is reduced to Exercise 3.16(1). As for (2), first note that

$$\left(1-\frac{\chi(p)}{p^s}\right)^{-1}=1+\frac{\chi(p)}{p^s}+\frac{\chi(p^2)}{p^{2s}}+\ldots,\qquad s>1.$$

Next, for a given natural number N, let $p_1,\ldots,p_l$ be all prime numbers $\leq N$. Then we have

$$\prod_{p\leq N}\left(1-\frac{\chi(p)}{p^s}\right)^{-1}=\sum_{e_1,\ldots,e_l=0}^{\infty}\frac{\chi(p_1^{e_1}\cdots p_l^{e_l})}{(p_1^{e_1}\cdots p_l^{e_l})^s}$$
$$=\sum_{\substack{n\text{ divisible}\\ \text{only by }p_i}}{}'\frac{\chi(n)}{n^s}.$$

Therefore

$$\left|L(s,\chi)-\prod_p\left(1-\frac{\chi(p)}{p^s}\right)^{-1}\right|=\left|\sum_{n=1}^{\infty}\frac{\chi(n)}{n^s}-\sum{}'\frac{\chi(n)}{n^s}\right|$$
$$=\left|\sum_{\substack{n\text{ at least one}\\ \text{prime factor }>N}}\frac{\chi(n)}{n^s}\right|<\sum{}''\frac{1}{n^s}<\sum_{n>N}\frac{1}{n^s}$$
$$\to 0\qquad (N\to\infty).$$

Hence, we get

$$\lim_{N\to\infty}\prod_{p\le N}\left(1-\frac{\chi(p)}{p^s}\right)^{-1}=\prod_p\left(1-\frac{\chi(p)}{p^s}\right)^{-1}=L(s,\chi). \qquad \text{Q.E.D.}$$

REMARK 3.10. Let $\chi=\epsilon$. By Proposition 3.12(2), we have $L(s,\epsilon)=\prod_{p\nmid m}(1-p^{-s})^{-1}$. On the other hand, by the proof of Proposition 3.12, $\zeta_{\mathbb{Q}}(s)=\prod_p(1-p^{-s})^{-1}$ and so we have $\zeta_{\mathbb{Q}}(s)=L(s,\epsilon)\prod_{p\mid m}(1-p^{-s})^{-1}$.

THEOREM 3.8. *Let $k=\mathbb{Q}(\zeta)$, $\zeta=e^{2\pi i/m}$, $m\ge 2$. Then we have*

$$\zeta_k(s)=\prod_{\mathfrak{p}\mid m}\left(1-\frac{1}{(N_k\mathfrak{p})^s}\right)^{-1}\prod_{\chi\in\widehat{(\mathbb{Z}/m\mathbb{Z})^\times}}L(s,\chi), \qquad s>1.$$

PROOF. First of all, as in the proof of Proposition 3.12(2), we obtain

$$\zeta_k(s)=\prod_{\mathfrak{p}}\left(1-\frac{1}{(N_k\mathfrak{p})^s}\right)^{-1}, \qquad s>1.\dagger$$

Therefore, putting

$$G(s)=\prod_{\mathfrak{p}\mid m}\left(1-\frac{1}{(N_k\mathfrak{p})^s}\right)^{-1},$$

we have

$$\zeta_k(s)=G(s)\prod_{p\nmid m}\prod_{\mathfrak{p}\mid p}\left(1-\frac{1}{(N_k\mathfrak{p})^s}\right)^{-1}=G(s)\prod_{p\nmid m}\left(1-\frac{1}{p^{sf_p}}\right)^{-g_p} \tag{3.58}$$

where $N_k\mathfrak{p}=p^{f_p}$ and $f_pg_p=\varphi(m)$. For the moment, put $f=f_p$, $g=g_p$, and $\eta=\eta_p=e^{2\pi i/f}$. Dividing both sides of the identity $x^f-1=\prod_{\nu=0}^{f-1}(X-\eta^\nu)$ by X^f and substituting $X=p^s$, we get

$$1-\frac{1}{p^{sf}}=\prod_{\nu=0}^{f-1}\left(1-\frac{\eta^\nu}{p^s}\right)$$

and hence (3.58) becomes

$$\zeta_k(s)=G(s)\prod_{p\nmid m}\prod_{\nu=0}^{f_p-1}\left(1-\frac{\eta_p^s}{p^s}\right)^{-g_p} \tag{3.59}$$

† In the proof of Proposition 3.12(2), replace $p_i\le N$ by $N_k\mathfrak{p}_i\le N$ and use the fact that $\zeta_k(s)$ converges for $s>1$ (Theorem 3.6).

On the other hand, for a fixed p, one verifies that

$$\prod_{\nu=0}^{f-1}\left(1-\frac{\eta^{\nu}}{p^{s}}\right)^{-g}=\prod_{\chi\in\widehat{(\mathbb{Z}/m\mathbb{Z})^{\times}}}\left(1-\frac{\chi(p)}{p^{s}}\right)^{-1}. \tag{3.60}$$

In fact, put $W_p=\langle\eta\rangle$, this being the group of fth roots of unity. Since $p^f\equiv 1\,(m)$ [see formula (3.52)], we have $\chi(p)^f=\chi(p^f)=1$ for any $\chi\in\widehat{(\mathbb{Z}/m\mathbb{Z})^{\times}}$ and so we have a homomorphism:

$$\pi_p\colon \widehat{(\mathbb{Z}/m\mathbb{Z})^{\times}}\to W_p,\qquad p\nmid m,\qquad \pi_p(\chi)=\chi(p).$$

This homomorphism π_p is surjective. To see this, let η^{ν} be any element of W_p. As $H=\langle p \bmod m\rangle$ is a cyclic subgroup of $(\mathbb{Z}/m\mathbb{Z})^{\times}$ of order f, there is a character $\psi\in\hat{H}$ such that $\psi(p)=\eta^{\nu}$; call χ a character of $\widehat{(\mathbb{Z}/m\mathbb{Z})^{\times}}$ which extends ψ [Exercise 3.23(2)], then $\chi(p)=\eta^{\nu}$, i.e., $\pi_p(\chi)=\eta^{\nu}$, which proves that π_p is surjective. From this we get $[\operatorname{Ker}\pi_p]=g$ and the equality (3.60). By (3.59) and (3.60) we obtain

$$\begin{aligned}\zeta_k(s)&=G(s)\prod_{p\nmid m}\prod_{\chi\in\widehat{(\mathbb{Z}/m\mathbb{Z})^{\times}}}\left(1-\frac{\chi(p)}{p^{s}}\right)^{-1}\\&=G(s)\prod_{\chi}\prod_{p\nmid m}\left(1-\frac{\chi(p)}{p^{s}}\right)^{-1}=G(s)\prod_{\chi}\prod_{p}\left(1-\frac{\chi(p)}{p^{s}}\right)^{-1}\\&=G(s)\prod_{\chi}L(s,\chi).\end{aligned}$$

Q.E.D.†

REMARK 3.11. From Remark 3.10 and Theorem 3.8, it follows that

$$\zeta_k(s)=T(s)\zeta_{\mathbb{Q}}(s)\prod_{\chi\neq\epsilon}L(s,\chi),\qquad \chi\in\widehat{(\mathbb{Z}/m\mathbb{Z})^{\times}}, \tag{3.61}$$

with

$$T(s)=\prod_{\mathfrak{p}\mid m}(1-(N_k\mathfrak{p})^{-s})^{-1}\prod_{p\mid m}(1-p^{-s})=\prod_{p\mid m}\{(1-p^{-sf_p})^{-g_p}(1-p^{-s})\}.$$

EXERCISE 3.25. Let $k=\mathbb{Q}(\zeta)$, $\zeta=e^{2\pi i/l}$, l an odd prime. Prove

† Note that $p\mid m\Rightarrow\chi(p)=0$. Since $\prod_p(1-\chi(p)/p^s)^{-1}$ converges absolutely for $s>1$, we can interchange the order of multiplication in the infinite product freely.

that

$$\zeta_k(s) = \zeta_{\mathbb{Q}}(s) \prod_{\chi \neq \epsilon} L(s, \chi), \qquad \chi \in \widehat{(\mathbb{Z}/l\mathbb{Z})^\times}.$$

EXERCISE 3.26. Let $k = \mathbb{Q}(i)$. Since $i = e^{2\pi i/4}$, k is the 4th cyclotomic field and $\widehat{(\mathbb{Z}/4\mathbb{Z})^\times} = \{\epsilon, \chi\}$, $\chi(3) = \chi(-1) = -1$. Prove that $\zeta_k(s) = \zeta_{\mathbb{Q}}(s)L(s, \chi)$.

EXERCISE 3.27. For a natural number n, denote by $r(n)$ the number of solutions $(x, y) \in \mathbb{Z}^2$ of $x^2 + y^2 = n$. Prove that

$$r(n) = 4 \sum_{d \mid n} \chi(d) = 4(d_1(n) - d_3(n))$$

where $d_1(n) = [\{d \in \mathbb{N};\ d \mid n,\ d \equiv 1 \quad (4)\}]$ and $d_3(n) = [\{d \in \mathbb{N};\ d \mid n,\ d \equiv 3 \quad (4)\}]$.

EXERCISE 3.28. For a series $f(s) = \sum_{n=1}^{\infty} a_n/n^s$, $a_n \in \mathbb{C}$, put $A_\nu = \sum_{n=1}^{\nu} a_n$, $\nu \geq 1$. If $|A_\nu| \leq B$ for all ν, $f(s)$ converges uniformly for $s \geq s_0 > 0$ for any $s_0 > 0$. [Hence $f(s)$ is continuous for $s > 0$.]

THEOREM 3.9. *For any character $\chi \neq \epsilon$ of $(\mathbb{Z}/m\mathbb{Z})^\times$, $L(s, \chi)$ represents a continuous function for $s > 0$.*

PROOF. Since $L(s, \chi) = \sum_{n=1}^{\infty} \chi(n)/n^s$, we have $A_\nu = \sum_{n=1}^{\nu} \chi(n)$. Writing $\nu = qm + r$, $0 \leq r < m$,

$$\begin{aligned} A_\nu &= \sum_{n=1}^{m} \chi(n) + \sum_{n=m+1}^{2m} \chi(n) + \cdots + \sum_{n=(q-1)m+1}^{qm} \chi(n) + \sum_{n=qm+1}^{qm+r} \chi(n) \\ &= \sum_{n=1}^{m} \chi(n) + \cdots + \sum_{n=1}^{m} \chi(n) + \sum_{n=1}^{r} \chi(n) = \sum_{n=1}^{r} \chi(n) \end{aligned}$$

[by the orthogonality (3.53)]. Hence $|A_\nu| \leq \sum_{n=1}^{r} |\chi(n)| \leq r < m$ and the assertion follows from Exercise 3.28. Q.E.D.

THEOREM 3.10. *Let $k = \mathbb{Q}(\zeta)$, $\zeta = e^{2\pi i/m}$, $m \geq 2$. Then we have*

$$(2^{r_1+r_2}\pi^{r_2}R_k h_k)/(w_k\, |\Delta_k|^{1/2}) = T(1) \prod_{\chi \neq \epsilon} L(1, \chi)$$

where

$$T(1) = \prod_{p \mid m} \{1 - p^{-f_p})^{-g_p}(1 - p^{-1})\}.$$

From this, we see that $L(1, \chi) \neq 0$ for $\chi \neq \epsilon$.

PROOF. Multiply $s-1$ on both sides of (3.61), take $\lim_{s\to 1+0}$, and use Theorem 3.6 and Theorem 3.9 for $\mathbb{Q}$, k. Q.E.D.

THEOREM 3.11. *Let l be an odd prime, $\zeta = e^{2\pi i/l}$, and $k = \mathbb{Q}(\zeta)$. We have*

$$(2^{(l-3)/2}\pi^{(l-1)/2}R_k h_k)l^{-l/2} = \prod_{\chi\neq\epsilon} L(1, \chi), \qquad \chi \in \widehat{(\mathbb{Z}/l\mathbb{Z})^\times}.$$

PROOF. Since $l \geq 3$, we have $r_1 = 0$ and $r_2 = (l-1)/2$. The equality in the theorem is obtained by substituting in the formula of Theorem 3.10 known data such as $w_k = 2l$ (Exercise 3.12), $|\Delta_k| = l^{l-2}$ (Exercise 2.18), and $T(1) = 1$ (Exercise 3.25). Q.E.D.

REMARK 3.12. Theorem 3.11 is the starting point of the calculation of h_k for the lth cyclotomic field $k = \mathbb{Q}(\zeta)$. (See Chapter 4.)

3.8. Dirichlet's Theorem on Arithmetical Progressions

THEOREM 3.12. *For $m \geq 2$, any coset of the group $(\mathbb{Z}/m\mathbb{Z})^\times$ contains infinitely many primes.*

PROOF. Since $|\chi(p)p^{-s}| < 1$ for $s > 1$, $\chi \in (\mathbb{Z}/m\mathbb{Z})^\times$, we have

$$\log\left(1 - \frac{\chi(p)}{p^s}\right)^{-1} = \frac{\chi(p)}{p^s} + \frac{\chi(p^2)}{2p^{2s}} + \frac{\chi(p^3)}{3p^{3s}} + \cdots \tag{3.62}$$

where we choose the principal value of the logarithm. Summing (3.62) with respect to all primes, we get

$$\sum_p \log\left(1 - \frac{\chi(p)}{p^s}\right)^{-1} = \sum_p \frac{\chi(p)}{p^s} + R(s, \chi) \tag{3.63}$$

where

$$R(s, \chi) = \sum_p (\chi(p^2)/(2p^{2s}) + \chi(p^3)/(3p^{3s}) + \cdots)$$

which converges for $s > 1$ because it is majorized by $\zeta_{\mathbb{Q}}(s)$. On the other hand, by Proposition 3.12, we have the infinite product

$$L(s, \chi) = \prod_p \left(1 - \frac{\chi(p)}{p^s}\right)^{-1}, \qquad s > 1.$$

This time, we choose the value for $\log L(s, \chi)$ so that

$$\log L(s, \chi) \stackrel{\text{def}}{=} \sum_p \log\left(1 - \frac{\chi(p)}{p^s}\right)^{-1}.\dagger \tag{3.64}$$

EXERCISE 3.29. Verify that there is a $b > 0$ so that $|R(s, \chi)| < b$ for $s \geq 1$.

We return to the proof of the theorem. For any coset C in $(\mathbb{Z}/m\mathbb{Z})^{\times}$, we have

$$\sum_p \frac{\chi(p)}{p^s} = \sum_{p \nmid m} \frac{\chi(p)}{p^s} = \sum_C \chi(C) f(s, C),$$

$$f(s, C) = \sum_{p \in C} \frac{1}{p^s}.$$

Therefore, by (3.63) and (3.64),

$$\log L(s, \chi) = \sum_C \chi(C) f(s, C) + R(s, \chi).$$

Take any $A \in (\mathbb{Z}/m\mathbb{Z})^{\times}$ and sum over all characters:

$$\begin{aligned} \sum_\chi \chi(A^{-1}) \log L(s, \chi) &= \sum_C \sum_\chi \chi(CA^{-1}) f(s, C) + R_A(s) \\ &= \varphi(m) f(s, A) + R_A(s) \end{aligned}$$

where

$$R_A(s) = \sum_\chi \chi(A^{-1}) R(s, \chi).$$

Thus we have

$$\log L(s, \epsilon) + \sum_{\chi \neq \epsilon} \chi(A^{-1}) \log L(s, \chi) = \varphi(m) f(s, A) + R_A(s). \tag{3.65}$$

By Exercise 3.29, notice that $|R_A(a)| \leq \varphi(m)\, |R(s, \chi)| < b\varphi(m)$. Suppose now that some A contains only finitely many primes. Then $f(s, A)$ must be bounded for $s \geq 1$. Since $R_A(s)$ is bounded

† When $\chi = \epsilon$, as no complex number appears, (3.64) follows from $L(s, \chi) = \prod_p (1 - \chi(p)/p^s)^{-1}$, $s > 1$.

for $s \geq 1$ and $L(1, \chi) \neq 0$ (Theorem 3.10) for $\chi \neq \epsilon$, the part $\sum_{\chi \neq \epsilon}$ of (3.65) is also bounded and hence $\log L(s, \epsilon)$ remains bounded when $s \to 1+0$. When $\chi = \epsilon$, (3.64) is an equality of ordinary logarithms in real analysis and it follows that $L(s, \epsilon)$ remains bounded for $s \to 1+0$. As $\zeta_{\mathbb{Q}}(s) = L(s, \epsilon) \prod_{p \mid m} (1 - p^{-s})^{-1}$ (Remark 3.10), $\zeta_{\mathbb{Q}}(s)$ must also remain bounded for $s \to 1+0$ which contradicts, however, the property $\lim_{s \to 1+0}(s-1)\zeta_{\mathbb{Q}}(s) = 1$ [Exercise 3.16(2)]. Q.E.D.

REMARK 3.13. Dirichlet's theorem says that the linear function $f(x) = a + mx$, $a, m \in \mathbb{N}$, $(a, m) = 1$, represents infinitely many primes. Although the theorem has been proved by making use of L-functions, it is somewhat strange that a linear function has something to do with the distribution of primes. What happens if one thinks of two linear functions? For example, the *twin-prime* conjecture asserts that there must be infinitely many x's in $\mathbb{N}$ such that $f(x) = x$ and $g(x) = x + 2$ are both prime. It is hoped that the conjecture is settled by introducing some type of "L-functions." Another pair of linear functions is $f(x) = x$ and $g(x) = 2x + 1$. One conjectures the same statement as above. Primes of the latter kind are called *Sophie Germain* primes, which are less well known but more academic than twin primes.† We owe her a beautiful theorem: if the Fermat equation $x^p + y^p = z^p$ has an integral solution x, y, z, $xyz \neq 0$, for a Sophie Germain prime p, then we have $xyz \equiv 0 \quad (p)$. In the context of this book, let us remark that if l and $(l-1)/2$ are both prime, i.e., if $(l-1)/2$ is a Sophie Germain prime, then the maximal real subfield $\mathbb{Q}(\zeta + \zeta^{-1})$ of the lth cyclotomic field $k = \mathbb{Q}(\zeta)$, $\zeta = e^{2\pi i/l}$ has a prime degree over $\mathbb{Q}$.

† Sophie Germain (1776–1831).

4

The lth Cyclotomic Field and Quadratic Fields

The main purpose of this chapter is to derive the class number formula for subfields of the lth cyclotomic field [formulas (4.21) and (4.22)] by making use of analytical results in Chapter 3. In particular, for the quadratic subfield $\mathbb{Q}(\sqrt{l^*})$ of the lth cyclotomic field, one reaches the final formula due to Dirichlet [formulas (4.30) and (4.33)] using the famous determination of the Gauss sums (Theorem 4.2).

In Section 4.5, we shall discuss how to determine the structure of ideal class groups of quadratic fields in an entirely elementary way, i.e., using only Minkowski constants. In Section 4.6, which may be considered as a foreshadow of the Gauss genus theory (Section 4.7), we study the cohomology of ideal class groups of quadratic fields. We shall not stick to the definition of genera. Opinion is divided how to define genera; one can use either cohomology or the system of characters χ_i, $1 \leq i \leq t$. However, the main theorem (Theorem 4.5) states that the two are essentially equivalent and so it is not a big issue to debate the definition of genera.

We shall conclude the book with continued fractions, which we introduced in Chapter 1 as an example of transformation groups. Here we shall study some connections between continued fractions and real quadratic fields (or quadratic irrationals). It is to be noted that special values of hypergeometric functions, such as $e = 2.718281828\ldots$, occasionally have continued fraction expansion with some regularity, although not periodic as in the case of quadratic irrationals:

$$e = [2, 1, 2, 1, 1, 4, 1, 1, 6, 1, 1, 8, 1, 1, \ldots].$$

4.1. Determination of Gauss Sums

Let G be a finite abelian group of order N. Denote by $L(G)$ the set of all $\mathbb{C}$-valued functions on G. Clearly, $L(G)$ makes up a vector space over $\mathbb{C}$. Let f_y be the characteristic function of one point set $\{y\}$, $y \in G$. By definition, we have

$$f_y(x) = \begin{cases} 1, & x = y \\ 0, & x \neq y \end{cases}$$

and since these functions form a basis for the vector space $L(G)$, we have $\dim_{\mathbb{C}} L(G) = N$.

EXERCISE 4.1. Verify that f_y, $y \in G$, form a basis of $L(G)$.

The group $\hat{G}$ of character of G is a subset of $L(G)$.

PROPOSITION 4.1. *$\hat{G}$ forms a basis of $L(G)$.*

PROOF. As $[\hat{G}] = N$ (Section 3.7), it is enough to show that N characters in $\hat{G}$ are linearly independent in $L(G)$. Suppose that

$$\sum_{\chi \in \hat{G}} a_\chi \chi = 0, \qquad a_\chi \in \mathbb{C}.$$

For any $\psi \in \hat{G}$,

$$0 = \sum_{\chi} a_\chi \chi(x) \bar{\psi}(x), \qquad x \in G.$$

By the orthogonality of characters, we have

$$0 = \sum_{x \in G} \sum_{\chi} a_\chi (\chi\bar{\psi})(x) = \sum_{\chi} a_\chi \sum_{x} (\chi\bar{\psi})(x) = N a_\psi$$

and so $a_\psi = 0$. Q.E.D.

We define an inner product in $L(G)$ by

$$\langle f, g \rangle = (1/N) \sum_{x \in G} f(x) \bar{g}(x), \qquad f, g \in L(G). \tag{4.1}$$

THEOREM 4.1. *With respect to the inner product* (4.1), *the vector space* $L(G)$ *makes up a Hilbert space and* G *forms an orthonormal basis.*

PROOF. The properties of the inner product such as $\langle f+g, h\rangle = \langle f, h\rangle + \langle g, h\rangle$, $\langle af, g\rangle = a\langle f, g\rangle$, $\langle f, ag\rangle = \bar{a}\langle f, g\rangle$, and $\overline{\langle f, g\rangle} = \langle g, f\rangle$ follow immediately from (4.1). Since

$$\langle f, f\rangle = (1/N)\sum_x f(x)\bar{f}(x) = (1/N)\sum_x |f(x)|^2 \geq 0,$$

we have $\langle f, f\rangle = 0 \Leftrightarrow f = 0$. Finally, for $\chi, \psi \in \hat{G}$,

$$\langle \chi, \psi\rangle = (1/N)\sum_x \chi(x)\bar{\psi}(x) = (1/N)\sum_x (\chi\bar{\psi})(x) = \begin{cases} 1, & \chi = \psi \\ 0, & \chi \neq \psi \end{cases}$$

which shows that $\hat{G}$ forms an orthonormal basis for $L(G)$ with respect to $\langle\ ,\ \rangle$. Q.E.D.

By Theorem 4.1, any function $f \in L(G)$ has the *Fourier expansion*

$$f = \sum_\chi c_\chi \chi, \qquad c_\chi = \langle f, \chi\rangle$$

where c_χ are Fourier coefficients of f.† On the other hand, if we consider the mapping $\chi \mapsto c_\chi$ as an element of $L(\hat{G})$, we obtain the Fourier transform $\hat{f} \in L(G)$ of f:

$$\hat{f}(\chi) = \langle f, \chi\rangle = (1/N)\sum_x f(x)\bar{\chi}(x).$$

From now on, let l be an odd prime and consider the additive group $\mathbb{F}_l = \mathbb{Z}/l\mathbb{Z}$. For each $\xi \in \mathbb{F}_l$, define a character $\psi_\xi \in \hat{\mathbb{F}}_l$ by $\psi_\xi(x) = \zeta^{\xi x}$, $\zeta = e^{2\pi i/l}$. One verifies easily that $\mathbb{F}_l \approx \hat{\mathbb{F}}_l$ by the mapping $\xi \mapsto \psi_\xi$. Identifying $\hat{\mathbb{F}}_l$ with $\mathbb{F}_l$ in this way, the Fourier transform $f \mapsto \hat{f}$ becomes an endomorphism of the vector space $L(\mathbb{F}_l)$. Similarly, we define an endomorphism T of $L(\mathbb{F}_l)$ by

$$(Tf)(\xi) \overset{\text{def}}{=} \sum_{x \in \mathbb{F}_l} f(x)\psi_\xi(x), \qquad f \in L(\mathbb{F}_l). \tag{4.2}$$

† Fourier (1768–1830).

Proposition 4.2. *$l^{-1/2}T$ is a unitary automorphism of the Hilbert space $L(\mathbb{F}_l)$.*

Proof. Observe that

$$(Tf)(\xi) = l\langle f, \psi_{-\xi}\rangle.\dagger$$

Clearly, T is an endomorphism of $L(\mathbb{F}_l)$. Furthermore, T is invertible. In fact, suppose that $Tf = 0$. Then $\langle f, \psi_{-\xi}\rangle = 0$ for all $\xi \in \mathbb{F}_l$ which implies that $f = 0$, because $\{\psi_\xi\}$ is an orthonormal basis for $L(\mathbb{F}_l)$. Finally, we must verify that $U = l^{-1/2}T$ leaves the inner product invariant: $\langle Uf, Ug\rangle = \langle f, g\rangle$, $f, g \in L(\mathbb{F}_l)$, i.e., $\langle Tf, Tg\rangle = l\langle f, g\rangle$. In fact, we have

$$\begin{aligned}\langle Tf, Tg\rangle &= l^{-1}\sum_\xi (Tf)(\xi)\overline{Tg}(\xi) = l^{-1}\sum_\xi \Big(\sum_x f(x)\psi_\xi(x)\Big)\Big(\sum_y \bar{g}(y)\bar{\psi}_\xi(y)\Big)\\ &= l^{-1}\sum_{x,y} f(x)\bar{g}(y)\sum_\xi \psi_\xi(x)\psi_\xi(-y)\\ &= \sum_{x,y} f(g)g(y)\Big(l^{-1}\sum_\xi \psi_\xi(x-y)\Big)\\ &= \sum_x f(x)\bar{g}(x) = l\langle f, g\rangle.\end{aligned}$$

Q.E.D.

For an endomorphism S of $L(\mathbb{F}_l)$, denote by S^* the adjoint of S defined by the relation $\langle Sf, g\rangle = \langle f, S^*g\rangle$. Note that

$$S \text{ is unitary} \Leftrightarrow SS^* = S^*S = 1.$$

Hence the endomorphism T satisfies the relation:

$$TT^* = T^*T = l.\ddagger \tag{4.3}$$

Exercise 4.2.

1. $(T^*f)(\xi) = (Tf)(-\xi)$.
2. $(T^2f)(\xi) = lf(-\xi)$.
3. $T^4 = l^2$.
4. $(\det T)^4 = l^{2l}$.

† Note that $\psi_{-\xi}(x) = \zeta^{-\xi x} = \overline{\zeta^{\xi x}} = \bar{\psi}_\xi(x)$.

‡ 1 in $SS^* = S^*S$ means the identity mapping $\mathbb{F}_l \to \mathbb{F}_l$. Similarly l on the right-hand side of (4.3) means the mapping $f \mapsto lf$, $f \in L(\mathbb{F}_l)$.

From Exercise 4.2(4), we see that $\det T = $ (a 4th root of $1) \times l^{l/2}$.

EXERCISE 4.3. Prove that $\det T = i^{l(l-1)/2} l^{l/2}$.

Now consider the group $\mathbb{F}_l^\times = (\mathbb{Z}/l\mathbb{Z})^\times$. As usual, we extend a character $\chi \in \hat{\mathbb{F}}_l^\times$ to a function $\chi \in L(\mathbb{F}_l)$ by setting $\chi(0) = 0$. For example, the unit character ϵ is a function:

$$\epsilon(x) = \begin{cases} 1, & x \neq 0, \\ 0, & x = 0. \end{cases}$$

If we denote by $\mathbf{1}$ the function taking the value 1 always, then we have $\epsilon = \mathbf{1} - f_0$ where f_0 is the characteristic function of $\{0\}$. Since the set $\{\chi\}$ of all characters forms a basis of $L(\mathbb{F}_l^\times)$ (Proposition 4.1), the set $\{\chi\}$ of extended functions is independent in $L(\mathbb{F}_l)$ and so it spans a hyperplane in $L(\mathbb{F}_l)$. As $\mathbf{1}$ is not in this hyperplane, the set $\{\mathbf{1}, \ldots, \chi, \ldots\}$ forms a basis of $L(\mathbb{F}_l)$. We shall determine the matrix B for T with respect to this basis.

EXERCISE 4.4.

1. $T\mathbf{1} = l\mathbf{1} - l\epsilon$.
2. $T\epsilon = (l-1)\mathbf{1} - l\epsilon$.

By Exercise 4.4, B must be of the form

$$(T\mathbf{1}, T\epsilon, \ldots) = (\mathbf{1}, \epsilon, \ldots) \begin{bmatrix} l & l-1 & \\ -l & -l & \\ 0 & 0 & * \\ \vdots & \vdots & \\ 0 & 0 & \end{bmatrix}.$$

In order to determine the rest entries of B, we need Gauss sums

$$\tau_\xi(\chi) = \sum_{x \in \mathbb{F}_l} \chi(x) \zeta^{\xi x}, \qquad \xi \in \mathbb{F}_l. \tag{4.4}$$

From (4.2) and (4.4), we see that

$$\tau_\xi(\chi) = (T\chi)(\xi). \tag{4.5}$$

Since $\tau(\chi) = \tau_1(\chi)$, we have

$$\tau(\chi) = (T\chi)(1). \tag{4.6}$$

By Proposition 2.8 and also (4.5) and (4.6),

$$T\chi = \tau(\chi)\bar{\chi}, \qquad \chi \neq \epsilon.$$

As we have, for $\chi \neq \epsilon$,

$$\chi = \bar{\chi} \Leftrightarrow \chi^2 = \epsilon, \tag{4.7}$$

and since $\hat{\mathbb{F}}_l^\times \approx \mathbb{F}_l^\times$ is a cyclic group of order $l-1$, we see that the Legendre character λ_l is the only character of order 2 and that all characters other than ϵ and λ_l appear as pairs $\{\chi, \bar{\chi}\}$. Thus, if we define B by the relation

$$(T\mathbf{1}, T\epsilon, T\lambda_l, T\chi, T\bar{\chi}, \ldots) = (\mathbf{1}, \epsilon, \lambda_l, \chi, \bar{\chi}, \ldots)B,$$

then B must be of the form

$$B = \left[\begin{array}{cc|c|cc|c} l & l-1 & & & & 0 \\ -l & -l & & & & \\ \hline & & \tau(\lambda_l) & & & \\ \hline & & & 0 & \tau(\bar{\chi}) & \\ & & & \tau(\chi) & 0 & \\ \hline 0 & & & & & \ddots \end{array}\right]$$

and we have

$$\det B = (-l)\tau(\lambda_l) \prod{}' (-\tau(\chi)\tau(\bar{\chi})) = (-1)^{(l-1)/2} l \tau(\lambda_l) \prod{}' (\tau(\chi)\tau(\bar{\chi})) \tag{4.8}$$

where $\prod'$ means the product over $\chi \neq \epsilon, \lambda_l$, choosing one χ out of a pair $\{\chi, \bar{\chi}\}$.

EXERCISE 4.5. $\tau(\chi)\tau(\bar{\chi}) = \chi(-1)l$.

From (4.8) and Exercise 4.5, we have

$$\det B = (-1)^{(l-1)/2} l^{(l-1)/2} \tau(\lambda_l) \prod{}' \chi(-1).\dagger$$

Note that there is a character χ^* such that $\chi^*(-1) \neq 1$ because $1 \neq -1$ in $\mathbb{F}_l$; we have $\chi^*(-1) = -1$, of course. Therefore, by the mapping $\chi \mapsto \chi\chi^*$ the sign of the value of characters at -1 must be

† Note that the number of factors in $\prod'$ is $(l-3)/2$.

changed. Since $\epsilon(-1)=1$, if $\lambda_l(-1)=1$, then among $\{\chi, \bar{\chi}, \ldots\}$ there are exactly $(l-1)/2$ characters whose values at -1 are -1. As $\chi(-1)=\bar{\chi}(-1)$, we see that in $\prod'$ there are exactly $(l-1)/4$ χ's for which $\chi(-1)=-1$. On the other hand, if $\lambda_l(-1)=-1$, then among $\{\chi, \bar{\chi}, \ldots\}$ there are exactly $(l-3)/2$ characters whose values at -1 are -1. Therefore in $\prod'$ there are exactly $(l-3)/4$ χ's for which $\chi(-1)=-1$. As $\lambda_l(-1)=(-1)^{(l-1)/2}$, we obtain

$$\det B=\begin{cases}(-1)^{(l-1)/2}(-1)^{(l-1)/4}l^{(l-1)/2}\tau(\lambda_l), & l\equiv 1\ (4),\\ (-1)^{(l-1)/2}(-1)^{(l-3)/4}l^{(l-1)/2}\tau(\lambda_l), & l\equiv 3\ (4).\end{cases} \tag{4.9}$$

On the other hand, since $\det B=\det T=i^{l(l-1)/2}l^{l/2}$ (Exercise 4.3), we get by (4.9) the following values:

$$\tau(\lambda_l)=\begin{cases}(-1)^{(l-1)/2}(-1)^{(l-1)/4}i^{l(l-1)/2}l^{1/2}, & l\equiv 1\quad (4),\\ (-1)^{(l-1)/2}(-1)^{(l-3)/4}i^{l(l-1)/2}l^{1/2}, & l\equiv 3\quad (4),\end{cases}$$

Or, after some computations, we obtain the following theorem.

THEOREM 4.2 (*Gauss*). *For an odd prime l, we have*

$$\tau(\lambda_l)=\begin{cases}\sqrt{l}, & l\equiv 1\quad (4),\\ i\sqrt{l}, & l\equiv 3\quad (4).\end{cases}$$

4.2. *L*-Functions and Gauss Sums

Let l be an odd prime and χ be a character $\neq\epsilon$ of the group $\mathbb{F}_l^\times$. As usual, we extend χ to a function on $\mathbb{F}_l$ by setting $\chi(0)=0$. For $s>1$, we have

$$L(s,\chi)=\sum_{n=1}^{\infty}\chi(n)/n^s=\sum_{x\in\mathbb{F}_l^\times}\chi(x)\sum_{n\equiv x(l)}1/n^s.$$

Since we can write

$$\sum_{n\equiv x(l)}1/n^s=\sum_{n=1}^{\infty}c(n)/n^s$$

with

$$c_n=\begin{Bmatrix}1, & n\equiv x\quad (l)\\ 0, & n\not\equiv x\quad (l)\end{Bmatrix}=l^{-1}\sum_{\xi\in\mathbb{F}_l}\zeta^{(x-n)\xi},$$

we have

$$L(s, \chi) = \sum_{x \in \mathbb{F}_l^\times} \chi(x) \sum_{n=1}^{\infty} 1/n^s \sum_{\xi \in \mathbb{F}_l} l^{-1} \zeta^{(x-n)\xi}$$
$$= l^{-1} \sum_{\xi \in \mathbb{F}_l} \left(\sum_{x \in \mathbb{F}_l^\times} \chi(x) \zeta^{x\xi} \right) \sum_{n=1}^{\infty} \zeta^{-n\xi}/n^s,$$

or

$$L(s, \chi) = l^{-1} \sum_{\xi \in \mathbb{F}_l^\times} \tau_\xi(\chi) \sum_{n=1}^{\infty} \zeta^{-n\xi}/n^s.\dagger \tag{4.10}$$

To study the inner sum, put $a_n = \zeta^{-n\xi}$. Then

$$A_\nu = \sum_{n=1}^{\nu} a_n = \sum_{n=1}^{l} a_n + \sum_{n=l+1}^{2l} a_n + \cdots + \sum_{n=(q-1)l+1}^{ql} a_n + \sum_{n=ql+1}^{\nu} a_n,$$
$$\nu = ql + r, \qquad 0 \le r < l,$$

and since $\sum_{n=1}^{l} \zeta^{-h\xi} = 0$ we have $|A_\nu| \le l$. Hence, by Exercise 3.28, the inner sum $\sum_{n=1}^{\infty} \zeta^{-n\xi}/n^s$ represents a continuous function for $s > 0$. On the other hand, as $\chi \neq \epsilon$, $L(s, \chi)$ represents also a continuous function for $s > 0$ (Theorem 3.9); thus one can evaluate both sides of (4.10) at $s = 1$ to get

$$L(1, \chi) = l^{-1} \sum_{\xi \in \mathbb{F}_l^\times} \tau_\xi(\chi) \sum_{n=1}^{\infty} \zeta^{-n\xi}/n. \tag{4.11}$$

Here, substituting the relation $\tau_\xi(\chi) = \bar{\chi}(\xi)\tau(\chi)$ (Proposition 2.8) in (4.11), we get

$$L(1, \chi) = l^{-1}\tau(\chi) \sum_{\xi \in \mathbb{F}_l^\times} \bar{\chi}(\xi) \sum_{n=1}^{\infty} \zeta^{-n\xi}/n. \tag{4.12}$$

Our next task is to obtain a finite expression for the infinite sum $\sum_{n=1}^{\infty} \zeta^{-n\xi}/n$. First of all, the above argument shows that one can evaluate the right-hand side of

$$-\log(1 - z) = \sum_{n=1}^{\infty} z^n/n$$

† Sine $\chi \neq \epsilon$, we removed $\xi = 0$ in $\sum_\xi$.

at $z=\zeta^{-\xi}\neq 1$ even though the point lies on the circle of convergence $|z|=1$. Hence, by a theorem of Abel,† we have

$$\lim_{z\to\zeta^{-\xi}}(-\log(1-z))=\sum_{n=1}^{\infty}\zeta^{-n\xi}/n. \tag{4.13}$$

On the other hand, since $\log(1-z)$ makes sense for $\mathbb{C}-[1,+\infty)$, we can evaluate the left-hand side of (4.13) at $z=\zeta^{-\xi}$ to get

$$-\log(1-\zeta^{-\xi})=\sum_{n=1}^{\infty}\zeta^{-n\xi}/n. \tag{4.14}$$

From (4.12) and (4.14), we conclude that

$$L(1,\chi)=l^{-1}\tau(\chi)\sum_{\xi\in\mathbb{F}_l^\times}\bar{\chi}(\xi)\log(1-\zeta^{-\xi}). \tag{4.15}$$

Let us put

$$S_\chi=\sum_{\xi\in\mathbb{F}_l^\times}\bar{\chi}(\xi)\log(1-\zeta^{-\xi})=\sum_{\xi=1}^{l-1}\bar{\chi}(\xi)\log(1-\zeta^{-\xi}).$$

Definition 4.1. A character $\chi\in\hat{\mathbb{F}}_l^\times$, $\chi\neq\epsilon$, is *even* if $\chi(-1)=1$ and *odd* if $\chi(-1)=-1$.

Exercise 4.6. Prove that

$$S_\chi=\begin{cases}\displaystyle\sum_{\xi=1}^{l-1}\bar{\chi}(\xi)\log(\sin(\pi\xi/l)), & \chi=\text{an even character},\\ \displaystyle-(\pi i/l)\sum_{\xi=1}^{l-1}\xi\bar{\chi}(\xi), & \chi=\text{an odd character}.\end{cases}$$

By (4.15) and Exercise 4.6, we obtain the following theorem.

Theorem 4.3. *Let l be an odd prime and χ be a character of $\mathbb{F}_l^\times$, $\chi\neq\epsilon$. Then*

$$L(1,\chi)=\begin{cases}\displaystyle-(\tau(\chi)/l)\sum_{\xi=1}^{l-1}\bar{\chi}(\xi)\log(\sin(\pi\xi/l)), & \chi\text{ even},\\ \displaystyle(\pi i/l^2)\tau(\chi)\sum_{\xi=1}^{l-1}\xi\bar{\chi}(\xi), & \chi\text{ odd}.\end{cases}$$

† Whittaker & Watson, *Modern Analysis* (4th ed.), p. 57.

4.3. Class Numbers of Subfields of the lth Cyclotomic Field

We have already studied these fields on several occasions (especially in Sections 2.6 and 2.16) The purpose of this section is to guide the reader to the gateway to the determination of class numbers of those field.

Let l be an odd prime, $\zeta = e^{2\pi i/l}$, $k = \mathbb{Q}(\zeta)$, and let $G(k/\mathbb{Q})$ be the Galois group of $k/\mathbb{Q}$. Since we have the isomorphism

$$\psi_{l-1}\colon \mathbb{F}_l^\times \xrightarrow{\sim} G(k/\mathbb{Q}), \qquad \psi_{l-1}(a) = \sigma_a, \qquad \zeta^{\sigma_a} = \zeta^a,$$

$G(k/\mathbb{Q})$ is cyclic of order $\varphi(l) = l - 1$ and so for any divisor n of $l - 1$ there is exactly one subfield k_n of k such that $[k_n\colon \mathbb{Q}] = n$.

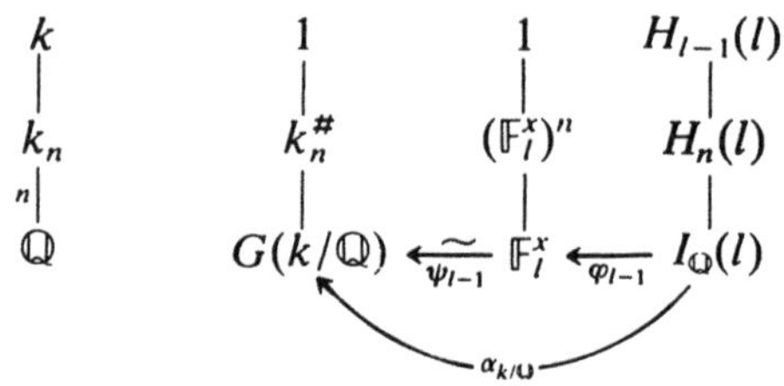

This figure is, as the reader can see, the same as in Section 2.16. We recall some basic facts. The mapping ψ_{l-1} induces naturally the isomorphism

$$\psi_n\colon \mathbb{F}_l^\times/(\mathbb{F}_l^\times)^n \xrightarrow{\sim} G(k_n/\mathbb{Q})$$

and the mapping φ_{l-1} induces the homomorphism

$$\varphi_n\colon I_\mathbb{Q}(l) \to \mathbb{F}_l^\times/(\mathbb{F}_l^\times)^n.$$

For the Artin map $\alpha_{k_n/\mathbb{Q}}$, we have

$$\psi_n \varphi_n = \alpha_{k_n/\mathbb{Q}}$$

and the exact sequence

$$1 \to H_n(l) \to I_\mathbb{Q}(l) \to G(k_n/\mathbb{Q}) \to 1.$$

In other words, for each divisor n of $l - 1$, we have the isomorphisms

$$G(k_n/\mathbb{Q}) \overset{\psi_n}{\underset{\sim}{\longleftarrow}} \mathbb{F}_l^\times/(\mathbb{F}_l^\times)^n \overset{\varphi_n}{\underset{\sim}{\longleftarrow}} I_\mathbb{Q}(l)/H_n(l) \qquad (4.16)$$

$$\alpha_{k_n/\mathbb{Q}}$$

where we have used the same notation φ_n and $\alpha_{k_n/\mathbb{Q}}$ after passing to quotient groups.

Now, we shall consider $\zeta_{k_n}(s)$, the Dedekind zeta function of k_n. First of all, we have the product expansion

$$\zeta_{k_n}(s) = \prod_{\mathfrak{p}} [1 - 1/(N\mathfrak{p})^s]^{-1}, \qquad s > 1$$

where $N = N_{k_n}$. Since l is the only prime which ramifies for $k/\mathbb{Q}$ and it ramifies completely, l has the same property for $k_n/\mathbb{Q}$ for each n and we have

$$\zeta_{k_n}(s) = (1 - 1/l^s)^{-1} \prod_{p \neq l} (1 - 1/p^{f_p s})^{-g_p} \qquad (4.17)$$

where $f_p g_p = n$, $N\mathfrak{p} = p^{f_p}$, and we shall occasionally put $f = f_p$ and $g = g_p$. Using the decomposition

$$X^f - 1 = \prod_{\nu=0}^{f-1} (X - \eta^\nu), \qquad \eta = e^{2\pi i/f},$$

we have

$$1 - 1/p^{sf} = \prod_{\nu=0}^{f-1} (1 - \eta^\nu/p^s)$$

and hence (4.17) becomes

$$\zeta_{k_n}(s) = (1 - 1/l^s)^{-1} \prod_{p \neq l} \prod_{\nu=0}^{f-1} (1 - \eta^\nu/p^s)^{-g}. \qquad (4.18)$$

On the other hand, for a fixed $p \neq l$, we claim that

$$\prod_{\nu=0}^{f-1} (1 - \eta^\nu/p^s)^{-g} = \prod_{\chi \in \widehat{\mathbb{F}_l^\times/(\mathbb{F}_l^\times)^n}} (1 - \chi(p)/p^s)^{-1}. \qquad (4.19)$$

To see this, we shall first show that there is a homomorphism

$$\pi_p: \widehat{\mathbb{F}_l^\times/(\mathbb{F}_l^\times)^n} \to W_p, \qquad \pi_p(\chi) = \chi(p)$$

where W_p is the cyclic group of order f generated by $\eta = e^{2\pi i/f}$. In fact, since

$$\alpha_{k_n/\mathbb{Q}}(p) = \left(\frac{k_n/\mathbb{Q}}{p}\right)$$

is of order f as being the Frobenius automorphism, we have $\alpha_{k_n/\mathbb{Q}}(p)^f = \alpha_{k_n/\mathbb{Q}}(p^f) = 1$, i.e., $p^f \in H_n(l)$ which means that p^f is an nth power residue modulo l. As χ is 1 on $(\mathbb{F}_l^\times)^n$, we have $\chi(p^f) = \chi(p)^f = 1$ and so $\chi(p) \in W_p$. It is clear that the mapping defined by $\pi_p(\chi) = \chi(p)$ is the required homomorphism. We next show that π_p is surjective. In fact, take any element $\eta^\nu \in W_p$. Since $H = \langle p \bmod (\mathbb{F}_l^\times)^n \rangle$ is a cyclic subgroup of order f of $\mathbb{F}_l^\times/(\mathbb{F}_l^\times)^n$,† there exist characters $\theta \in \hat{H}$ such that $\theta(p \bmod (\mathbb{F}_l^\times)^n) = \chi(p) = \eta^\nu$, i.e., $\pi_p(\chi) = \eta^\nu$ which proves that π_p is surjective. Therefore we have $[\operatorname{Ker} \pi_p] = g_p$ and equality (4.19) is proved. From (4.18) and (4.19) we have

$$\begin{aligned}
\zeta_{k_n}(s) &= (1 - 1/l^s)^{-1} \prod_{p \neq l} \prod_{\chi} (1 - \chi(p)/p^s)^{-1} \\
&= (1 - 1/l^s)^{-1} \prod_{\chi} \prod_{p} (1 - \chi(p)/p^s)^{-1} \\
&= (1 - 1/l^s)^{-1} \prod_{\chi} L(s, \chi).
\end{aligned}$$

[Note that $\chi(l) = 0$.] On the other hand, since

$$\begin{aligned}
L(s, \epsilon) &= \prod_{p} (1 - \epsilon(p)/p^s)^{-1} = \prod_{p \neq l} (1 - 1/p^s)^{-1} \\
&= (1 - 1/l^s) \prod_{p} (1 - 1/p^s)^{-1} = (1 - 1/l^s) \zeta_{\mathbb{Q}}(s)
\end{aligned}$$

we get

$$\zeta_{k_n}(s) = \zeta_{\mathbb{Q}}(s) \prod_{\chi \in \widehat{\mathbb{F}_l^\times/(\mathbb{F}_l^\times)^n},\ \chi \neq \epsilon} L(s, \chi). \tag{4.20}$$

By Theorem 3.9, $L(s, \chi)$, $\chi \neq \epsilon$, is continuous for $s > 0$ and we

† Note that $\alpha_{k_n/\mathbb{Q}}(p)$ is not of order f.

obtain, taking residues at $s = 1$ of both sides of (4.20), the equality

$$\lim_{s\to 1+0}(s-1)\zeta_{k_n}(s) = \prod_{\chi\in\widehat{\mathbb{F}_l^\times/(\mathbb{F}_l^\times)^n},\ \chi\neq\epsilon} L(1,\chi). \tag{4.21}$$

On the other hand, we know that (Theorem 3.6)

$$\lim_{s\to 1+0}(s-1)\zeta_{k_n}(s) = \frac{2^{r_1+r_2}\pi^{r_2}R_{k_n}h_{k_n}}{W_{k_n}\sqrt{|\Delta_{k_n}|}} \tag{4.22}$$

and that

$$L(1,\chi) = \begin{cases} -(\tau(\chi)/l)\sum_{\xi=1}^{l-1}\bar{\chi}(\xi)\log(\sin(\pi\xi/l)), & \chi \text{ even},\\ (\pi i/l^2)\tau(x)\sum_{\xi=1}^{l-1}\xi\bar{\chi}(\xi), & \chi \text{ odd}. \end{cases} \tag{4.23}$$

By substituting (4.22) and (4.23) in (4.21), one determines the class number h_{k_n}.

REMARK 4.1. We have just said that "one determines the class number h_{k_n}"; but to do this we must know R_{k_n}, $\tau(\chi)$, etc., and it is not an easy matter to determine h_{k_n} all the way to the end for any divisor n of $l-1$. In the next section, we shall treat in detail the case $n = 2$, i.e., the case $k_2 = \mathbb{Q}(\sqrt{l^*})$.

4.4. Class Number of $\mathbb{Q}(\sqrt{l^*})$

With the notation as in Section 4.3, we put $n = 2$; thus $k = k_2 = \mathbb{Q}(\sqrt{l^*})$, $l^* = (-1)^{(l-1)/2}l$. Since $n = 2$ the group $\mathbb{F}_l^\times/(\mathbb{F}_l^\times)^2$ is of order 2 and so its nontrivial character is the Legendre character λ_l. As $\lambda_l(-1) = (-1)^{(l-1)/2}$, we have

$$\begin{cases} \lambda_l \text{ is even} \Leftrightarrow l \equiv 1\ (4) \Leftrightarrow k \text{ is real quadratic},\\ \lambda_l \text{ is odd} \Leftrightarrow l \equiv 3\ (4) \Leftrightarrow k \text{ is imaginary quadratic}. \end{cases}$$

As for the Gauss sum, by Theorem 4.2, we have

$$\tau(\lambda_l) = \begin{cases} \sqrt{l}, & l \equiv 1 \quad (4),\\ i\sqrt{l}, & l \equiv 3 \quad (4). \end{cases}$$

After substituting this in (4.23), we get

$$L(1, \lambda_l) = \begin{cases} -1/\sqrt{l} \sum_{\xi=1}^{l-1} \lambda_l(\xi) \log(\sin(\pi\xi/l)), & l \equiv 1 \quad (4), \\ -\pi/(l\sqrt{l}) \sum_{\xi=1}^{l-1} \xi\lambda_l(\xi), & l \equiv 3 \quad (4). \end{cases} \tag{4.24}$$

As for the right-hand side of (4.22),

$$r_1 = \begin{cases} 2, \\ 0, \end{cases} \quad r_2 = \begin{cases} 0, \\ 1, \end{cases} \quad r = r_1 + r_2 - 1 = \begin{cases} 1, & l \equiv 1 \quad (4), \\ 0, & l \equiv 3 \quad (4); \end{cases} \tag{4.25}$$

$$w_{k_2} = \begin{cases} 2, & l \equiv 1 \quad (4), \\ 6, & l = 3, \\ 2, & l \equiv 3 \quad (4), l \geq 7; \end{cases} \tag{4.26}$$

$$\Lambda_{k_2} = l^* = \begin{cases} l, & l \equiv 1 \quad (4), \\ -l, & l \equiv 3 \quad (4); \end{cases} \tag{4.27}$$

$$R_{k_2} = \begin{cases} \log \epsilon, & l \equiv 1 \quad (4), \\ 1, & l \equiv 3 \quad (4). \end{cases} \tag{4.28}$$

When $l \equiv 1 \quad (4)$, i.e., when k_2 is real, $\mathfrak{o}_{k_2}^{\times} = \langle \pm\epsilon \rangle$, $\epsilon > 1$; here ϵ is uniquely determined and is called the *fundamental unit* of k_2. Since (4.21) reduces to

$$\lim_{s \to 1+0} (s-1)\zeta_{k_2}(s) = L(1, \lambda_l).$$

we obtain from (4.24)–(4.28), with $k = k_2$,

$$h_k = \begin{cases} -(2\log\epsilon)^{-1} \sum_{\xi=1}^{l-1} \lambda_l(\xi) \log(\sin(\pi\xi/l)), & l \equiv 1 \quad (4), \\ -l^{-1} \sum_{\xi=1}^{l-1} \xi\lambda_l(\xi), \quad l \equiv 3 \quad (4), & l \geq 7. \end{cases} \tag{4.29}$$

Remark 4.2. When $l = 3$, we know that $h_k = 1$ (Exercise 3.8). This can also be written as

$$h_k = -\sum_{\xi=1}^{l-1} \xi\lambda_l(\xi) = -(\lambda_l(1) + 2\lambda_l(2)) = -(1-2) = 1.$$

Real case. $k = \mathbb{Q}(\sqrt{l}),\ l \equiv 1 \pmod 4$.

In the equality

$$\sum_{\xi=1}^{l-1} \lambda_l(\xi)\log(\sin(\pi\xi/l)) = \sum_{\xi=1}^{(l-1)/2} \lambda_l(\xi)\log(\sin(\pi\xi/l)) + \sum_{\xi=(l+1)/2}^{l-1} \lambda_l(\xi)\log(\sin(\pi\xi/l)),$$

as $\lambda_l(l-\xi) = \lambda_l(-\xi) = \lambda_l(\xi)$ and $\sin(\pi-\theta) = \sin\theta$, we see that the two summations on the right-hand side are equal. Therefore (4.29) implies

$$h_k = -(1/\log\epsilon)\sum_{0<\xi<1/2} \lambda_l(\xi)\log(\sin(\pi\xi/l)).$$

If we denote by a and b those natural numbers $<\frac{l}{2}$ such that $\lambda_l(a) = 1$ and $\lambda_l(b) = -1$, then we have

$$h_k = -(1/\log\epsilon)\sum_a \log(\sin(\pi a/l)) + (1/\log\epsilon)\sum_b \log(\sin(\pi b/l))$$

and hence

$$h_k \log\epsilon = \log\left(\prod_b \sin(\pi b/l) \Big/ \prod_a \sin(\pi a/l)\right).$$

In other words, we find the following interesting relation between the fundamental unit ϵ and the class number h_k of a real quadratic field k:

$$\epsilon^{h_k} = \prod_b \sin(\pi b/l) \Big/ \prod_a \sin(\pi a/l). \tag{4.30}$$

Imaginary case. $k = \mathbb{Q}(\sqrt{-l}),\ l \equiv 3 \pmod 4$.

From the following equalities

$$\begin{aligned}\sum_{\xi=1}^{l-1} \xi\lambda_l(\xi) &= \sum_{\xi=1}^{(l-1)/2} \xi\lambda_l(\xi) + \sum_{\xi=(l+1)/2}^{l-1} \xi\lambda_l(\xi)\\ &= \sum_{\xi=1}^{(l-1)/2} \xi\lambda_l(\xi) + \sum_{\xi=1}^{(l-1)/2} (l-\xi)\lambda_l(l-\xi)\\ &= \sum_{0<\xi<l/2} \xi\lambda_l(\xi) + \sum_{0<\xi<l/2} (\xi-l)\lambda_l(\xi)\\ &= 2\sum_{0<\xi<l/2} \xi\lambda_l(\xi) - l\sum_{0<\xi<l/2} \lambda_l(\xi)\end{aligned}$$

we have

$$h_k l = -2 \sum_{0<\xi<l/2} \xi\lambda_l(\xi) + l \sum_{0<\xi<l/2} \lambda_l(\xi). \tag{4.31}$$

On the other hand, we have

$$h_k l = -\sum_{\xi=1}^{l-1} \xi\lambda_l(\xi) = -\sum_{\xi \text{ even}} \xi\lambda_l(\xi) - \sum_{\eta \text{ odd}} \eta\lambda_l(\eta).$$

If we put $\xi = l - \eta$ in the second term of the right-hand side, we get

$$\begin{aligned} h_k l &= -\sum_{\xi \text{ even}} \xi\lambda_l(\xi) - \sum_{\xi \text{ even}} (l-\xi)\lambda_l(l-\xi) \\ &= -\sum_{\xi \text{ even}} \xi\lambda_l(\xi) + \sum_{\xi \text{ even}} (l-\xi)\lambda_l(\xi) \\ &= -2 \sum_{\substack{1\le\xi\le l-1 \\ \xi \text{ even}}} \xi\lambda_l(\xi) + l \sum_{\substack{1\le\xi\le l-1 \\ \xi \text{ even}}} \lambda_l(\xi) \\ &= -2 \sum_{0<\xi<l/2} (2\xi)\lambda_l(2\xi) + l \sum_{0<\xi<l/2} \lambda_l(2\xi) \\ &= -4 \sum_{0<\xi<l/2} \xi\lambda_l(2\xi) + l \sum_{0<\xi<l/2} \lambda_l(2\xi) \end{aligned}$$

and hence

$$h_k l\lambda_l(2) = -4 \sum_{0<\xi<l/2} \xi\lambda_l(\xi) + l \sum_{0<\xi<l/2} \lambda_l(\xi). \tag{4.32}$$

It follows from (4.31) and (4.32) that

$$h_k l\lambda_l(2) = 2\Big(h_k l - l \sum_{0<\xi<l/2} \lambda_l(\xi)\Big) + l \sum_{0<\xi<l/2} \lambda_l(\xi)$$

or

$$h_k(2 - \lambda_l(2)) = \sum_{0<\xi<l/2} \lambda_l(\xi).$$

Now, we have

$$\lambda_l(2) = \begin{cases} 1, & l \equiv 7 \quad (8) \\ -1, & l \equiv 3 \quad (8) \end{cases}$$

and we obtain

$$h_k = \begin{cases} \sum_{0<\xi<l/2} \lambda_l(\xi), & l \equiv 7 \quad (8), \\ \frac{1}{3} \sum_{0<\xi<l/2} \lambda_l(\xi), & l \equiv 3 \quad (8), \qquad l \neq 3. \end{cases} \tag{4.33}$$

REMARK 4.3. In (4.33) the summation Σ is nothing but the excess of the number of quadratic residues mod l between 0 and $l/2$ over the number of quadratic nonresidues in the same interval. Since $h_k > 0$, the formula shows that when $l \equiv 3 \quad (4)$ the excess in that interval is always positive and it is a multiple of 3 if $l \equiv 3 \quad (8)$. On the other hand, when $l \equiv 1 \quad (4)$, since $\lambda_l(-1) = 1$ we find that the numbers of quadratic residues and nonresidues in the interval are equal; it is $(l-1)/4$.

EXERCISE 4.7. Verify that $h_k = 9$ for $k = \mathbb{Q}(\sqrt{-199})$.

4.5. Ideal Class Groups of Quadratic Fields

The Dirichlet's formula (4.33) is beautiful. However, when h_k is a composite number such as $h_k = 9$ as in Exercise 4.7, the formula cannot tell whether the group H_k is a cyclic group of order 9 or a product of two cyclic groups of order 3. So, in this section, let us think out an elementary method (using only Minkowski constants) by which we can determine not only the number h_k but the group structure of H_k.

First, we need some facts on ideals in quadratic fields. Let $k = \mathbb{Q}(\sqrt{m})$ be a quadratic field where m is a square-free integer. For a number $z = x + y\sqrt{m} \in k$, we denote its conjugate for $k/\mathbb{Q}$ by $z' = x - y\sqrt{m}$; when $m < 0$, z' is the complex conjugate of z. The canonical basis for $\mathfrak{o}_k$ is given by $\mathfrak{o}_k = [1, \omega]$ where (Exercise 2.17)

$$\omega = \begin{cases} \sqrt{m}, & m \equiv 2, 3 \quad (4), \\ (1+\sqrt{m})/2, & m \equiv 1 \quad (4), \end{cases}$$

and

$$\Delta_k = \begin{cases} 4m, & m \equiv 2, 3 \quad (4), \\ m, & m \equiv 1 \quad (4). \end{cases}$$

Let u and v be numbers in $\mathfrak{o}_k$ which are independent over $\mathbb{Q}$ and let $\mathfrak{m} = [u, v]$ be the free module of rank 2 in $\mathfrak{o}_k$ generated by u and v. If we determine a matrix $A \in M_2(\mathbb{Z})$ by the relation $(u, v) = (1, \omega)A$, then $[\mathfrak{o}_k: \mathfrak{m}] = |\det A|$ (Proposition 2.18). Since $1 \in \mathfrak{o}_k$, we have $[\mathfrak{o}_k: m] \in \mathfrak{m}$ (using Proposition 1.7 for the finite group $\mathfrak{o}_k/\mathfrak{m}$ of order $[\mathfrak{o}_k: \mathfrak{m}]$ and the element $1 \bmod \mathfrak{m}$ in it). Therefore we have $\mathfrak{m} \cap \mathbb{Z} = \mathbb{Z}a$, $a \geq 1$; a is the smallest natural number in $\mathfrak{m}$. Similarly, since $[\mathfrak{o}_k: \mathfrak{m}]\omega \in \mathfrak{m}$, $\mathfrak{m}$ contains an element of the form $b + c\omega$, $b, c \in \mathbb{Z}$, $c \geq 1$. Let us assume that the number c is the smallest one for those elements. obviously, we may vary b modulo a. The above observation shows that $\mathfrak{m}$ contains the module $[a, b + c\omega]$. In fact, we have the following stronger result:

Exercise 4.8. Prove that $\mathfrak{m} = [a, b + c\omega]$.

Thus, given $\mathfrak{m}$, the natural numbers a and c are naturally determined by $\mathfrak{m}$. As for b, assume that $m = [a, b + c\omega] = [a, b' + c\omega]$. Then $b' + c\omega = \lambda a + \mu(b + c\omega) = (\lambda a + \mu b) + \mu c\omega$ and so $c = \mu c$, or $\mu = 1$, i.e., $b' = \lambda a + b$ or $b' \equiv b \pmod{a}$. Conversely, if $b' \equiv b \pmod{a}$, then we obtain the same $\mathfrak{m}$. In other words, the triple $\{a, c, b \bmod a\}$ is uniquely determined by $\mathfrak{m}$. We call $\mathfrak{m} = [a, b + c\omega]$ the *canonical basis* expression for $\mathfrak{m}$.

Exercise 4.9. For $a, b \in \mathbb{N}$, $b \in \mathbb{Z}$, put $\mathfrak{m} = [a, b + c\omega]$. Prove that $\mathfrak{m}$ is a free module of rank 2 in $\mathfrak{o}_k$ and that $\{a, b + c\omega\}$ is the canonical basis for $\mathfrak{m}$.

In the following, when we speak of a module in $\mathfrak{o}_k$ we shall mean always one of rank 2. We shall list basic facts on ideals in $\mathfrak{o}_k$ as Exercises.

Exercise 4.10. If $\mathfrak{m} = [a, b + c\omega]$ is an ideal of $\mathfrak{o}_k$, then $c \mid a$ and $c \mid b$.

Exercise 4.11. $\mathfrak{m} = [a, b + c\omega]$ is an ideal $\Leftrightarrow a \mid N(b + \omega)$.

Exercise 4.12. An ideal $\mathfrak{a} \subset \mathfrak{o}_k$ is called a *primitive ideal* if $d = 1$ is the only natural number such that $d \mid \mathfrak{a}$. A module $\mathfrak{m} = [a, b + c\omega]$ is a primitive ideal $\Leftrightarrow c = 1$ and $a | N(b + \omega)$.

Exercise 4.13. For an ideal $\mathfrak{a} = [\alpha, \beta]$, show that $\mathfrak{a} = (\alpha, \beta) = \mathfrak{o}_k\alpha + \mathfrak{o}_k\beta$.

Exercise 4.14. Let $\mathfrak{a}'$ be the ideal consisting of all conjugates of elements of an ideal $\mathfrak{a}$ in $\mathfrak{o}_k$. Prove that $\mathfrak{a}\mathfrak{a}' = (N\mathfrak{a})$.

EXERCISE 4.15. Let $\mathfrak{a}=[a, b+\omega]$ be a primitive ideal. If $a=a_1a_2$ is a factorization of a then $\mathfrak{a}_i=[a_i, b+\omega]$, $i=1, 2$, are primitive ideals and we have $\mathfrak{a}=\mathfrak{a}_1\mathfrak{a}_2$.

Here, we shall summarize basic facts on quadratic fields scattered mainly in Chapter 2.

For each quadratic field k, there is associated the Kronecker character χ_k whose values at primes are given by

$$\chi_k(p)=\begin{cases} 0, & p=\mathfrak{p}^2 & & \text{(ramified)},\\ 1, & p=\mathfrak{p}\mathfrak{p}', & \mathfrak{p}\neq\mathfrak{p}' & \text{(splits completely)},\\ -1, & p=\mathfrak{p} & & \text{(remains prime)}.\end{cases}$$

These values are computed by the rule

$$\chi_k(p)=\begin{cases} 0, & p\mid\Delta_k, & \\ \left(\dfrac{\Delta_k}{p}\right). & p\nmid\Delta_k, & p\neq 2,\\ (-1)^{(\Delta_k^2-1)/8}, & p=2 & \text{[this happens only for } m\equiv 1 \quad (4)].\end{cases}$$

Since $\mathfrak{o}_k=\mathbb{Z}[\omega]$ as $\mathfrak{o}_k=[1, \omega]$, we can determine the decomposition of p in k by decomposing the minimal polynomial $f_\omega(X)$ modulo p (Theorem 2.17). We shall denote by $\mathfrak{p}_p$ and $\mathfrak{p}_p'$ the prime factors of p in k; it may happen that $\mathfrak{p}_p=\mathfrak{p}_p'$, of course.

The Minkowski constant M_k is defined by

$$M_k=\begin{cases}(1/2)\sqrt{\Delta_k}, & \Delta_k>0 \quad (k \text{ real}),\\ (2/\pi)\sqrt{-\Delta_k}, & \Delta_k<0 \quad (k \text{ imaginary}).\end{cases}$$

Any ideal class in H_k contains an ideal $\mathfrak{a}$ in $\mathfrak{o}_k$ such that $N_k\mathfrak{a}\leq M_k$ (Theorem 3.3).

PROPOSITION 4.3. *For a quadratic field k, put*

$$S_k \overset{\text{def}}{=} \{p; p\leq M_k, \chi_k(p)\neq -1\}.$$

Then the ideal class group H_k is generated by the classes of $\mathfrak{p}_p$, $p\in S_k$.

PROOF. In each class of H_k there is an ideal $\mathfrak{a}$ such that $N\mathfrak{a}\leq M_k$. Any prime factor of such an ideal $\mathfrak{a}$ satisfies $N\mathfrak{p}\leq M_k$. Therefore H_k is generated by classes of prime ideals $\mathfrak{p}$ such that

$N\mathfrak{p} \le M_k$. For such $\mathfrak{p}$, let p be the prime number below $\mathfrak{p}$. If $\chi_k(p) = -1$, then $\mathfrak{p} = p \sim 1$. If $\chi_k(p) \neq -1$, then $N\mathfrak{p} = p$ and so $p = \mathfrak{p}_p$ or $p = \mathfrak{p}'_p$; in the latter case, we have $\mathfrak{p}'_p \sim \mathfrak{p}_p^{-1}$ (Exercise 4.14); hence to generate H_k it is enough to take classes of $\mathfrak{p}_p$. As $p = \mathbb{N}\mathfrak{p} \le M_k$ in case $\chi_k(p) = -1$, we see that H_k is generated by classes of $\mathfrak{p}_p$ with $p \in S_k$. Q.E.D.

In the sequel, we put

$$l = \begin{cases} -m, & m \equiv 2, 3 \quad (4), \\ (1-m)/4, & m \equiv 1 \quad (4), \end{cases}$$

$$P(X) \stackrel{\text{def}}{=} N(x+\omega) = \begin{cases} x^2 + l, & m \equiv 2, 3 \quad (4), \\ x^2 + x + l, & m \equiv 1 \quad (4). \end{cases}$$

Since H_k is generated by classes of $\mathfrak{p}_p$, $p \in S_k$, the structure of H_k will be determined by relations among these generators.

Now, let b be a nonnegative integer such that all prime factors of the integer $P(b) = N(b+\omega)$ belong to the set S_k:

$$|P(b)| = |N(b+\omega)| = p_1^{e_1} \cdots p_n^{e_n}, \qquad p_i \in S_k \tag{4.34}$$

By Exercise 4.11 and Exercise 4.12, we see that

$$(b+\omega) = [|N(b+\omega)|, b+\omega]$$

is the canonical basis expression of the primitive principal ideal $(b+\omega)$. By (4.34) and Exercise 4.15, we have

$$1 \sim (b+\omega) = [p_1, b+\omega]^{e_1} \cdots [p_n, b+\omega]^{e_n}. \tag{4.35}$$

Since each factor on the right-hand side of (4.35) is of the form $[p, b+\omega]$ which is a prime ideal of norm p, we must have $[p, b+\omega] = \mathfrak{p}_p$ or $\mathfrak{p}'_p$ and hence $[p, b+\omega] \sim \mathfrak{p}_p$ or $\mathfrak{p}_p^{-1}$. Thus we obtain, by (4.35), the following relation among generators:

$$\mathfrak{p}_{p_1}^{\pm e_1} \cdots \mathfrak{p}_{p_n}^{\pm e_n} \sim 1 \tag{4.36}$$

where we take + (resp. −) when $[p, b+\omega] \sim \mathfrak{p}_p$ (resp. $\sim \mathfrak{p}_p^{-1}$). Doing this for b's satisfying (4.34), we obtain relations like (4.36) and will eventually be able to determine the structure of H_k. Let us try this method for $m = \pm 199$ in view of Exercise 4.7.

For $m = 199$ we have $199 \equiv 3\,(4)$, $\omega = \sqrt{199}$, $\Delta_k = 4.199$, k real, $M_k = (\frac{1}{2})\sqrt{\Delta_k} = \sqrt{199} \fallingdotseq 14.1067$, $p = 2, 3, 5, 7, 11, 13 \leq M_k$.

p	2	3	5	7	11	13
$\chi_p(p)$	0	1	1	-1	1	1

Hence $S_k = \{2, 3, 5, 11, 13\}$. For $p \in S_k$, we determine $\mathfrak{p}_p$ and $\mathfrak{p}'_p$. The minimal polynomial $f_\omega(X) = X^2 - 199$.

When $p = 2$, $X^2 - 199 \equiv X^2 + 1 \equiv (X+1)^2 \quad (2) \Rightarrow 2 = \mathfrak{p}_2^2$, $\mathfrak{p}_2 = (2, 1+\omega) = [2, 1+\omega]$,† $\mathfrak{p}'_2 = \mathfrak{p}_2$.

When $p = 3$, $X^2 - 199 \equiv X^2 - 1 = (X-1)(X+1) \quad (3) \Rightarrow 3 = \mathfrak{p}_3\mathfrak{p}'_3$, $\mathfrak{p}_3 = (3, \omega - 1) = (3, 2+\omega) = [3, 2+\omega]$, $p'_3 = (3, \omega+1) = [3, 1+\omega]$.

When $p = 5$, $X^2 - 199 \equiv X^2 + 1 \equiv (X+2)(X+3) \quad (5) \Rightarrow 5 = \mathfrak{p}_5\mathfrak{p}'_5$, $\mathfrak{p}_5 = (5, 3+\omega) = [5, 3+\omega]$, $\mathfrak{p}'_5 = (5, 2+\omega) = [5, 2+\omega]$.

When $p = 11$, $X^2 - 199 \equiv X^2 - 1 = (X+1)(X-1) \quad (11) \Rightarrow 11 = \mathfrak{p}_{11}\mathfrak{p}'_{11}$, $\mathfrak{p}_{11} = (11, \omega - 1) = (11, \omega + 10) = [11, 10+\omega]$, $\mathfrak{p}'_{11} = (11, 1+\omega) = [11, 1+\omega]$.

When $p = 13$, $X^2 - 199 \equiv X^2 - 4 = (X+2)(X-2) \quad (13) \Rightarrow 13 = \mathfrak{p}_{13}\mathfrak{p}'_{13}$, $\mathfrak{p}_{13} = (13, \omega - 2) = [13, 11+\omega]$, $\mathfrak{p}'_{13} = (11, 2+\omega) = [13, 2+\omega]$.

Next, for $P(b) = N(b+\omega) = b^2 - 199$, $b \geq 0$, we decompose $|P(b)|$ into prime factors and show them in boldface type if all factors of $|P(b)|$ belong to $S_k = \{2, 3, 5, 11, 13\}$.

b	0	1	2	3	4	5	6
$P(b)$	-199	-198	-195	-190	-183	-174	-163
$\lvert P(b)\rvert$	199	$\mathbf{2 \cdot 3^2 \cdot 11}$	$\mathbf{3 \cdot 5 \cdot 13}$	$2 \cdot 5 \cdot 19$	$3 \cdot 61$	$2 \cdot 87$	163

b	7	8	9	10	11	12	13	14
$P(b)$	-150	-135	-118	-99	-78	-55	-30	-3
$\lvert P(b)\rvert$	$\mathbf{2 \cdot 3 \cdot 5^2}$	$\mathbf{3^3 \cdot 5}$	$2 \cdot 59$	$\mathbf{3^2 \cdot 11}$	$\mathbf{2 \cdot 3 \cdot 13}$	$\mathbf{5 \cdot 11}$	$\mathbf{2 \cdot 3 \cdot 5}$	$\mathbf{3}$

† The inclusion $(2, 1+\omega) \supset [2, 1+\omega]$ is O.K. The right-hand side is also an ideal and we get an equality by comparing norms of ideals on both sides. We put $\mathfrak{p}_3 = [3, 2+\omega]$ and $\mathfrak{p}'_3 = [3, 1+\omega]$; we can, of course, interchange $\mathfrak{p}_3$ and $\mathfrak{p}'_3$.

The ideal class group H_k is generated by classes of $\mathfrak{p}_2$, $\mathfrak{p}_3$, $\mathfrak{p}_5$, $\mathfrak{p}_{11}$, and $\mathfrak{p}_{13}$. Using the table above, we determine the relations (4.36). Since $P(14) = -3$ contains the least number of prime factors among $|P(b)|$'s in boldface type, let us start with $b = 14$.† By (4.35), we get

$$1 \sim [|P(b)|, b+\omega] = [3, 14+\omega] = [3, 2+\omega] = \mathfrak{p}_3, \qquad \text{i.e., } \mathfrak{p}_3 \sim 1.$$

Next, we choose $b = 8$ because $|P(b)|$ contains 3 which has already been taken and the number of prime factors is only two. We have

$$\begin{aligned} 1 \sim [3^3 \cdot 5, 8+\omega] &= [3, 8+\omega]^3[5, 8+\omega] = [3, 2+\omega]^3[5, 3+\omega] \\ &= \mathfrak{p}_3^3\mathfrak{p}_5 \sim \mathfrak{p}_5, \\ &\text{i.e.,} \quad \mathfrak{p}_5 \sim 1. \end{aligned}$$

Next, consider the case $b = 10$, $|P(b)| = 3^2 \cdot 11$. We have

$$\begin{aligned} 1 \sim [3^2 \cdot 11, 10+\omega] &= [3, 10+\omega]^2[11, 10+\omega] = [3, 1+\omega]^2[11, 10+\omega] \\ &= (\mathfrak{p}_3')^2\mathfrak{p}_{11} \sim \mathfrak{p}_{11}, \qquad \text{i.e, } \mathfrak{p}_{11} \sim 1. \end{aligned}$$

Next, $b = 1$, $|P(b)| = 2 \cdot 3^2 \cdot 11$. Then

$$\begin{aligned} 1 \sim [2 \cdot 3^2 \cdot 11, 1+\omega] = [2, 1+\omega][3, 1+\omega]^2[11, 1+\omega] = \mathfrak{p}_2(\mathfrak{p}_3')\mathfrak{p}_{11}' \sim \mathfrak{p}_2, \\ \text{i.e., } \mathfrak{p}_2 \sim 1. \end{aligned}$$

Finally, $b = 2$, $|P(b)| = 3 \cdot 5 \cdot 13$. Then

$$\begin{aligned} 1 \sim [3 \cdot 5 \cdot 13, 2+\omega] = [3, 2+\omega][5, 2+\omega][13, 2+\omega] = \mathfrak{p}_3\mathfrak{p}_5'\mathfrak{p}_{13}' \sim \mathfrak{p}_{13}', \\ \text{i.e., } \mathfrak{p}_{13} \sim 1. \end{aligned}$$

Therefore, we find that $h_k = 1$ for $k = \mathbb{Q}(\sqrt{199})$.

For $m = -119$ we have $-199 \equiv 1 \pmod 4$, $\omega = (1+\sqrt{-199})/2$, $\Delta_k = -199$, k imaginary, $M_k = (2/\pi)\sqrt{-\Delta_k} < 0.63663\sqrt{199} \doteqdot 8.98077$, $p = 2, 3, 5, 7 \leq M_k$.

p	2	3	5	7
$\chi_k(p)$	1	-1	1	1

$S_k = \{2, 5, 7\}$

† There is no big reason why we stopped at $b = 14$ in making the table. Perhaps the fact that $P(b)$ is prime might be one of reasons. If you stop too soon, you will not come to a settlement.

$f_\omega(X) = X^2 - X - (m-1)/4 = X^2 - X + 50.$

When $p = 2$, $X^2 - X + 50 \equiv X(X+1) \pmod{2} \Rightarrow 2 = \mathfrak{p}_2\mathfrak{p}_2'$, $\mathfrak{p}_2 = (2, \omega) = [2, \omega]$, $\mathfrak{p}_2' = (2, \omega + 1) = [2, 1 + \omega]$.

When $p = 5$, $X^2 - X + 50 \equiv X(X-1) \pmod{5} \Rightarrow 5 = \mathfrak{p}_5\mathfrak{p}_5'$, $\mathfrak{p}_5 = (5, \omega - 1) = [5, 4 + \omega]$, $\mathfrak{p}_5' = (5, \omega) = [5, \omega]$.

When $p = 7$, $X^2 - X + 50 \equiv X^2 - X + 1 \equiv (X+4)(X+2) \pmod{7} \Rightarrow \mathfrak{p}_7 = (7, 4 + \omega) = [7, 4 + \omega]$, $\mathfrak{p}_7' = (7, 2 + \omega) = [7, 2 + \omega]$.

$P(b) = N(b + \omega) = b^2 + b + 50.$

b	0	1	2	3	4
$P(b)$	50	52	56	62	70
$\lvert P(b)\rvert$	$\mathbf{2 \cdot 5^2}$	$2^2 \cdot 13$	$\mathbf{2^3 \cdot 7}$	$2 \cdot 31$	$\mathbf{2 \cdot 5 \cdot 7}$

The group H_k is generated by classes of $\mathfrak{p}_2$, $\mathfrak{p}_5$, and $\mathfrak{p}_7$.

For $b = 0$, $P(b) = 2 \cdot 5^2$, we have

$$1 \sim [2 \cdot 5^2, \omega] = [2, \omega][5, \omega]^2 = \mathfrak{p}_2(\mathfrak{p}_5')^2, \qquad \text{i.e., } \mathfrak{p}_2 \sim \mathfrak{p}_5^5.$$

Next, $b = 2$, $P(b) = 2^3 \cdot 7$. Then

$$1 \sim [2^3 \cdot 7, 2 + \omega] = [2, 2 + \omega]^3[7, 2 + \omega] = [2, \omega]^3[7, 2 + \omega] = \mathfrak{p}_2^3\mathfrak{p}_7',$$
$$\text{i.e., } \mathfrak{p}_7 \sim \mathfrak{p}_2^3 \sim \mathfrak{p}_5^6.$$

Therefore we find that H_k is generated by the class of $\mathfrak{p}_5$, i.e., H_k is a cyclic group. To determine the order of H_k, we argue as follows. Let $b = 4$. So $P(b) = 2 \cdot 5 \cdot 7$. We have

$$\begin{aligned} 1 \sim [2 \cdot 5 \cdot 7, 4 + \omega] &= [2, 4 + \omega][5, 4 + \omega][7, 4 + \omega] \\ &= [2, \omega][5, 4 + \omega][7, 4 + \omega] \\ &= \mathfrak{p}_2\mathfrak{p}_5\mathfrak{p}_7 \sim \mathfrak{p}_5^2\mathfrak{p}_5\mathfrak{p}_5^6 \sim \mathfrak{p}_5^9, \qquad \text{i.e., } \mathfrak{p}_5^9 \sim 1. \end{aligned}$$

Therefore H_k is a cyclic group generated by the class of $\mathfrak{p}_5$ of order at most 9. We claim that $\mathfrak{p}_5^3 \not\sim 1$. In fact, suppose, on the contrary, that $\mathfrak{p}_5^3 \sim 1$. Then we have $\mathfrak{p}_5^3 = (x + y\omega)$, $x, y \in \mathbb{Z}$. Taking norms of both sides, we get $125 = N(x + y\omega) = x^2 + xy + 50y^2$, or $500 = (2x + y)^2 + (199y^2)$. Thus $u^2 + 199v^2 = 500$ must have integral solution (u, v). Clearly $v = 0, \pm 1$ are only possibilities; however,

$v = 0 \Rightarrow u^2 = 500$ and $v = \pm 1 \Rightarrow u^2 = 301$, and so there are no such (u, v). Therefore H_k is a cyclic group of order 9, a result better than just $h_k = 9$ (Exercise 4.7).

Long ago (Exercise 1.10), we verified that the values of quadratic polynomial $P(x) = x^2 + x + 41$ are all prime for $x = 0, 1, 2, \ldots, 38, 39$. When $k = \mathbb{Q}(\sqrt{-163})$ we observe that $P(x)$ (in the sense of this section) $= N(x + \omega) = x^2 + x + l = x^2 + x + 41$ as $l = (1 - m)/4 = (1 + 163)/4 = 41$. This kind of phenomenon (discovered by Euler) is a special case of the following exercise.

Exercise 4.16 (Rabinovitch). Let $k = \mathbb{Q}(\sqrt{m})$ be an imaginary quadratic field, $m \neq -1, -3$. Let

$$l = N\omega = \begin{cases} -m & m \equiv 2, 3 \quad (4), \\ (1 - m)/4, & m \equiv 1 \quad (4), \end{cases}$$

$$P(n) = N(x + \omega) = \begin{cases} x^2 + l, & m \equiv 2, 3 \quad (4), \\ x^2 + x + l, & m \equiv 1 \quad (4). \end{cases}$$

Then we have

$$P(x),\ 0 \le x \le l - 2, \text{ are all prime} \Leftrightarrow h_k = 1.$$

Remark 4.4. For any natural number ν, we denote by $\deg \nu$ the number of prime factors in ν (counted with multiplicities). Thus $\deg 10 = 2$, $\deg 8 = 3$, etc., and ν is prime $\Leftrightarrow \deg \nu = 1$. For an imaginary quadratic field k, using notation of this section, put

$$p_k \overset{\text{def}}{=} \max_{0 \le x \le l-2} (\deg P(x)).$$

Thus Exercise 4.16 states that $p_k = 1 \Leftrightarrow h_k = 1$. It is known that $p_k = 2 \Leftrightarrow h_k = 2$ (R. Sasaki). However, we have $p_k = 3 < h_k = 4$ when $k = \mathbb{Q}(\sqrt{-21})$. Actually we know that $p_k \le h_k$ for any imaginary quadratic field k (R. Sasaki).

4.6. Cohomology of Quadratic Fields

In Section 2.7, for a finite cyclic group $G = \langle \sigma \rangle$ and a G-module A, we defined two cohomology groups

$$H^0(A) = A^G/NA \qquad \text{and} \qquad H^1(A) = {}_NA/A^{1-\sigma}.$$

We also defined the Herbrand quotient $Q(A)=[H^1(A)]/[H^0(A)]$ when both $H^0(A)$ and $H^1(A)$ are finite and, for three G-modules $A\subset B$, $C=B/A$ with C finite, we proved that $Q(A)=Q(B)$ whenever at least one of $Q(A)$ and $Q(B)$ is finite. If, in particular, $A=0$ and B is finite, then $C=B$ and we have $[H^0(B)]=[H^1(B)]$ because $Q(A)=Q(B)=1$. Summarizing, we get the following proposition.

PROPOSITION 4.4. *Let $G=\langle\sigma\rangle$ be a finite cyclic group and A be a finite G-module. Then $[H^0(A)]=[H^1(A)]$.*

In the sequel, let $k/\mathbb{Q}$ be a quadratic field with the Galois group $G=G(k/\mathbb{Q})=\langle\sigma\rangle$. We shall denote by α^σ and $\mathfrak{a}^\sigma$ the conjugates of a number α and a fractional ideal $\mathfrak{a}$ of k instead of α' and $\mathfrak{a}'$ as before. Among G-modules associated with k, the most important one in Gauss genus theory (in the next section) is the group H_k^+ of ideal classes of k in the narrow sense. So let us begin with some definitions. As before, I_k denotes the group of fractional ideals of k and P_k denotes the subgroup of I_k consisting of principal fractional ideals of k. Put

$$P_k^+=\{(\alpha)\in P_k;\, N_{k/\mathbb{Q}}\alpha>0\}.$$

Clearly, P_k^+ is a subgroup of P_k.

EXERCISE 4.17. When k is real, let ϵ be the fundamental unit of k. Prove that

$$[P_k\colon P_k^+]=\begin{cases}1, & \Delta_k<0 \text{ or } \Delta_k>0 \text{ and } N\epsilon=-1,\\ 2, & \Delta_k>0 \text{ and } N\epsilon=1.\end{cases}$$

The quotient group $H_k^+=I_p/P_k^+$ is the *ideal class group in the narrow sense* and its order h_k^+ is the *class number in the narrow sense.* Hence

$$h_k^+=\begin{cases}h_k, & \Delta_k<0 \text{ or } \Delta_k>0 \text{ and } N\epsilon=-1,\\ 2h_k, & \Delta_k>0 \text{ and } N\epsilon=1.\end{cases}$$

We shall denote $\mathfrak{a}\pm\mathfrak{b}$ the equivalence of fractional ideals modulo P_k^+. If $\mathfrak{a}\sim\mathfrak{b}$, then $\mathfrak{a}^\sigma\sim\mathfrak{b}^\sigma$ and so the Galois group $G=\langle\sigma\rangle$ acts on H_k. Similarly, if $\mathfrak{a}\pm\mathfrak{b}$, then $\mathfrak{b}=(\alpha)\mathfrak{a}$, $N\mathfrak{a}>0$, so $\mathfrak{b}^\sigma=(\alpha^\sigma)\mathfrak{a}^\sigma$ with

$N\alpha^\sigma = N\alpha > 0$, i.e., $\mathfrak{a} \pm \mathfrak{b}$. Hence G acts on H_k^+, too. We shall study the cohomology groups

$$H^0(H_k^+) = (H_k^+)^G / NH_k^+ \tag{4.37}$$

and

$$H^1(H_k^+) = {}_N H_k^+ / (H_k^+)^{1-\sigma}. \tag{4.38}$$

Here is a remark about the norm of ideals. For an ideal $\mathfrak{a} \subset \mathfrak{o}_k$, we defined the norm by $N_k\mathfrak{a} = [\mathfrak{o}_k : \mathfrak{a}]$. On the other hand, the norm in the sense of cohomology is defined by the action of the Galois group G, i.e., $N\mathfrak{a} = \mathfrak{a}\mathfrak{a}^\sigma = \mathfrak{a}^{1+\sigma}$. However, these two kinds of norms are related by the equality $N\mathfrak{a} = (N_k\mathfrak{a})$. (See Exercise 4.14 for quadratic fields and Exercise 4.18 for Galois extension $k/\mathbb{Q}$.)

Exercise 4.18. Let $k/\mathbb{Q}$ be a Galois extension. For $\mathfrak{a} \subset \mathfrak{o}_k$, put $N\mathfrak{a} = \prod_{\sigma \in G} \mathfrak{a}^\sigma$, $G = G(k/\mathbb{Q})$. Prove that $N\mathfrak{a} = (N_k\mathfrak{a})$. The equality can naturally be extended for fractional ideals.

Proposition 4.5. *For a quadratic field $k/\mathbb{Q}$, we have* $H^0(H_k^+) = {}_2H_2^+$,† $H^1(H_k^+) = H_k^+/(H_k^+)^2$.

Proof. For a class $[\mathfrak{a}] \in H_k^+$, since $N\mathfrak{a} = (N_k\mathfrak{a}) \pm 1$, we have $N[\mathfrak{a}] = 1 \in H_k^+$ and so $H^0(H_k^+) = (H_k^+)^G$. Now, $[\mathfrak{a}] \in (H_k^+)^G \Leftrightarrow [\mathfrak{a}]^\sigma = [\mathfrak{a}] \Leftrightarrow \mathfrak{a}^\sigma \pm \mathfrak{a} \Leftrightarrow \mathfrak{a}^{-1} \pm \mathfrak{a} \Leftrightarrow \mathfrak{a}^2 \pm 1 \Leftrightarrow [\mathfrak{a}]^2 = 1$, i.e., $H^0(H_k^+) = {}_2H_k^+$. Next, we have ${}_NH_k^+ = H_k^+$ because $N\mathfrak{a} \pm 1$. Furthermore, if $\mathfrak{a}\mathfrak{a}^\sigma \pm 1$, then $\mathfrak{a}^{-\sigma} \pm \mathfrak{a}$, or $\mathfrak{a}^{1-\sigma} \pm \mathfrak{a}^2$ and so $(H_k^+)^{1-\sigma} = (H_k^+)^2$, i.e., $H^1(H_k^+) = H_k^+/(H_k^+)^2$. Q.E.D.

Remark 4.5. From Exercise 3.23 it follows that $H_k^+/(H_k^+)^2 \approx {}_2H_k^+$. Therefore Proposition 4.5 implies that

$$H^0(H_k^+) \approx H^1(H_k^+), \qquad k \text{ quadratic.} \tag{4.39}$$

Exercise 4.19. For a cyclic extension $k/\mathbb{Q}$, prove that $H^1(I_k) = 1$.

Let us return to a quadratic field k and start with the exact sequence

$$1 \to P_k^+ \xrightarrow{f} I_k \xrightarrow{g} H_k^+ \to 1 \tag{4.40}$$

By Theorem 2.7, we obtain the following exact hexagonal

† For a group A, ${}_2A \overset{\text{def}}{=} \{a \in A; a^2 = e\}$.

sequence:

$$
\begin{array}{ccccc}
 & H^0(P_k^+) & \xrightarrow{f_0} & H^0(I_k) & \\
{\scriptstyle \delta_1} \nearrow & & & & \searrow {\scriptstyle g_0} \\
H^1(H_k^+) & & & & H^0(H_k^+) \\
{\scriptstyle g_1} \nwarrow & & & & \swarrow {\scriptstyle \delta_0} \\
 & H^1(I_k) & \xleftarrow[f_1]{} & H^1(P_k^+) & \\
 & \| & & \| & \\
 & 1 & & 1 &
\end{array}
\tag{4.41}
$$

Proposition 4.6. $H^1(P_k^+) = 1$.

Proof. For any $(\alpha) \in {}_NP_k^+$, we have $N(\alpha) = (N\alpha) = (1)$ and so $N\alpha = 1$ because $N\alpha > 0$. Furthermore, we may assume that $\alpha \neq -1$ because if $\alpha = -1$ we can replace it by $\alpha = 1$. Then from $N\alpha = \alpha\alpha^\sigma = 1$ it follows that $\alpha + 1 = \alpha + \alpha\alpha^\sigma = \alpha(1 + \alpha\alpha^\sigma)$ or $\beta = \alpha\beta^\sigma$ with $\beta = 1 + \alpha \neq 0$, i.e., $\alpha = \beta^{1-\sigma}$. Hence $\beta\alpha^{-1} = \beta^\sigma$, i.e., $N\beta = \beta\beta^\sigma = \beta^2\alpha^{-1} > 0$ even when k is real. If we write $(\alpha) = (\beta)^{1-\sigma}$ then we find that $(\alpha) \in (P_k^+)^{1-\sigma}$. Q.E.D.

Remark 4.6. If we only consider the full group H_k instead of H_k^+ we cannot prove that $H^1(P_k) = 1$ because $(\alpha) \in {}_NP_k$ does not imply $N\alpha = 1$.

Exercise 4.20. Find a quadratic field such that $H^1(P_k) \neq 1$.

Proposition 4.7. *The sequence*

$$1 \to (P_k^+)^G \to I_k^G \to (H_k^+)^G \to 1$$

is exact.

Proof. Since $H^1(P_k^+) = 1$ (Proposition 4.6) the mapping

$$g_0\colon H^0(I_k) = I_k^G/NI_k \to H^0(H_k^+) = (H_k^+)^G$$

in the sequence (4.41) is surjective. From this we see that the natural mapping $I_k^G \to (H_k^+)^G$ is surjective and the exactness of the sequence in the proposition follows immediately. Q.E.D.

PROPOSITION 4.8. *Let k be a quadratic field, $l_1, \ldots, l_t$ be the distinct prime factors of Δ_k, and $\mathfrak{l}_1, \ldots, \mathfrak{l}_t$ be prime ideals in k lying above $l_1, \ldots, l_t$: $l_i = \mathfrak{l}_i^2$, $\mathfrak{l}_i^\sigma = \mathfrak{l}_i$, $1 \le i \le t$.* When $2 \mid \Delta_k$, *we agree to put $l_1 = 2$. Then the group $I_k^G/(P_k^+)^G$ is generated by classes* $\mathrm{mod}(P_k^+)^G$ *of 2^t ideals $\mathfrak{l}_1^{\epsilon_1} \cdots \mathfrak{l}_t^{\epsilon_t}$, $\epsilon_i = 0$ or* 1.

PROOF. Note first that $(P_k^+)^G \supset I_{\mathbb{Q}} = \mathbb{Q}_+^\times$ where we have embedded $I_{\mathbb{Q}}$ naturally in I_k (Remark 2.18). A prime divisor $\mathfrak{p}$ of $\mathfrak{a} \in I_k^G$ of the form $\mathfrak{p} = (p)$ does not contribute to the quotient group $I_k^G/(P_k^+)^G$. If $\mathfrak{p}$ is such that $\mathfrak{p}\mathfrak{p}^\sigma = (p)$ and $\mathfrak{p} \ne \mathfrak{p}^\sigma$, then $\mathfrak{p}^\alpha(\mathfrak{p}^\sigma)^\beta \mid \mathfrak{a}$ implies $(\mathfrak{p}^\sigma)^\alpha\mathfrak{p}^\beta \mid \mathfrak{a}$; hence

$$\mathfrak{a} = \mathfrak{a}^\sigma \Rightarrow \alpha = \beta \Rightarrow \mathfrak{p}^\alpha(\mathfrak{p}^\sigma)^\beta = (\mathfrak{p}\mathfrak{p}^\sigma)^\alpha = (p)^\alpha \in I_{\mathbb{Q}}.$$

Therefore only divisors which contribute to the quotient group $I_k^G/(P_k^+)$ are of the form $\mathfrak{l}_1^{\epsilon_1} \cdots \mathfrak{l}_t^{\epsilon_t}$, $\epsilon_i = 0$ or 1 because $\mathfrak{l}_i^2 = l_i \sim 1$.
Q.E.D.

PROPOSITION 4.9. $[I_k^G/(P_k^+)^G] \le 2^{t-1}$.

PROOF. By Proposition 4.8 $I_k^G/(P_k^+)^G$ is generated by the classes of 2^t ideals $\mathfrak{l}_1^{\epsilon_1} \cdots \mathfrak{l}_t^{\epsilon_t}$, $\epsilon_i = 0$ or 1. To prove the assertion, it is enough to show that there is at least one ideal $\mathfrak{l}_1^{\epsilon_1} \cdots \mathfrak{l}_t^{\epsilon_t} \pm 1$ with $(\epsilon_1, \ldots, \epsilon_t) \ne (0, \ldots, 0)$.

1. $\Delta_k < 0$ or $\Delta_k > 0$ and $N\epsilon = -1$. (a) $m \equiv 1 \pmod 4$. In this case, since $\Delta_k = m = \pm l_1 \cdots l_t$ we have $(\sqrt{m}) = \mathfrak{l}_1 \cdots \mathfrak{l}_t$ and so $\mathfrak{l}_1 \cdots \mathfrak{l}_t \pm 1$. [Note that in the case (1) there is no distinction between $\pm$ and $\sim$.] (b) $m \equiv 2 \pmod 4$. In this case, since $\Delta_k = 4m = \pm 8 l_2 \cdots l_t$, $m = \pm 2 l_2 \cdots l_t$ we have $(\sqrt{m}) = \mathfrak{l}_1\mathfrak{l}_2 \cdots \mathfrak{l}_t$, $2 = \mathfrak{l}_1^2$, and $\mathfrak{l}_1 \cdots \mathfrak{l}_t \pm 1$. (c) $m \equiv 3 \pmod 4$. In this case, since $\Delta_k = 4m = \pm 2^2 l_2 \cdots l_t$ we have $m = \pm l_2 \cdots l_t$ and $(\sqrt{m}) = \mathfrak{l}_2 \cdots \mathfrak{l}_t \pm 1$.†

2. $\Delta_k > 0$, $N\epsilon = 1$. Let $(1 + \epsilon) = (a)\mathfrak{a}$, $a \in \mathbb{N}$, and $\mathfrak{a}$ be a primitive ideal. Note that $\epsilon(1 + \epsilon^\sigma) = 1 + \epsilon$ as $N\epsilon = 1$. Hence $\mathfrak{a}^\sigma = \mathfrak{a}$ and by the argument in the proof of Proposition 4.8 we find that $\mathfrak{a} = \mathfrak{l}_1^{\epsilon_1} \cdots \mathfrak{l}_t^{\epsilon_t}$, $\epsilon_i = 0$ or 1. If $(\epsilon_1, \ldots, \epsilon_t) = (0, \ldots, 0)$ then $\mathfrak{a} = (1)$ and so $1 + \epsilon = a\eta$, $\eta \in \mathfrak{o}_k^\times$. From $\epsilon(1 + \epsilon^\sigma) = 1 + \epsilon$, it follows that $\epsilon N(1 + \epsilon) > 0$ and so $N\eta > 0$, i.e., $N\eta = 1$. Since $1 + \epsilon^\sigma = a\eta^\sigma = a\eta^{-1}$, we have $\epsilon = (1 + \epsilon)/(1 + \epsilon^\sigma) = \eta^2$ which contradicts that ϵ is the fundamental unit. Therefore $1 \pm \mathfrak{a} = \mathfrak{l}_1^{\epsilon_1} \cdots \mathfrak{l}_t^{\epsilon_t}$ as $N(1 + \epsilon) > 0$.‡
Q.E.D.

† If $m = -1$ in (c), we have $t = 1$. In this case, the matter is trivial because $h_k = h_k^+ = 1$.

‡ As $N\epsilon = 1$, we have $N(1 + \epsilon) = 2 + (\epsilon + \epsilon^\sigma)$, where $\epsilon + \epsilon^\sigma > 0$ (cf. the argument at the beginning of Section 4.8).

THEOREM 4.4. *Two cohomology groups* $H^0(H_k^+)$ *and* $H^1(H_k^+)$ *are isomorphic and have order* 2^{t-1}, *where* t *is the number of distinct prime factors of* Δ_k.

PROOF. By Proposition 4.5 and Remark 4.5 we already know that the two groups are isomorphic. Since $[(H_k^+)^G]=[I_k^G/(P_k^+)^G]$ (Proposition 4.7), to prove our theorem it is enough to prove that the equality holds in Proposition 4.9, i.e., that there are no more relations of the form $\mathfrak{l}_1^{\epsilon_1}\cdots\mathfrak{l}_t^{\epsilon_t}\approx 1$ than we found in the proof of Proposition 4.9. Thus suppose that

$$\mathfrak{l}_1^{\epsilon_1}\cdots\mathfrak{l}_t^{\epsilon_t}=(\alpha),\qquad N\alpha>0. \tag{4.42}$$

Since $(\alpha)^\sigma=(\alpha)$, we have $\alpha^\sigma=\eta\alpha$ and so $N\eta=1$, $\eta\in\mathfrak{o}_k^\times$.

1. $\Delta_k<0$. Since we can exclude the trivial cases $k=\mathbb{Q}(\sqrt{-1})$, $\mathbb{Q}(\sqrt{-3})$ where $h_k=h_k^+=1$, we may assume that $\eta=\pm 1$. If $\eta=1$, then $\alpha^\sigma=\alpha$, i.e., $\alpha\in\mathbb{Z}$. As the ideal on the left-hand side of (4.42) is primitive, we must have $\alpha=\pm 1$ which implies that $\epsilon_1=\cdots=\epsilon_t=0$. On the other hand, if $\eta=-1$, then $\alpha^\sigma=-\alpha$, i.e, α is purely imaginary: $\alpha=a\sqrt{m}$, $a\in\mathbb{Z}$. Thus (4.42) becomes $\mathfrak{l}_1^{\epsilon_1}\cdots\mathfrak{l}_t^{\epsilon_t}=(\sqrt{m})$; the case occurred in (1) of the proof of Proposition 4.9.

2. $\Delta_k>0$, $N\epsilon=-1$. Since $N\eta=1$, we have $\eta=\epsilon^{2v}$. Hence $\alpha^\sigma=\epsilon^{2v}\alpha$, or we have $\epsilon^{\sigma v}\alpha^\sigma=\epsilon^{\sigma v}\epsilon^{2v}\alpha=(N\epsilon)^v\epsilon^v\alpha=(-1)^v\epsilon^v\alpha$. If v is even, then $\epsilon^v\alpha\in\mathbb{Z}$ and since the left-hand side of (4.42) is an primitive ideal we have $\epsilon_1=\cdots=\epsilon_t=0$. On the other hand, if v is odd, then $\epsilon^v\alpha$ is purely imaginary: $\epsilon^v\alpha=a\sqrt{m}$, $a\mathbb{Z}$. Thus (4.42) again becomes the case that occurred in (1) of the proof of Proposition 4.9.

3. $\Delta_k>0$, $N\epsilon=1$. Since we have $\epsilon=(1+\epsilon)/(1+\epsilon^\sigma)$ in this case, there is an integer $n\in\mathbb{Z}$ such that $\eta=\epsilon^{-n}=(1+\epsilon^\sigma)^n/(1+\epsilon)^n$. Then $\alpha^\sigma=\eta\alpha=((1+\epsilon^\sigma)^n/(1+\epsilon)^n)\alpha$ and so $\alpha=a(1+\epsilon)^n$, $a\in\mathbb{Q}^\times$. Thus we have $\mathfrak{l}_1^{\epsilon_1}\cdots\mathfrak{l}_t^{\epsilon_t}=(a)(1+\epsilon)^n$. By the argument in (2) of the proof of Proposition 4.9, we have $(1+\epsilon)=(b)\mathfrak{l}_1^{\eta_1}\cdots\mathfrak{l}_t^{\eta_t}$, $\eta_i=0$ or 1, $(\eta_1,\ldots,\eta_t)\neq(0,\ldots,0)$, $b\in\mathbb{N}$. Therefore, we can find $A, B\in\mathbb{N}$, $(A,B)=1$, such that

$$A\mathfrak{l}_1^{\epsilon_1}\cdots\mathfrak{l}_t^{\epsilon_t}=B\mathfrak{l}_1^{n\eta_1}\cdots\mathfrak{l}_t^{n\eta_t}. \tag{4.43}$$

If n is even here, then the right-hand side of (4.43) is an ideal generated by an element of $\mathbb{Z}$; from this it follows that $(\epsilon_1,\ldots,\epsilon_t)=(0,\ldots,0)$ because $\mathfrak{l}_1^{\epsilon_1}\cdots\mathfrak{l}_t^{\epsilon_t}$ is a primitive ideal. On the other hand, if n is odd, we may assume $n=1$ because we can adjust n mod 2. Then we have

$$A\mathfrak{l}_1^{\epsilon_1}\cdots\mathfrak{l}_t^{\epsilon_t}=B\mathfrak{l}_1^{\eta_1}\cdots\mathfrak{l}_t^{\eta_t},\qquad (A,B)=1.$$

Again, by the primitivity of ideals $\mathfrak{l}_1^{\epsilon_1}\cdots\mathfrak{l}_t^{\epsilon_t}$, $\mathfrak{l}_1^{\eta_1}\cdots\mathfrak{l}_t^{\eta_t}$, we find that $A=B=1$ and so $(\epsilon_1,\ldots,\epsilon_t)=(\eta_1,\ldots,\eta_t)$, which means that (4.42) becomes the case that already occurred in (2) of the proof of Proposition 4.9. Q.E.D.

THEOREM 4.5. *Let k be a quadratic field such that Δ_k is a prime. Then h_k^+ is odd. If $\Delta_k>0$, then $N\epsilon=-1$, where ϵ is the fundamental unit of k.*

PROOF. Since $t=1$, we have $[H^0(H_k^+)]=[{}_2H_k^+]=1$ (Theorem 4.4), i.e., there are no elements of even order in H_k^+. In case $\Delta_k>0$, if $N\epsilon=1$, then $h_k^+=2h_k$ which contradicts the first half of our assertion. Q.E.D.

REMARK 4.7. For $k=\mathbb{Q}(\sqrt{m})$ we have $t=1$ if and only if $k=\mathbb{Q}(i)$, $\mathbb{Q}(\sqrt{2})$, $\mathbb{Q}(\sqrt{l^*})$, l is odd prime, $l^*=(-1)^{(l-1)/2}l$.

EXERCISE 4.21. If $k=\mathbb{Q}(\sqrt{m})$ with $m=1-2^{225}$, then h_k $(=h_k^+)$ is a multiple of 1826816.

THEOREM 4.6. *Let k be a quadratic field and let ϵ be the fundamental unit of k when $\Delta_k>0$. Then we have*:

	$H^0(\mathfrak{o}_k^\times)$	$H^1(\mathfrak{o}_k^\times)$	$Q(\mathfrak{o}_k^\times)$
$\Delta_k<0$	2	2	1
$\Delta_k>0$, $N\epsilon=1$	2	4	2
$\Delta_k>0$, $N\epsilon=-1$	1	2	2

PROOF. The first column follows immediately from the equalities $H^0(\mathfrak{o}_k^\times)=(\mathfrak{o}_k^\times)^G/N\mathfrak{o}_k^\times=\{\pm1\}/N\mathfrak{o}_k^\times$. The rest of the table will be checked as follows:

1. $\Delta_k<0$, $k\neq\mathbb{Q}(i)$, $\mathbb{Q}(\rho)$, $\rho=e^{2\pi i/3}$. In this case, $\mathfrak{o}_k^\times=\{\pm1\}\Rightarrow N\mathfrak{o}_k^\times=\mathfrak{o}_k^\times=\{\pm1\}$, $(\mathfrak{o}_k^\times)^{1-\sigma}=\{1\}\Rightarrow[H^1(\mathfrak{o}_k^\times)]=2$.

2. $k=\mathbb{Q}(i)$. $\mathfrak{o}_k^\times=\{\pm1,\pm i\}={}_N\mathfrak{o}_k^\times$, $(\mathfrak{o}_k^\times)^{1-\sigma}=\{\pm1\}\Rightarrow[H^1(\mathfrak{o}_k^\times)]=2$.

3. $k=\mathbb{Q}(\rho)$. $\mathfrak{o}_k^\times=\{\pm1,\pm\rho,\pm\rho^2\}={}_N\mathfrak{o}_k^\times$, $(\mathfrak{o}_k^\times)^{1-\sigma}=\{1,\rho,\rho^2\}\Rightarrow[H^1(\mathfrak{o}_k^\times)]=2$.

4. $\Delta_k > 0$, $N\epsilon = 1$. $\mathfrak{o}_k^\times = \{\pm 1\} \times \langle \epsilon \rangle = {}_N\mathfrak{o}_k^\times$, $(\mathfrak{o}_k^\times)^{1-\sigma} = \{1\} \times \langle \epsilon^{1-\sigma} \rangle$. $1 = N\epsilon = \epsilon\epsilon^\sigma \Rightarrow \epsilon/\epsilon^\sigma = \epsilon^2$, i.e., $(\mathfrak{o}_k^\times)^{1-\sigma} = \langle 1 \rangle \times \langle \epsilon^2 \rangle \Rightarrow [H^1(\mathfrak{o}_k^\times)] = 4$.

5. $\Delta_k > 0$, $N\epsilon = -1$. $\mathfrak{o}_k^\times = \{\pm 1\} \times \langle \epsilon \rangle$, ${}_N\mathfrak{o}_k^\times = \{\pm 1\} \times \langle \epsilon^2 \rangle$. $\epsilon^{1-\sigma} = -\epsilon^2 \Rightarrow (\mathfrak{o}_k^\times)^{1-\sigma} = \{1\} \times \langle -\epsilon^2 \rangle \Rightarrow [H^1(\mathfrak{o}_k^\times)] = 2$. Q.E.D.

4.7. Gauss Genus Theory

Let $k = \mathbb{Q}(\sqrt{m})$ be a quadraic field. In Section 2.17, we defined the Kronecker character χ_k as the unique solution which makes the following diagram commutative:

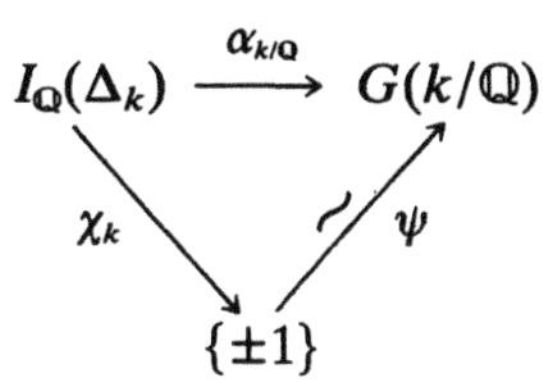

where $\alpha_{k/\mathbb{Q}}$ is the Artin map and ψ is the obvious isomorphism. Values of χ_k at generators, i.e., primes p, $p \nmid \Delta_k$, are determined by

$$\chi_k(p) = \begin{cases} \lambda_p(\Delta_k), & p \neq 2 \\ (-1)^{(\Delta_k^2-1)/8}, & p = 2 \end{cases}$$

where $\lambda_p(a) = (a/p)$, $p \nmid a$, is the Legendre character.

If $\Delta_k \equiv 1\ (4)$, then $\Delta_k = m = \pm l_1 l_2 \cdots l_t$. In this case, we know that [Exercise 2.33(1)]

$$\chi_k(n) = \lambda_{l_1}(n) \cdots \lambda_{l_t}(n), \qquad n \in \mathbb{Z}, \qquad (n, \Delta_k) = 1. \qquad (4.44)\dagger$$

When $\Delta_k \equiv 0\ (4)$, using notation in Exercise 2.33, put

$$\Delta_k = 4m = 2^\beta \Delta_k', \qquad 2 \nmid \Delta_k', \qquad \beta = \begin{cases} 3, & m \equiv 2\ (4), \\ 2, & m \equiv 3\ (4). \end{cases}$$

† In Exercise 2.33(1), we have $n > 0$. However, we can extend the variable n of χ_k for $n < 0$, $(n, \Delta_k) = 1$, By Exercise 2.33(4) or by the fact that Legendre characters on the right-hand side of (4.44) make sense for $n < 0$. In this context, it can be shown that $\chi_k(-1) = \text{sgn}(\Delta_k)$.

Then, by Exercise 2.32(4) and Exercise 2.33(2), we get

$$\chi_k(n)=(-1)^{((n-1)/2)((\Delta_k'-1)/2)+((n^2-1)/8)\beta}\lambda_{l_2}(n)\cdots\lambda_{l_t}(n),$$

$$n\in\mathbb{Z},\qquad (n,\Delta_k)=1.\dagger\quad (4.45)$$

EXERCISE 4.22. Verify that

$$(-1)^{((n-1)/2)((\Delta_k'-1)/2)+((n^2-1)/8)\beta}=\begin{cases}(-1)^{(n-1)/2}, & m\equiv 3\ (4),\\ (-1)^{(n^2-1)/8}, & m\equiv 2\ (8),\\ (-1)^{(n^2-1)/8+(n-1)/2}, & m\equiv 6\ (8).\end{cases}$$

Summarizing we get the following theorem.

THEOREM 4.7. *Let $l_1, l_2, \ldots, l_t$ be distinct prime factors of Δ_k, $k=\mathbb{Q}(\sqrt{m})$. We agree to put $l_1=2$ when $2\mid\Delta_k$. For each i, $1\le i\le t$, define a character χ_i of the group $(\mathbb{Z}/\Delta_k\mathbb{Z})^\times$ by*

$$\chi_1(n)=\left.\begin{cases}(-1)^{(n-1)/2}, & m\equiv 3\ (4)\\ (-1)^{(n^2-1)/8}, & m\equiv 3\ (8)\\ (-1)^{(n^2-1)/8+(n-1)/2}, & m\equiv 6\ (8)\end{cases}\right\},\qquad 2\mid\Delta_k,$$

$$\chi_i(n)=\lambda_{l_i}(n)=\left(\frac{n}{l_i}\right),\qquad \begin{matrix}1\le i\le t, & 2\nmid\Delta_k,\\ 2\le i\le t, & 2\mid\Delta_k,\end{matrix}$$

then the Kronecker character χ_k can be written

$$\chi_k(n)=\chi_1(n)\chi_2(n)\cdots\chi_t(n),\qquad n\in\mathbb{Z},\qquad (n,\Delta_k)=1.\ddagger$$

THEOREM 4.8. $\chi_k(n)=1\Leftrightarrow\exists\mathfrak{a}\subset\mathfrak{o}_k$, $(\mathfrak{a},\Delta_k)=1$, *such that* $n\equiv N\mathfrak{a}\ (\Delta_k)$.

PROOF. ($\Leftarrow$) Immediate from Proposition 2.42. ($\Rightarrow$) This is essentially Theorem 2.21. Here we go again (without fractional ideals)! By the theorem of arithmetical progressions (Theorem 3.12), there is a prime $p\ne 2$ such that $n\equiv p\ (\Delta_k)$. Since $\chi_k(p)=\chi_k(n)=1$ and $p\nmid\Delta_k$, we have $p=N\mathfrak{p}$ and so $n\equiv N\mathfrak{p}\ (\Delta_k)$.
Q.E.D.

PROPOSITION 4.10. *Each class of H_k^+ contains an ideal which is prime to a given ideal $\mathfrak{m}\subset\mathfrak{o}_k$.*

† Compare Footnote † on p. 173.

‡ That χ_i is a character mod Δ_k is obvious for $l_i\ne 2$. For $l_1=2$, one verifies it directly using (4.46) and (4.47), or by Exercise 2.33(4).

PROOF. Take any $C \in H_k^+$ and take any ideal $\mathfrak{a}$ in the class C^{-1}. It is enough to find an ideal $\mathfrak{b}$ such that $\mathfrak{ab} \pm 1$, $(\mathfrak{b}, \mathfrak{m}) = 1$. Put $\mathfrak{p}_1 \cdots \mathfrak{p}_l$ be distinct prime ideals dividing $\mathfrak{m}$. So we must find $\mathfrak{b}$ such that $\mathfrak{ab} \pm 1$ and $(\mathfrak{b}, \mathfrak{p}_1 \cdots \mathfrak{p}_l) = 1$. To do this, for each i, $1 \leq i \leq l$, take $\alpha_i \in \mathfrak{a}\mathfrak{p}_1 \overset{i}{\smile} \mathfrak{p}_l$, $\alpha_i \notin \mathfrak{a}\mathfrak{p}_1 \cdots \mathfrak{p}_l$, and put $\alpha = \alpha_1 + \cdots + \alpha_l$. Hence $\alpha \in \mathfrak{a}$ but $\alpha \notin \mathfrak{a}\mathfrak{p}_j$, $1 \leq j \leq l$. Now, let $a = N\mathfrak{a}$, $b = N\mathfrak{m}$, and put $\beta = \alpha + abt$, $t \in \mathbb{N}$. Then one verifies that $\beta \in \mathfrak{a}$ and $\beta \notin \mathfrak{a}\mathfrak{p}_j$, $1 \leq j \leq l$. In fact, since $a \in \mathfrak{a}$ we have $\beta \in \mathfrak{a}$; next, if for some j_0, $\beta \in \mathfrak{a}\mathfrak{p}_{j_0}$, then we have $abt \in \mathfrak{a}\mathfrak{p}_{j_0}$ which implies that $\alpha \in \mathfrak{a}\mathfrak{p}_{j_0}$ against the assumption on α_{j_0}. Furthermore, since $N(\alpha + abt) = N\alpha + (t_{k/\mathbb{Q}}\alpha)abt + a^2b^2t^2$ we can choose β such that $N\beta > 0$ by taking sufficient large t. Finally, as $\beta \in \mathfrak{a}$, we find $\mathfrak{b} \subset \mathfrak{o}_k$ such that $\mathfrak{ab} = (\beta)$, i.e., $\mathfrak{ab} \pm 1$, $(\mathfrak{b}, \mathfrak{p}_1 \cdots \mathfrak{p}_l) = 1$. Q.E.D.

PROPOSITION 4.11. $\alpha \in \mathfrak{o}_k$, $(\alpha, \Delta_k) = 1 \Rightarrow \chi_i(N\alpha) = 1$, $1 \leq i \leq t$.

PROOF. Suppose first that $l = l_i \neq 2$. Since l is ramified for $k/\mathbb{Q}$ we have $[\mathfrak{o}_k/\mathfrak{l}: \mathbb{Z}/l\mathbb{Z}] = 1$, $\mathfrak{l} \mid l$, and so there is $a \in \mathbb{Z}$ such that $(a, \Delta_k) = 1$ and $\alpha \equiv a \quad (\mathfrak{l})$. As $\mathfrak{l}^\sigma = \mathfrak{l}$, $\alpha\alpha^\sigma \equiv a^2 \quad (\mathfrak{l})$ or $N\alpha \equiv a^2 \quad (l)$; thus $N\alpha$ is a quadratic residue mod l, i.e., $\chi_i(N\alpha) = \lambda_{l_i}(N\alpha) = 1$. Next, when $l = 2$, we have $\Delta_k = 4m$, $m \equiv 3, 2 \quad (4)$, $\omega = \sqrt{m}$, and so $\alpha = x + y\sqrt{m}$, $x, y \in \mathbb{Z}$, $N\alpha = x^2 - my^2$. If $m \equiv 3 \quad (4)$, then $N\alpha \equiv x^2 + y^2 \equiv 1 \quad (4)$ because $\Delta_k = 4m$ implies $N\alpha$ is odd. Therefore $\chi_1(N\alpha) = (-1)^{(N\alpha-1)/2} = 1$. If $m \equiv 2 \quad (4)$, then, as m is even, x must be odd when $N\alpha$ is odd. If, in particular, $m \equiv 2 \quad (8)$, then $N\alpha \equiv x^2 - 2y^2 \equiv 1, 7 \quad (8)$ and hence $\chi_1(N\alpha) = (-1)^{((N\alpha)^2-1)/8} = 1$. On the other hand, if $m \equiv 6 \quad (8)$, then $N\alpha \equiv x^2 + 2y^2 \equiv 1, 3 \quad (8)$ and so $\chi_1(N\alpha)^{((N\alpha)^2-1)/8 + (N\alpha-1)/2} = 1$.

Q.E.D.

Any class in H_k^+ can be written $[\mathfrak{a}]$, $\mathfrak{a} \subset \mathfrak{o}_k$, $(\mathfrak{a}, \Delta_k) = 1$ (Proposition 4.10). If $[\mathfrak{a}] = [\mathfrak{b}]$, then there are $\alpha, \beta \in \mathfrak{o}_k$ such that $(\beta)\mathfrak{a} = (\alpha)\mathfrak{b}$, $(\alpha, \Delta_k) = (\beta, \Delta_k) = 1$, and $N\alpha \cdot N\beta > 0$.† Taking norms of both sides of the equality, we have $|N\beta|\, N\mathfrak{a} = |N\alpha|\, N\mathfrak{b}$ and hence $N\beta \cdot N\mathfrak{a} = N\alpha \cdot N\mathfrak{b}$ in view of the condition $N\alpha \cdot N\beta > 0$. Then Proposition 4.11 implies $\chi_i(N\alpha) = \chi_i(N\beta) = 1$ and so

† By Exercise 2.21, there is an ideal $\mathfrak{c}$ such that $\mathfrak{ac} = (\alpha)$ and $(\mathfrak{c}, \Delta_k) = 1$. If $[\mathfrak{a}] = [\mathfrak{b}]$, then $\mathfrak{b} = (\gamma)\mathfrak{a}$, with $N\gamma > 0$. Hence

$$(\gamma) = \frac{\mathfrak{b}}{\mathfrak{a}} = \frac{\mathfrak{bc}}{\mathfrak{ac}} = \frac{\mathfrak{bc}}{(\alpha)} \qquad \text{and} \qquad \mathfrak{bc} = (\alpha\gamma), \qquad \beta = \alpha\gamma.$$

$\chi_i(N\mathfrak{a}) = \chi_i(N\mathfrak{b})$. Therefore a mapping

$$\chi_i^+: H_k^+ \to \{\pm 1\}$$

can be defined by

$$\chi_i^+([\mathfrak{a}]) \stackrel{\text{def}}{=} \chi_i(N\mathfrak{a}), \qquad 1 \le i \le t. \tag{4.46}$$

THEOREM 4.9. *The mapping $\chi_i^+: H_k^+ \to \{\pm 1\}$ defined by* (4.46) *is a character of the group H_k^+. We have*

$$\prod_{i=1}^{t} \chi_i^+ = 1 \qquad (\textit{the trivial character}) \tag{4.47}$$

and

$$\bigcap_{i=1}^{t} \operatorname{Ker} \chi_i^+ = (H_k^+)^2. \tag{4.48}$$

PROOF. It follows from (4.46) that

$$\begin{aligned}\chi_i^+([\mathfrak{a}][\mathfrak{b}]) &= \chi_i^+([\mathfrak{a}\mathfrak{b}]) = \chi_i(N(\mathfrak{a}\mathfrak{b})) = \chi_i(N\mathfrak{a}N\mathfrak{b}) = \chi_i(N\mathfrak{a})\chi_i(N\mathfrak{b}) \\ &= \chi_i^+([\mathfrak{a}])\chi_i^+([\mathfrak{b}]);\end{aligned}$$

hence χ_i^+ is a character. Next, since

$$\prod_{i=1}^{t} \chi_i([\mathfrak{a}]) = \prod_{i=1}^{t} \chi_i(N\mathfrak{a}) = \chi_k(N\mathfrak{a}) = 1 \qquad \text{(Theorem 4.8)}$$

equality (4.47) is proved. As for equality (4.48) since $\chi_i^2 = 1$ the inclusion $\bigcap_{i=1}^{t} \operatorname{Ker} \chi_i^+ \supset (H_k^+)^2$ is trivial. To complete the proof of (4.48), consider a homomorphism $\psi: H_k^+ \to \{\pm 1\}^t$ defined by

$$\psi([\mathfrak{a}]) = (\chi_1^+([\mathfrak{a}]), \ldots, \chi_t^+([\mathfrak{a}])).$$

In view of (4.47) ψ is not surjective; however, we wish to verify that (4.47) is the only relation among χ_i^+'s. So let $\epsilon = (\epsilon_1, \ldots, \epsilon_t)$ be any element of the group $\{\pm 1\}^t$ which satisfies the relation $\epsilon_1 \cdots \epsilon_t = 1$. We wish to find an ideal class $[\mathfrak{a}] \in H_k^+$ such that $\psi([\mathfrak{a}]) = \epsilon$. First of all, for each i, since χ_i is a nontrivial character mod Δ_k there is $n_i \in \mathbb{Z}$, $(n_i, \Delta_k) = 1$, such that $\chi_i(n_i) = \epsilon_i$. For $\Delta_k = l_1 l_2 \cdots l_t$, $m \equiv 1 \pmod 4$ [resp. $\Delta_k = 2^\beta l_2 \cdots l_t$, $m \equiv 2, 3 \pmod 4$], find an integer n such that

$$\begin{cases} n \equiv n_1 & (l_1) \\ n \equiv n_2 & (l_2) \\ \quad\vdots \\ n \equiv n_t & (l_t) \end{cases} \qquad \left(\text{resp.} \begin{cases} n \equiv n_1 & (2^\beta) \\ n \equiv n_2 & (l_2) \\ \quad\vdots \\ n \equiv n_t & (l_t) \end{cases}\right)$$

Finally, by the theorem of arithmetical progressions, find an odd prime p such that $p \equiv n \quad (\Delta_k)$. Then $\chi_i(p)=\chi_i(n)=\chi_i(n_i)=\epsilon_i$, $1 \leq i \leq t$. Since

$$\chi_k(p)=\prod_{i=1}^{t}\chi_i(p)=\prod_{i=1}^{t}\epsilon_i=1$$

we have $p=N\mathfrak{p}$ and hence $\chi_i^+([\mathfrak{p}])=\chi_i(N\mathfrak{p})=\chi_i(p)=\epsilon_i$, or $\psi([\mathfrak{p}])=\epsilon$. Therefore we have proved that $[H_k^+ : \operatorname{Ker}\psi]=2^{t-1}$. On the other hand, we have

$$\operatorname{Ker}\psi=\bigcap_{i=1}^{t}\operatorname{Ker}\chi_i^+ \supset (H_k^+)^2 \qquad \text{and} \qquad [H_k^+ : (H_k^+)^2]=2^{t-1}$$

(Proposition 4.5, Theorem 4.4); hence $\ker\psi=(H_k^+)^2$. Q.E.D.

REMARK 4.8. So far, we have not met the term *genus* although the section is entitled *Gauss genus theory*. This is a term in biology which Gauss employed when he classified quadratic forms. As we outline in Note D, there is a parallelism between quadratic forms and quadratic fields. Hence we shall explain here how these terms in biology are employed to classify ideals.

First of all, for $\mathfrak{a}, \mathfrak{b} \in I_k$, we put $\mathfrak{a} \sim \mathfrak{b}$ when $\mathfrak{b}=(\alpha)\mathfrak{a}$, $\alpha \in k^\times$; a class defined by this equivalence relation is the *class* of (fractional) ideals and the set of all classes forms the ideal class group H_k. Next, for $\mathfrak{a}, \mathfrak{b} \in I_k$, we put $\mathfrak{a} \pm \mathfrak{b}$ when $\mathfrak{b}=(\alpha)\mathfrak{a}$, $N\alpha>0$; a class defined by this equivalence relation is the *class in the narrow sense* and the set of all classes in the narrow sense forms the group H_k^+. Lastly, for $[\mathfrak{a}]$, $[\mathfrak{b}]$, $\mathfrak{a}, \mathfrak{b} \subset \mathfrak{o}_k$, $(\mathfrak{a}, \Delta_k)=(\mathfrak{b}, \Delta_k)=1$, we define the following equivalence relation (in H_k^+):

$$[\mathfrak{a}] \approx [\mathfrak{b}] \Leftrightarrow \chi_i^+([\mathfrak{a}])=\chi_i^+([\mathfrak{b}]), \qquad 1 \leq i \leq t \tag{4.49}$$

and a class defined by this equivalence relation in H_k^+ is the *genus* of ideal classes in the narrow sense. In view of definitions in this section and Theorem 4.9, condition (4.49) is equivalent to either of the following conditions:

$$[\mathfrak{a}] \approx [\mathfrak{b}] \Leftrightarrow \chi_i(N\mathfrak{a})=\chi_i(N\mathfrak{b}), \qquad 1 \leq i \leq t, \tag{4.50}$$

$$[\mathfrak{a}] \approx [\mathfrak{b}] \Leftrightarrow [\mathfrak{a}] \equiv [\mathfrak{b}] \bmod (H_k^+)^2. \tag{4.51}$$

From (4.51), we see that the equivalence $\approx$ is nothing else than the classification of H_k^+ modulo $(H_k^+)^2$; hence a genus is a coset of $H_k^+ \bmod (H_k^+)^2$ and so the number of classes (in the narrow sense) in a genus is independent of the choice of genera and equals $[(H_k^+)^2] = [H_k^+ : {}_2H_k^+]$. Therefore if we put $h_k^* = [(H_k^+)^2] = [H_k^+ : {}_2H_k^+]$, then we obtain

$$h_k^+ = 2^{t-1} h_k^* \qquad \text{(Theorem 4.4, Theorem 4.9)},$$

which describes one of the important results of Gauss's *Disquisitiones arithmeticae.*

As an example, let $m = -14$ and $k = \mathbb{Q}(\sqrt{-14})$. Then $m = -14 \equiv 2 \pmod 4$, $\omega = \sqrt{-14}$, $\Delta_k = -2^3 \cdot 7$, and $M_k = (2/\pi)\sqrt{|\Delta_k|} \fallingdotseq 0.63662 \times 2\sqrt{14} \fallingdotseq 4.76$. Since $p = 2, 3 \le M_k$ the group $H_k = H_k^+$ is generated by the classes of $\mathfrak{p}_2$ and $\mathfrak{p}_3$. We have $f_\omega(X) = X^2 + 14$.

1. $p = 2$, $f(X) = X^2 + 14 \equiv X^2 \pmod 2$. Hence $2 = \mathfrak{p}_2^2$, $\mathfrak{p}_2 = (2\omega) = [2, \omega]$.

2. $p = 3$, $f_\omega(X) = X^2 + 14 \equiv X^2 - 1 \equiv (X + 1)(X - 1) \pmod 3$. Hence $3 = \mathfrak{p}_3\mathfrak{p}_3'$, $\mathfrak{p}_3 = (3, 1 + \omega) = [3, 1 + \omega]$, and $\mathfrak{p}_3' = (3, -1 + \omega) = [3, 2 + \omega]$ (Figure 4.1). Now, as $N(2 + \omega) = 2^2 + 14 = 18 = 2 \cdot 3^2$, we have

$$\begin{aligned} 1 \sim (2 + \omega) &= [N(2 + \omega), 2 + \omega] = [2 \cdot 3^2, 2 + \omega] \\ &= [2, 2 + \omega][3, 2 + \omega]^2 = \mathfrak{p}_2(\mathfrak{p}_3')^2 \sim \mathfrak{p}_2\mathfrak{p}_3^{-2}, \end{aligned}$$

and so $\mathfrak{p}_2 \sim \mathfrak{p}_3^2$. Therefore H_k^+ is a cyclic group generated by the class of $\mathfrak{p}_3$. On the other hand, we find $\mathfrak{p}_2^2 = 2 \sim 1$ but $\mathfrak{p}_2 \not\sim 1$; in

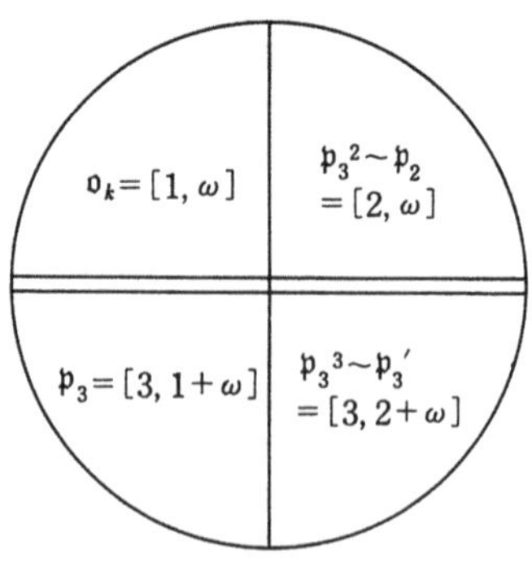

Figure 4.1

fact, if $\mathfrak{p}_2 = (x + y\omega)$, then $2 = N(x + y\omega) = x^2 + 14y^2$ but obviously it is impossible to find $x, y \in \mathbb{Z}$. Hence H_k^+ is a cyclic group of order 4 $(=h_k^+)$ generated by the class of $\mathfrak{p}_3$. And so $(H_k^+)^2 = \{[\mathfrak{o}_k], [\mathfrak{p}_3^2]\}$; hence $t = 2$, $h_k^* = 2$, and there are two genera.

4.8. Quadratic Irrationals

Let $k = \mathbb{Q}(\sqrt{m})$ be a real quadratic field. By Dirichlet's unit theorem, we have $\mathfrak{o}_k^\times = \{\pm\epsilon^n\}$, $n \in \mathbb{Z}$. There are four ways to choose such an ϵ; if an ϵ is chosen, then four numbers $\pm\epsilon^{\pm 1}$ have the same property.

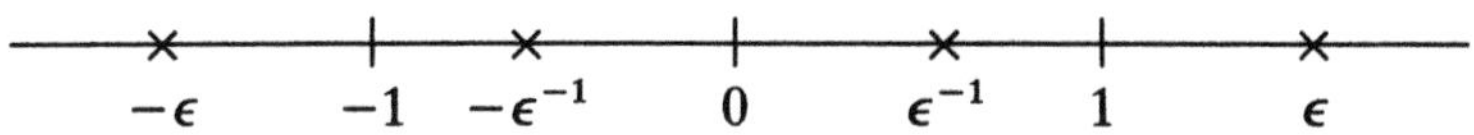

Among the four ϵ's there is a unique ϵ such that $\epsilon > 1$; this is the *fundamental unit* of k. If $u \in \mathfrak{o}_k^\times$ is such that $u > 1$, then we have $u = \epsilon^n$, $n \geq 1$, and so $u \geq \epsilon$. In other words, ϵ is the smallest number in $\mathfrak{o}_k^\times$ such that $u > 1$.

For $m \equiv 2, 3 \pmod 4$ let $\epsilon = a + b\sqrt{m}$, $a, b \in \mathbb{Z}$, be the fundamental unit. Since $\epsilon' = a - b\sqrt{m}$ and $\pm 1 = N\epsilon = \epsilon\epsilon'$, we have $\epsilon^{-1} = \pm\epsilon'$ and $\pm\epsilon^{\pm 1} = \pm a \pm b\sqrt{m}$. As ϵ is the largest among the four numbers, we must have $a, b \geq 1$. For a unit $u > 1$, write $u = \epsilon^n$, $n \geq 1$, and $\epsilon^n = a_n + b_n\sqrt{m}$ ($a_1 = a$, $b_1 = b$). Then

$$\begin{aligned} \epsilon^{n+1} &= a_{n+1} + b_{n+1}\sqrt{m} = \epsilon\epsilon^n = (a + b\sqrt{m})(a_n + b_n\sqrt{m}) \\ &= aa_n + bb_n m + (ab_n + ba_n)\sqrt{m} \end{aligned}$$

and

$$\begin{cases} a_{n+1} = aa_n + bb_n m, \\ b_{n+1} = ab_n + ba_n. \end{cases}$$

From this, it follows that $0 < a_1 < a_2 < \cdots$ and $0 < b_1 < b_2 < \cdots$. Therefore, (a, b) is the solution $(x, y) \in \mathbb{Z}^2$, $x \geq 1$, $y \geq 1$, to the equation

$$x^2 - my^2 = \pm 1 \tag{4.52}$$

such that x (and hence y) is the smallest one.

Exercise 4.23. Find the fundamental unit of $k = \mathbb{Q}(\sqrt{2}), \mathbb{Q}(\sqrt{3}), \mathbb{Q}(\sqrt{6}), \mathbb{Q}(\sqrt{7})$.

For $m \equiv 1 \pmod 4$ let $\epsilon = (a + b\sqrt{m})/2$, $a \equiv b \pmod 2$ be the fundamental unit. As above, we have $\epsilon^{-1} = \pm\epsilon'$, $\pm\epsilon^{\pm 1} = (\pm a \pm$

$b\sqrt{m})/2$, and $a, b \geq 1$. In view of the denominator 2, (a, b) is the solution $(x, y) \in \mathbb{Z}^2$, $x \geq 1$, $y \geq 1$, to the equation

$$x^2 - my^2 = \pm 4 \tag{4.53}$$

such that x (and hence y) is the smallest one.

Exercise 4.24. Find the fundamental unit of $k = \mathbb{Q}(\sqrt{5}), \mathbb{Q}(\sqrt{13})$.

Remark 4.9. It is not always easy to determine ϵ, i.e., to solve equations (4.52) and (4.53) in $\mathbb{Z}$. For example, consider $k = \mathbb{Q}(\sqrt{199})$ which has $h_k = 1$ (Section 4.5). Can you find a solution of

$$x^2 - 199y^2 = \pm 1 \tag{4.54}$$

by a naive method? If we consult Mr. Wada's table, we learn that

$$(x, y) = (16266196520, 1153080099) \tag{4.55}$$

is the smallest solution to equation (4.54).† By the way, $m = 199$ is one of small numbers for which the smallest solution is large. In order to get (4.55) from (4.54) mechanically, we need the help of the continued-fraction expansion of $\sqrt{199}$.

Let us summarize the basic facts studied in Chapter 1.

$$C: X = \mathbb{R} - \mathbb{Q} \xrightarrow{\sim} \mathbb{Z} \times \prod_2^\infty \mathbb{N}$$

sending α to $\{q_1, q_2, \ldots, q_n, \ldots\}$ where

$$\begin{cases} \alpha = q_1 + \dfrac{1}{\alpha_2}, & q_1 = [\alpha], \\ \alpha_2 = q_2 + \dfrac{1}{\alpha_3}, & q_2 = [\alpha_2], \\ \alpha_3 = q_3 + \dfrac{1}{\alpha_4}, & q_3 = [\alpha_3], \\ \quad\vdots & \\ \alpha_n = q_n + \dfrac{1}{\alpha_{n+1}}, & q_n = [\alpha_n], \\ \quad\vdots & \end{cases}$$

† $N\epsilon = 1$.

The inverse D of C sends $s=\{q_1, q_2, \dots, q_n, \dots\} \in \mathbb{Z}\times\prod_2^\infty \mathbb{N}$ to $\alpha=\lim_{n\to\infty} a_n \overset{\text{def}}{=} [q_1, q_2, \dots, q_n, \dots]$ where

$$a_n=[q_1, q_2, \dots, q_n]=q_1+\cfrac{1}{q_2+\cfrac{1}{q_3+\ddots\cfrac{1}{q_{n-1}+\cfrac{1}{q_n}}}}$$

We have, for each $n\geq 0$, polynomials P_n and Q_n in $(q_1, \dots, q_n)$ defined inductively:

$$\begin{cases} P_0=1, & P_1=q_1, \quad P_n=q_nP_{n-1}+P_{n-2}, \\ Q_0=0, & Q_1=1, \quad Q_n=q_nQ_{n-1}+Q_{n-2}, \end{cases} \qquad n\geq 2.$$

Here are some basic formulas:

$$\begin{aligned} &[q_1, \dots, q_n]=\frac{P_n}{Q_n}, \qquad n\geq 1 \\ &P_nQ_{n-1}-P_{n-1}Q_n=(-1)^n, \qquad n\geq 1 \\ &P_nQ_{n-2}-P_{n-2}Q_n=(-1)^{n-1}q_n, \qquad n\geq 2 \\ &\alpha=[q_1, \dots, q_{n-1}, q_n, \dots]=[q_1, \dots, q_{n-1}, \alpha_n] \\ &\quad=\frac{\alpha_nP_{n-1}+P_{n-2}}{\alpha_nQ_{n-1}+Q_{n-2}}, \qquad \text{where } \alpha_n=[q_n, q_{n+1}, \dots]. \end{aligned} \tag{4.56}$$

As for the action of the group $G=\mathrm{GL}_2(\mathbb{Z})$ on X (Section 1.5), for

$$\sigma=\begin{pmatrix} a & b \\ c & d\end{pmatrix}\in G, \qquad \alpha\in X,$$

we define

$$\sigma\alpha=\frac{a\alpha+b}{c\alpha+d}$$

and consider the following two equivalence relations:

$$\alpha,\ \alpha'\in X, \qquad \alpha \underset{G}{\sim} \alpha' \overset{\text{def}}{\Leftrightarrow} \alpha'=\sigma\alpha, \qquad \sigma\in G,$$

$$\alpha,\ \alpha'\in X, \qquad \alpha \underset{C}{\sim} \alpha' \overset{\text{def}}{\Leftrightarrow} \exists \quad n, m\in\mathbb{N}, \qquad \alpha_n=\alpha'_m,$$

where $\alpha=[q_1, \ldots, q_{n-1}, \alpha_n]$ and $\alpha'=[q_1', \ldots, q_{m-1}', \alpha_m']$. Then we have (Theorem 1.9)

$$\alpha \underset{G}{\approx} \alpha' \Leftrightarrow \alpha \underset{C}{\approx} \alpha'.$$

In what follows, we shall consider a subset $X_2 \subset X$ consisting of quadratic irrationals (Definition 4.3). Hence it is natural to find a characterization of these numbers in terms of continued fractions; Theorem 4.10 will be the answer to this question.

EXERCISE 4.25. Verify that

$$\sqrt{199}=[14, \overline{9, 2, 1, 2, 2, 5, 4, 1, 1, 13, 1, 1, 4, 5, 2, 2, 1, 2, 9, 28}]$$

where the bar indicates the period. (According to Definition 4.2 below, for $\sqrt{199}$, $n_0=2$ and the period $m=20$.)

DEFINITION 4.2. Let $\alpha \in X$ and let $\alpha=[q_1, q_2, \ldots, q_n, \ldots]$ be the continued-fraction expansion of α. Then α is *periodic* with period $m \geq 1$ if there is an integer $n_0 \geq 1$ such that $q_n=q_{n'}$ whenever $n' \geq n \geq n_0$ and $n' \equiv n\ (m)$. If, in particular, $n_0=1$, then α is said to be *purely periodic.*

DEFINITION 4.3. Define a subset X_2 of $X=\mathbb{R}-\mathbb{Q}$ by

$$X_2=\{\alpha \in X \cap \bar{\mathbb{Q}}; \deg f_\alpha=2\},$$

where f_α is the minimal polynomial of α over $\mathbb{Q}$. Numbers in X_2 are called *quadratic irrationals.*

REMARK 4.10. If $\alpha \in X_2$, then the conjugate α' is different from α because $\alpha \notin \mathbb{Q}$. α is a root of the equation

$$ax^2+bx+c=0, \qquad a, b, c \in \mathbb{Z}, \qquad (a, b, c)=1, \qquad a>0$$

and the ordered set $\{a, b, c\}$ is uniquely determined by α. Hence the *discriminant* $D(\alpha)=b^2-4ac$ is well defined by α.

EXERCISE 4.26. Prove that the ordered set $\{a, b, c\}$ above is uniquely determined by α.

DEFINITION 4.4. For $D \in \mathbb{N}$, put

$$X_2(D) \overset{\text{def}}{=} \{\alpha \in X_2; D(\alpha)=D\}.\dagger$$

† Of course, it may happen that $X_2(D)=\varnothing$. Since α is irrational, $D(\alpha)$ cannot be square. As $D(\alpha')=D(\alpha)$, $\alpha \in X_2(D) \Rightarrow \alpha' \in X_2(D)$.

EXERCISE 4.27. Verify that $G = \mathrm{GL}_2(\mathbb{Z})$ acts on $X_2(D)$.

DEFINITION 4.5. $\alpha \in X_2(D)$ is *reduced* if $\alpha > 1$, $-1 < \alpha' < 0$ (i.e., $-1/\alpha' > 1$). We denote by $R_2(D)$ the set of all reduced irrationals in $R_2(D)$.

PROPOSITION 4.12. *$R_2(D)$ is a finite set. For any $\alpha \in X_2(D)$ there is a $\beta \in R_2(D)$ such that $\alpha \underset{G}{\sim} \beta$.*

PROOF.

1. For $\alpha \in R_2(D)$, let $\{a, b, c\}$ be the ordered set of integers determined by α as in Remark 4.10. Therefore

$$\alpha = \frac{-b + \epsilon\sqrt{D}}{2a} > 1, \qquad -1 < \alpha' = \frac{-b - \epsilon\sqrt{D}}{2a} < 0, \qquad \epsilon = \pm 1.$$

If $\epsilon = -1$ then $-b - \sqrt{D} > 0$ and $-b + \sqrt{D} < 0$, which is impossible; we must have $\epsilon = 1$. From $-b + \sqrt{D} > 2a > b + \sqrt{D}$, we find $b < 0$ and, as $-b - \sqrt{D} < 0$, we have $0 < -b < \sqrt{D}$. Therefore, $|b| < \sqrt{D}$ and we see that all of a, b, c have only finite possibilities, i.e., that $R_2(D)$ is a finite set.

2. Let α be any number in $X_2(D)$ and let $\alpha = [q_1, \ldots, q_n, \alpha_{n+1}]$ with $\alpha_{n+1} = [q_{n+1}, \ldots] > 1$ be its continued-fraction expansion. According to (4.56), we write

$$\alpha = \frac{\alpha_{n+1}P_n + P_{n-1}}{\alpha_{n+1}Q_n + Q_{n-1}} = \sigma_n \alpha_{n+1}, \qquad \sigma_n = \begin{pmatrix} P_n & P_{n-1} \\ Q_n & Q_{n-1} \end{pmatrix} \in G,$$

then

$$\alpha' = \frac{\alpha'_{n+1}P_n + P_{n-1}}{\alpha'_{n+1}Q_n + Q_{n-1}} \quad \text{and} \quad \alpha'_{n+1} = -\frac{Q_{n-1}}{Q_n}\frac{\alpha' - a_{n-1}}{\alpha' - a_n}, \qquad a_n = \frac{P_n}{Q_n}.$$

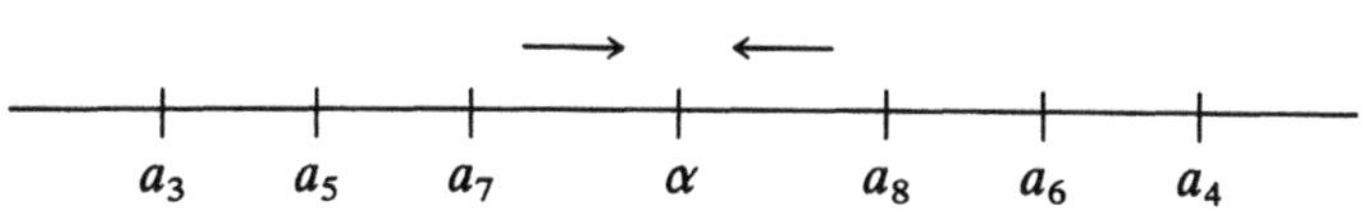

Since a_n approaches α from both sides as shown above (Theorem 1.8), $(\alpha' - a_{n-1})/(\alpha' - a_n)$ approaches 1 from both sides alternately. On the other hand, for any n, $0 < Q_{n-1} < Q_n$, i.e., $0 < Q_{n-1}/Q_n < 1$; therefore, for sufficiently large n, we have $-1 < \alpha'_{n+1} < 0$. As $\alpha_{n+1} > 1$, we have $\beta = \alpha_{n+1} \in R_2(D)$ and $\beta \underset{G}{\sim} \alpha$.

Q.E.D.

Proposition 4.13. *For an* $\alpha \in R_2(D)$, *if we determine* β *by* $\alpha = [\alpha] + 1/\beta$, *then* $\beta \in R_2(D)$.

Proof. Since

$$\beta = \frac{1}{\alpha - q} = \begin{pmatrix} 0 & 1 \\ 1 & -q \end{pmatrix} \alpha \quad \text{with} \begin{pmatrix} 0 & 1 \\ 1 & -q \end{pmatrix} \in G,$$

we have $\beta \in X_2(D)$ (Exercise 4.27). By definition of $[\alpha]$, we have $0 < 1/\beta < 1$, i.e., $\beta > 1$. On the other hand, $\alpha' = [\alpha] + 1/\beta'$ and so $-1/\beta' = [\alpha] - \alpha' > [\alpha] \geq 1$, i.e., $\beta \in R_2(D)$. Q.E.D.

Theorem 4.10 (*Euler–Lagrange*).† *For* $\alpha \in X = \mathbb{R} - \mathbb{Q}$, *we have*

1. $\alpha \in X_2 \Leftrightarrow$ *the continued-fraction expansion of* α *is periodic,*
2. $\alpha \in R_2(D(\alpha)) \Leftrightarrow$ *the continued-fraction expansion of* α *is purely periodic.*

Proof.

1. ($\Rightarrow$). By the proof of Proposition 4.12, for $\alpha \in X_2$, we have of course $\alpha \in X_2(D(\alpha))$ and so, for a sufficiently large n, $\alpha_n \in R_2(D(\alpha))$ where α_n is determined as

$$\alpha = [q_1, \ldots, q_{n-1}, \alpha_n], \; \alpha_n = [q_n, q_{n+1}, \ldots].$$

Call n_0 the smallest n of this property; hence, e.g., $n_0 = 1$ means that $\alpha = \alpha_1 \in R_2(D(\alpha))$. Now, $\alpha_{n_0} \in R_2(D(\alpha))$ implies $\alpha_{n_0+1} \in R_2(D(\alpha))$ because $\alpha_{n_0} = [q_{n_0}, \alpha_{n_0+1}] = q_{n_0} + 1/\alpha_{n_0+1}$ (Proposition 4.13) and hence $\alpha_{n_0+k} \in R_2(D(\alpha))$ for all $k \geq 1$. However, as $R_2(D(\alpha))$ is a finite set (Proposition 4.13), there must be $n_1 \geq n_0$ and $k \geq 1$ such that $\alpha_{n_1} = \alpha_{n_1+k}$. Then we have

$$\alpha_{n_1} = [\alpha_{n_1}] + \frac{1}{\alpha_{n_1+1}} = \alpha_{n_1+k} = [\alpha_{n_1+k}] = \frac{1}{\alpha_{n_1+k+1}}$$

which implies $\alpha_{n_1+1} = \alpha_{n_1+k+1}$ and, in turn, $\alpha_{n_1+j} = \alpha_{n_1+j+k}$ for all j. Therefore we have $q_{n_1} = q_{n_1+k}, \ldots, q_{n_1+k-1} = q_{n_1+2k-1}, \ldots$ and so

$$\alpha = [q_1, \ldots, q_{n_1-1}, \overline{q_{n_1}, \ldots, q_{n_1+k+1}}].$$

† Lagrange (1736–1813).

2 (⇒). If $\alpha = \alpha_1 \in R_2(D(\alpha))$, then $\alpha_n \in R_2(D(\alpha))$ for all n. As above, there is n_1 such that $\alpha_{n_1} = \alpha_{n_1+k}$. Now, for $\alpha \in R_2(D(\alpha))$, write $\alpha = [\alpha] + 1/\beta$. Then we have $\beta \in R_2(D(\alpha))$ (Proposition 4.13) and $\alpha' = [\alpha] + 1/\beta'$. As $-1/\beta' = [\alpha] - \alpha' = [\alpha] + (-\alpha')$ and $0 < -\alpha' < 1$, we have $[\alpha] = [-1/\beta']$. Thus

$$\alpha_{n_1-1} = [\alpha_{n_1-1}] + \frac{1}{\alpha_{n_1}} \Rightarrow [\alpha_{n_1-1}] = \left[-\frac{1}{\alpha'_{n_1}}\right]$$

and

$$\alpha_{n_1+k-1} = [\alpha_{n_1+k-1}] + \frac{1}{\alpha_{n_1+k}} \Rightarrow [\alpha_{n_1+k-1}] = \left[-\frac{1}{\alpha'_{n_1+k}}\right].$$

Therefore we obtain $\alpha_{n1-1} = \alpha_{n_1+k-1}$ and in turn $\alpha_j = \alpha_{j+k}$ for all $j \geq 1$. This means that $\alpha = [\overline{q_1, \ldots, q_k}]$, i.e., α is purely periodic.

1 (⇐). Suppose that α is periodic, i.e.,

$$\alpha = [q_1, \ldots, q_{n_0-1}, \alpha_{n_0}] \qquad \text{and} \qquad \alpha_{n_0} = [\overline{q_{n_0}, \ldots, q_{n_0+m-1}}].$$

From (4.56), we have

$$\alpha = \frac{\alpha_{n_0} P_{n_0-1} + P_{n_0-2}}{\alpha_{n_0} Q_{n_0-1} + Q_{n_0-2}}. \tag{4.57}$$

On the other hand,

$$\begin{aligned}\alpha_{n_0} &= [\overline{q_{n_0}, \ldots, q_{n_0+m-1}}] = [q_{n_0}, \ldots, q_{n_0+m-1}, \alpha_{n_0}] \\ &= \frac{\alpha_{n_0} P_m(q_{n_0}, \ldots, q_{n_0+m-1}) + P_{m-1}(q_{n_0}, \ldots, q_{n_0+m-2})}{\alpha_{n_0} Q_m(q_{n_0}, \ldots, q_{n_0+m-1}) + Q_{m-1}(q_{n_0}, \ldots, q_{n_0+m-2})}.\end{aligned} \tag{4.58}$$

Now $\alpha_{n_0} \in X$ as $\alpha \in X$, and since values of P_m and Q_m are in $\mathbb{Z}$, we have $\alpha_{n_0} \in X_2$ by (4.58) and hence $\alpha \in X_2$ by (4.57).

2 (⇐). Finally, suppose that α is purely periodic, i.e., $n_0 = 1$. Then

$$\alpha = [q_1, \ldots, q_m, \alpha] = [q_1, \ldots, q_m, q_1, \ldots, q_m, \alpha] = \cdots$$

and so $\alpha = \alpha_{m+1} = \alpha_{2m+1} = \cdots$. On the other hand, as $\alpha_n \in R_2(D(\alpha))$ for sufficiently large n (Proof of Proposition 4.12), we get $\alpha \in R_2(D(\alpha))$. Q.E.D.

4.9. Real Quadratic Fields and Continued Fractions

In this last section of the book, we explain how to describe the group $\mathfrak{o}_k^\times$ of units and the ideal class group H_k of a real quadratic field k in terms of continued-fraction expansions of quadratic irrationals belonging to Δ_k, the discriminant of k. Throughout the section, let $\Delta = \Delta_k$ and consider the sets (Definition 4.4, Definition 4.5),

$$X_2(\Delta) = \{\alpha \in X_2; D(\alpha) = \Delta\} = \{\alpha \in k - \mathbb{Q}; D(\alpha) = \Delta\}$$

and

$$R_2(\Delta) = \{\alpha \in X_2(\Delta); \alpha > 1, -1 < \alpha' < 0\}.$$

By Exercise 4.25, for each $\alpha \in X_2(\Delta)$, an ordered triple $\{a, b, c\}$ of integers is uniquely determined by α so that

$$a > 0, \qquad (a, b, c) = 1, \qquad D(\alpha) = b^2 - 4ac = \Delta. \tag{4.59}$$

If, in particular, α is reduced, i.e., if $\alpha \in R_2(\Delta)$, then we know that

$$\alpha = \frac{-b + \sqrt{\Delta}}{2a} \quad \text{and} \quad \alpha' = \frac{-b - \sqrt{\Delta}}{2a} \tag{4.60}$$

(See the proof of Propostion 4.12.) By Exercise 4.26, the group $G = \mathrm{GL}_2(\mathbb{Z})$ acts on $X_2(\Delta)$. We shall denote by $\tilde{X}_2(\Delta)$ the quotient set $X_2(\Delta)/G$, the set of equivalence classes in $X_2(\Delta)$ defined by the relation $\alpha \underset{G}{\sim} \beta$, $\alpha, \beta \in X_2(\Delta)$. We shall also denote by $\tilde{R}_2(\Delta)$ the set of equivalence classes in $R_2(\Delta)$ defined by the same relation $\alpha \underset{G}{\sim} \beta$, $\alpha, \beta \in R_2(\Delta)$. By Proposition 4.12, there is a natural bijection $\tilde{R}_2(\Delta) \approx \tilde{X}_2(\Delta)$.

Proposition 4.14. *Notation being as in* (4.59) *and* (4.60), *the mapping*

$$R_2(\Delta) \ni \alpha = \frac{-b + \sqrt{\Delta}}{2a} \mapsto \mathfrak{a}_\alpha = [2a, -b + \sqrt{\Delta}]$$

induces naturally a bijection $\tilde{R}_2(\Delta) \xrightarrow{\sim} H_k$.

Proof. In view of the natural bijection $\tilde{R}_2(\Delta) \approx \tilde{X}_2(\Delta)$, it is enough to verify that the mapping in Proposition 4.14 induces a bijection $\tilde{X}_2(\Delta) \rightrightarrows H_k$. First of all, for an

$$\alpha = \frac{-b + \sqrt{\Delta}}{2a} \in X_2(\Delta),$$

one verifies easily that $\mathfrak{a}_\alpha = [2a, -b + \sqrt{\Delta}]$ is an ideal which is equal to $2[a, -h + \omega]$, with

$$h = \begin{cases} \dfrac{b}{2}, & m \equiv 2, 3 \quad (4) \\[2ex] \dfrac{b+1}{2}, & m \equiv 1 \quad (4) \end{cases}$$

(Exercise 4.28). Next, we verify that for

$$\alpha = \frac{-b + \sqrt{\Delta}}{2a} \quad \text{and} \quad \beta = \frac{-b_1 + \sqrt{\Delta}}{2a_1}$$

in $X_2(\Delta)$ we have

$$\alpha \underset{G}{\sim} \beta \Leftrightarrow \mathfrak{a}_\alpha \sim \mathfrak{a}_\beta.$$

In fact, if $\alpha = \sigma\beta$,

$$\sigma = \begin{pmatrix} p & q \\ r & s \end{pmatrix} \in G,$$

put $\rho = a/a_1(r\beta + s)$. Then we have

$$a\alpha = \frac{a(p\beta + q)}{r\beta + s} = \frac{aa_1(p\beta + q)}{a_1(r\beta + s)} = \rho a_1(p\beta + q)$$

and so

$$\begin{aligned} \tfrac{1}{2}\mathfrak{a}_\alpha &= [a, a\alpha] = [\rho a_1(r\beta + s), \rho a_1(p\beta + q)] = \rho a_1[r\beta + s, p\beta + q] \\ &= \rho a_1[1, \beta] = \rho[a_1, a_1\beta] = \rho(\tfrac{1}{2}\mathfrak{a}_\beta), \end{aligned}$$

i.e., $\mathfrak{a}_\alpha \sim \mathfrak{a}_\beta$. Conversely, if $\mathfrak{a}_\alpha = \rho\mathfrak{a}_\beta$, $\rho \in k^\times$, then we have

$$\tfrac{1}{2}\mathfrak{a}_\alpha = [a, a\alpha] = [a, -h + \omega] = \rho(\tfrac{1}{2}\mathfrak{a}\beta) = [\rho a_1, \rho(-h_1 + \omega)] = [\rho a_1, \rho a_1\beta].$$

Hence, there is a

$$\sigma = \begin{pmatrix} p & q \\ r & s \end{pmatrix} \in G$$

such that

$$\begin{pmatrix} a\alpha \\ a \end{pmatrix} = \begin{pmatrix} p & q \\ r & s \end{pmatrix} \begin{pmatrix} \rho a_1 \beta \\ \rho a_1 \end{pmatrix},$$

or $\alpha = \sigma\beta$, which means that $\alpha \underset{\tilde{G}}{\sim} \beta$. The above argument shows that the set $\tilde{R}_2(\Delta)$ can be embedded in H_k. Therefore, to complete the proof, it remains to show that the mapping $\alpha \mapsto \mathfrak{a}_\alpha$ induces a surjective mapping from $\tilde{X}_2(\Delta)$ to H_k. Thus let C be an ideal class in H_k. Let $\mathfrak{a} = [a, r + \omega]$ be a primitive ideal in C. Then $N(r + \omega) = ac$, $c \in \mathbb{Z}$. Then we have

$$-b = T(r + \omega) = \begin{cases} 2r & m \equiv 2, 3 \pmod 4, \\ 2r + 1, & m \equiv 1 \pmod 4. \end{cases}$$

Therefore $r + \omega$ satisfies the quadratic equation $x^2 + bx + ac = 0$. On the other hand,

$$\begin{aligned} \Delta &= (\omega - \omega')^2 = ((r + \omega) - (r + \omega'))^2 \\ &= ((r + \omega) + (r + \omega'))^2 - 4(r + \omega)(r + \omega') \\ &= (T(r + \omega))^2 - 4N(r + \omega) = b^2 - 4ac. \end{aligned}$$

From these, we see that

$$\alpha = \frac{r + \omega}{a} = \frac{-b + \sqrt{\Delta}}{2a} \in X_2(\Delta) \quad \text{and} \quad \mathfrak{a}_\alpha = [2a, -b + \sqrt{\Delta}] = 2\mathfrak{a} \sim \mathfrak{a} \in C$$

which proves that $\alpha \mapsto \mathfrak{a}_\alpha$ induces a surjection from $\tilde{X}_2(\Delta)$ to H_k.
Q.E.D.

Exercise 4.28. Notation being as in the proof of Proposition 4.14, prove that $\mathfrak{a}_\alpha = [2a, -b + \sqrt{\Delta}] = 2[a, -h + \omega]$.

Now, let α be any reduced quadratic irrational in $R_2(\Delta)$ and let $\alpha = [\overline{q_1, \ldots, q_N}]$ be the continued-fraction expansion of α with (minimal) period N, i.e.,

$$\begin{cases} \alpha = \alpha_1 = q_1 + \dfrac{1}{\alpha_2}, & q_1 = [\alpha_1], \\ \alpha_2 = q_2 + \dfrac{1}{\alpha_3}, & q_2 = [\alpha_2], \\ \vdots & \\ \alpha_N = q_N + \dfrac{1}{\alpha}, & q_N = [\alpha_N]. \end{cases} \tag{4.61}$$

Since

$$\alpha_i = q_i + \frac{1}{\alpha_{i+1}} = \begin{pmatrix} q_i & 1 \\ 1 & 0 \end{pmatrix} \alpha_{i+1}, \qquad 1 \le i \le N, \qquad \alpha_{N+1} = \alpha,$$

we have $\alpha_1 \underset{G}{\approx} \alpha_2 \underset{G}{\approx} \cdots \underset{G}{\approx} \alpha_N$. On the other hand, if $\alpha \underset{G}{\approx} \beta = [\overline{r_1, \ldots, r_M}] \in R_2(\Delta)$ with

$$\begin{cases} \beta = \beta_1 = r_1 + \dfrac{1}{\beta_2}, & r_1 = [\beta_1], \\ \beta_2 = r_2 + \dfrac{1}{\beta_3}, & r_2 = [\beta_2], \\ \quad \vdots & \\ \beta_M = r_M + \dfrac{1}{\beta}, & r_M = [\beta_M], \end{cases}$$

then, by Theorem 1.9, there are $n, m \in \mathbb{N}$ such that $\alpha_n = \beta_m$ in the notation of Definition 1.10. By the periodicity, however, we may take n and m so that $n \le N$ and $m \le M$ and we have $\alpha = \alpha_1 \underset{G}{\approx} \alpha_n = \beta_m \underset{G}{\approx} \beta_1 = \beta$. Therefore the equivalence class containing $\alpha \in R_2(\Delta)$ is precisely the set $A_\alpha = \{\alpha_1, \alpha_2, \ldots, \alpha_N\}$. If $A_\alpha = R_2(\Delta)$, we get $h_k = 1$; if not, take $\beta \in R_2(\Delta) - A_\alpha$ and get the second equivalence class $A_\beta = \{\beta_1, \ldots, \beta_M\}$. Since $R_2(\Delta)$ is a finite set (Proposition 4.12), we reach the class number h_k as the number of classes A_α in $R_2(\Delta)$. This is the procedure how to calculate h_k by the use of continued fractions. Here the complete determination of the finite set $R_2(\Delta)$ must precede everything (Exercise 4.29).

Exercise 4.29. Let a and b be integers such that $a > 0$, $1 \le -b < \sqrt{\Delta}$, $(b + \sqrt{\Delta})/2a < 1 < (-b + \sqrt{\Delta})/2a$, and $4a \mid (b^2 - \Delta)$. Then we have $\alpha = (-b + \sqrt{\Delta})/2a \in R_2(\Delta)$. (Note that $(a, b, c) = 1$, with $c = (b^2 - \Delta)/4a$, follows from the conditions.) Conversely, any $\alpha \in R_2(\Delta)$ is obtained from some a and b in this way.

Let us now move to the group $\mathfrak{o}_k^\times$ of units. For $\alpha \in R_2(\Delta)$, put $G_\alpha = \{\sigma \in G, \sigma\alpha = \alpha\}$, this being obviously a subgroup of G.

Proposition 4.15. *Let $\alpha \in R_2(\Delta)$. Then the mapping*

$$\sigma = \begin{pmatrix} p & q \\ r & s \end{pmatrix} \mapsto r\alpha + s$$

gives an isomorphism: $G_\alpha \xrightarrow{\sim} \mathfrak{o}_k^\times$.

Proof. Let $\{a, b, c\}$ be the triple of integers (4.59) associated with $\alpha \in R_2(\Delta)$. If

$$\begin{pmatrix} p & q \\ r & s \end{pmatrix} \in G_\alpha,$$

then $\alpha = (p\alpha + q)/(r\alpha + s)$ and we have $r\alpha^2 + (s - p)\alpha - q = 0$. This implies that

$$\frac{r}{a} = \frac{s-p}{b} = -\frac{q}{c} = u \in \mathbb{Z} \tag{4.62}$$

(Exercise 4.30). Put $s + p = t$. Then we have

$$\begin{cases} p = \dfrac{t - bu}{2}, & q = -cu, \\ r = au, & s = \dfrac{t + bu}{2}. \end{cases} \tag{4.63}$$

Since $\det \sigma = \pm 1 = ps - rq = (t^2 - b^2u^2)/4 + acu^2$ and $\Delta = b^2 - 4ac$, we get

$$t^2 - \Delta u^2 = \pm 4.\dagger \tag{4.64}$$

Conversely, for a solution $(t, u) \in \mathbb{Z}^2$ of equation (4.64), define $p, q, r,$ and s by (4.63). Then one verifies easily that

$$\sigma = \begin{pmatrix} p & q \\ r & s \end{pmatrix} \in G_\alpha$$

and that the mapping $\sigma \mapsto (t, u)$ gives a bijection

$$G_\alpha \xrightarrow{\sim} \{(t, u) \in \mathbb{Z}^2;\, t^2 - \Delta u^2 = \pm 4\}. \tag{4.65}$$

On returning to the mapping in Proposition 4.15, we have

$$r\alpha + s = \frac{au(-b + \sqrt{\Delta})}{2a} + \frac{t + bu}{2} = \frac{t + u\sqrt{\Delta}}{2} \tag{4.66}$$

and since any unit in $\mathfrak{o}_k^\times$ can be written as $(t + u\sqrt{\Delta})/2$ with a

† (4.64) is usually called Pell's equation.

solution of (4.64) (Exercise 4.31), we find from (4.65) and (4.66) that the mapping $f: \sigma \mapsto r\alpha + s$ is a bijection $G_\alpha \approx \mathfrak{o}_k^\times$ of sets. It remains to check that f is a homomorphism. In fact, for

$$\sigma = \begin{pmatrix} p & q \\ r & s \end{pmatrix},\ \tau = \begin{pmatrix} p_1 & q_1 \\ r_1 & s_1 \end{pmatrix} \in G_\alpha$$

we have

$$f(\sigma\tau) = f\begin{pmatrix} * & * \\ rp_1 + sr_1 & rq_1 + ss_1 \end{pmatrix} = (rp_1 + sr_1)\alpha + (rq_1 + ss_1).$$

On the other hand,

$$\begin{aligned} f(\sigma)f(\tau) &= (r\alpha + s)(r_1\alpha + s_1) = rr_1\alpha^2 + (rs_1 + sr_1)\alpha + ss_1 \\ &= r((p_1 - s_1)\alpha + q_1) + (rs_1 + sr_1)\alpha + ss_1 \end{aligned}$$

as $r_1\alpha^2 + (s_1 - p_1)\alpha - q_1 = 0$ (for $\tau \in G_\alpha$). Therefore we obtain $f(\sigma\tau) = f(\sigma)f(\tau)$. Q.E.D.

Exercise 4.30. Verify (4.62).

Exercise 4.31. Prove that any unit of $\mathfrak{o}_k^\times$ can be written as $(t + u\sqrt{\Delta})/2$ with $t^2 - \Delta u^2 = \pm 4$.

Proposition 4.16. *Let $\alpha \in R_2(\Delta)$. Using the notation in* (4.61), *we have $\alpha = [q_1, \ldots, q_N, \alpha] = \sigma_0 a$, with*

$$\sigma_0 = \begin{pmatrix} P_N & P_{N-1} \\ Q_N & Q_{N-1} \end{pmatrix} \in G_\alpha.$$

In this situation $\epsilon = Q_N\alpha + Q_{N-1}$ is the fundamental unit of k.

Proof. Since $\epsilon = Q_N\alpha + Q_{N-1} > 1$, to show that ϵ is the fundamental unit it suffices to prove that for any

$$\sigma = \begin{pmatrix} p & q \\ r & s \end{pmatrix} \in G_\alpha$$

such that $r\alpha + s > 1$ we have $r\alpha + s = \epsilon^k$ for some $k \in \mathbb{N}$. Here we used the fact that any unit of $\mathfrak{o}_k^\times$ has the form $r\alpha + s$ for some $\sigma \in G_\alpha$ (Proposition 4.15). Put $\eta = r\alpha + s$; hence $\eta' = r\alpha' + s$. Since $\eta > 1$ by the assumption and $N\eta = \eta\eta' = \pm 1$, we have

$|\eta'|<1$ and hence $\eta-\eta'>0$, or $r(\alpha-\alpha')>0$. As $\alpha\in R_2(\Delta)$, we have $\alpha>1$, $0>\alpha'>-1$, and so $r>0$. We find also that $s>r\alpha'+s>-r+s$, or $s>\eta'>-r+s$. Now, we have

$$N\eta=1\Rightarrow 1>\eta'>0\Rightarrow r\geq s>0, \tag{4.67}$$

$$N\eta=-1\Rightarrow 0>\eta'>-1\Rightarrow r>s\geq 0. \tag{4.68}$$

Case 1: $r>s>0$. In this case, by Lemma in the proof of Theorem 1.9, we have $\alpha=\alpha_m$ for some $m\in\mathbb{N}$. As $\alpha=[q_1,\ldots,q_{m-1},\alpha_m]=[q_1,\ldots,q_{m-1},\alpha]$ we must have $m-1=kN$ for some $k\in\mathbb{N}$ and so $\alpha=\sigma\alpha=\sigma_0^k\alpha$ with

$$\sigma_0=\begin{pmatrix}P_N & P_{N-1}\\ Q_N & Q_{N-1}\end{pmatrix}.$$

Call f the isomorphism $G_\alpha\xrightarrow{\sim}\mathfrak{o}_k^\times$ in Proposition 4.15. Then we have $\eta=f(\sigma)=f(\sigma_0^k)=f(\sigma_0)^k=\epsilon^k$ and our assertion is proved in this case.

Case 2: $r=s$ in (4.67). In this case, $r=s=1$. Since $1=N\eta=ps-qr$ (Exercise 4.32), we have $p-q=1$. Hence

$$\alpha=\frac{(q+1)\alpha+q}{\alpha+1}=[q,1,\alpha]$$

and $\alpha=\sigma\alpha=\sigma_0^k\alpha$ with some $k\in\mathbb{N}$, again.

Case 3: $s=0$ in (4.68). In this case, since $-1=N\eta=ps-qr$ (Exercise 4.32), we have $qr=1$. As $r>0$, we must have $q=1$, $r=1$, and so $\alpha=(p\alpha+1)/\alpha=[p,\alpha]$, i.e., $\alpha=\sigma\alpha=\sigma_0\alpha$, again.

Q.E.D.

EXERCISE 4.32. For $\alpha\in R_2(\Delta)$ and

$$\sigma=\begin{pmatrix}p & q\\ r & s\end{pmatrix}\in G_\alpha,$$

prove that $N(r\alpha+s)=\det\sigma$.

EXAMPLE 1. $k=\mathbb{Q}(\sqrt{15})$, $m=15\equiv 3\ (4)$, $\Delta=4m=60$, $\sqrt{\Delta}\doteqdot 7.746$, $\omega=\sqrt{15}$. The set $R_2(\Delta)$ consists of $\alpha=(-b+\sqrt{\Delta})/2a$ with conditions $a>0$, $1\leq -b<\sqrt{\Delta}$, $(b+\sqrt{\Delta})/2a<1<(-b+\sqrt{\Delta})/2a$, and $4a\mid(b^2-\Delta)$ (Exercise 4.29). In our case, we find that $-b=6$ and $a=1, 2, 3, 6$ are the only possibilities. Therefore $R_2(\Delta)$ consists of $3+\omega$, $(3+\omega)/2$, $(3+\omega)/3$, and $(3+\omega)/6$. Let

$\alpha = 3 + \omega$. Then

$$\alpha = \alpha_1 = 6 + \frac{1}{\alpha_2} \qquad \text{with } \alpha_2 = \frac{3+\omega}{6},$$

$$\alpha_2 = 1 + \frac{1}{\alpha}.$$

Hence $\alpha = [6, 1, \alpha] = [\overline{6, 1}] = (7\alpha + 6)/(\alpha + 1)$. From this we see that the fundamental unit is $\epsilon = \alpha + 1 = 4 + \omega$ and that

$$A_\alpha = \left\{3 + \omega, \frac{3+\omega}{6}\right\}.$$

Next, let $\beta = (3 + \omega)/2$. Then

$$\beta = \beta_1 = 3 + \frac{1}{\beta_2} \qquad \text{with } \beta_2 = \frac{3+\omega}{3},$$

$$\beta_2 = 2 + \frac{1}{\beta}.$$

From this we see that $\varepsilon = 1 + 2\beta = 4 + \omega$ (again) and that

$$A_\beta = \left\{\frac{3+\omega}{2}, \frac{3+\omega}{3}\right\}.$$

Since $R_2(\Delta)$ is the union of two orbits A_α and A_β, we see that $h_k = 2$. The nontrivial ideal class is represented by $[2, 1 + \omega]$.

Example 2. $k = \mathbb{Q}(\sqrt{199})$, $m = 199 \equiv 3 \pmod 4$, $\Delta = 4m = 796$, $\sqrt{\Delta} \doteqdot 28.213$, $\omega = \sqrt{199}$. With the same notation as in Example 1, we find that

$-b$	a
4	13, 15
14	10, 15
16	9, 15
20	9, 11
22	6, 13
24	5, 11
26	2, 3, 5, 6, 10, 15
28	1, 3

are all possibilities for a and b. There are 20 pairs (a, b) in the table; hence $R_2(\Delta)$ has 20 numbers. Put $\alpha = (28 + \sqrt{\Delta})/2 = 14 + \omega$. Then, by Exercise 4.25, we get

$$\alpha = [\overline{28, 9, 2, 1, 2, 2, 5, 4, 1, 1, 13, 1, 1, 4, 5, 2, 2, 1, 2, 9}]. \qquad (4.69)$$

On the other hand, it is seen that one obtains 20 distinct numbers $\alpha_i \sim \alpha$, $1 \leq i \leq 20$, by taking all cyclic permutations of (4.69): hence $h_k = 1$. (Compare Section 4.5.) Finally, computing $Q_N\alpha + Q_{N-1}$ $(N = 20)$ from (4.69) (Proposition 4.16), we get $\epsilon = 16266196520 + 1153080099\omega$. [Compare formula (4.55)].

Answers and Hints to Exercises

Chapter 1

EXERCISE 1.1. $0a = (0+0)a = 0a + 0a$.

EXERCISE 1.2. $0 = 1 \Rightarrow 0a = 1a = a,\ \forall a \in K \Rightarrow K = \{0\}$ which contradicts the definition of the field (Definition 1.1).

EXERCISE 1.3. $0 = (1 + (-1))(-1) = 1(-1) + (-1)(-1) = (-1) + (-1)(-1)$. Add 1 on both ends.

EXERCISE 1.4. For $k \in \mathbb{N}$, put $S_k(n) = 1^k + 2^k + \cdots + n^k$. Use $S_k(n+1) = S_k(n) + (n+1)^k$ and $S_1(n) = n(n+1)/2$.

EXERCISE 1.5. Think of the definition of the baldness.

EXERCISE 1.6. If $a > 1$ is not prime, then $a = bc,\ 1 < b \leq c < a$. Hence
$a = bc \geq b^2$.

EXERCISE 1.7. $\mathbb{Z}a + \mathbb{Z}b + \mathbb{Z}c = (\mathbb{Z}a + \mathbb{Z}b) + \mathbb{Z}c = \mathbb{Z}a + (\mathbb{Z}b + \mathbb{Z}c)$.

EXERCISE 1.8. $a_{n+2} = a_n + a_{n+1} \Rightarrow (a_n, a_{n+1}) = (a_{n+1}, a_{n+2})$. Induction works because $(a_0, a_1) = (0, 1) = 1$. $a_{n+1} = a_{n-1} + a_n \leq 2a_n$.

EXERCISE 1.9. Note first that the series $f(x)$ converges for every x; hence one can differentiate $f(x)$ as many times as one wants. From $a_{n+2} = a_n + a_{n+1}$ we find that $y = f(x)$ satisfies the differential equation $y'' = y + y'$; hence $y = Ae^{\lambda e} + Be^{\mu x}$. As λ and μ are solutions of $X^2 = 1 + X$, we have $\lambda = (1+\sqrt{5})/2$ and $\mu = (1-\sqrt{5})/2$. Putting $x = 0$ in $f(x) = Ae^{\lambda x} + Be^{\mu x}$ and $f'(x) = A\lambda e^{\lambda x} + B\mu e^{\mu x}$, we find $0 = A + B$ and $1 = A\lambda + B\mu$; hence $A = 1/\sqrt{5}$ and $B = -1/\sqrt{5}$. Values of a_n are obtained by the equality $e^x = \sum_{n=0}^{\infty} (x^n/n!)$.

EXERCISE 1.10. Use Exercise 1.6 to check each $P(x)$, $0 \leq x \leq 39$, is prime.

EXERCISE 1.11. One proves $a_{n+1} = a_1 a_2 \cdots a_n + 1$ by induction. Then we find $(a_{n+1}, a_i) = 1$ for all i, $1 \leq i \leq n$; hence $(a_i, a_j) = 1$ for $i \neq j$. Since

$a_i = 2$, $i \geq 1$, if we take any prime p_i dividing a_i, we get infinitely many primes.

EXERCISE 1.12. Similar to Exercise 1.11.

EXERCISE 1.13. $(95, 432) = 1$, $(1986, 304) = 2$.

EXERCISE 1.14. $(-42, 191)$.

EXERCISE 1.15. $(-93, 62, -1)$.

EXERCISE 1.16. $\sqrt{2} = [\sqrt{2}] + (\sqrt{2})_0 = 1 + (\sqrt{2} - 1)$, $q_1 = 1$. $1/(\sqrt{2} - 1) = \sqrt{2} + 1 = 2 + (\sqrt{2} - 1)$, $q_2 = 2$. Then $q_i = 2$, $i \geq 2$.

EXERCISE 1.17. For $n = 1, 2, 3$, see text. For general n, the induction goes as follows:

$$[\lambda_1, \ldots, \lambda_{n+1}] = [\lambda_1, \ldots, \lambda_{n-1}, \lambda_n + 1/\lambda_{n+1}]$$
$$= \frac{P_n\left(\lambda_1, \ldots, \lambda_{n-1}, \lambda_n + \dfrac{1}{\lambda_{n+1}}\right)}{Q_n\left(\lambda_1, \ldots, \lambda_{n-1}, \lambda_n + \dfrac{1}{\lambda_{n+1}}\right)} = \frac{\left(\lambda_n + \dfrac{1}{\lambda_{n+1}}\right)P_{n-1} + P_{n-2}}{\left(\lambda_n + \dfrac{1}{\lambda_{n+1}}\right)Q_{n-1} + Q_{n-2}}$$
$$= \frac{\lambda_{n+1}(\lambda_n P_{n-1} + P_{n-2}) + P_{n-1}}{\lambda_{n+1}(\lambda_n Q_{n-1} + Q_{n+2}) + Q_{n-1}} = \frac{\lambda_{n+1} P_n + P_{n-1}}{\lambda_{n+1} Q_n - Q_{n-1}} = \frac{P_{n+1}}{Q_{n+1}}.$$

EXERCISE 1.18. $P_n Q_{n-1} - P_{n-1} Q_n = (\lambda_n P_{n-1} + P_{n-2}) Q_{n-1} - P_{n-1}(\lambda_n Q_{n-1} + Q_{n-2}) = P_{n-2} Q_{n-1} - P_{n-1} Q_{n-2} = -(P_{n-1} Q_{n-2} - Q_{n-1} P_{n-2}) = \cdots = (-1)^{n-1}(P_1 Q_0 - P_0 Q_1) = (-1)^n$. From this we find that polynomials P_n and Q_n are mutually prime.

EXERCISE 1.19. $P_n Q_{n-2} - P_{n-2} Q_n = (\lambda_n P_{n-1} + P_{n-2}) Q_{n-2} - P_{n-2}(\lambda_n Q_{n-1} + Q_{n-2}) = \lambda_n (P_{n-1} Q_{n-2} - P_{n-2} Q_{n-1}) = \lambda_n (-1)^{n-1}$.

EXERCISE 1.20. $\alpha = \lim_{N \to \infty} [q_1, \ldots, q_{n-1}, q_N] = \lim_{N \to \infty} [q_1, \ldots, q_{n-1}, [q_n, \ldots q_N]]$

$$= \lim_{N\to\infty} q_1 + \cfrac{1}{q_2 + \cfrac{1}{\ddots \; q_{n-1} + \cfrac{1}{[q_n, \ldots, q_N]}}} = q_1 + \cfrac{1}{q_2 + \cfrac{1}{\ddots \; q_{n-1} + \cfrac{1}{\lim_{N\to\infty} [q_n, \ldots, q_N]}}}$$

$$= [q_1, \ldots, q_{n-1}, \alpha_n].$$

EXERCISE 1.21. $$\alpha = \frac{1}{2} + \frac{\sqrt{5}}{2} = 1 + \frac{\sqrt{5} - 1}{2} = 1 + \cfrac{1}{\cfrac{2}{\sqrt{5} - 1}} = 1 + \cfrac{1}{\cfrac{\sqrt{5} + 1}{2}}$$
$$= 1 + \cfrac{1}{1 + \cfrac{\sqrt{5} - 1}{2}} = \cdots = [1, 1, 1, \cdots].$$

Since $Q_0 = 0$, $Q_1 = 1$, and $Q_n = q_n Q_{n-1} + Q_{n-2} = Q_{n-1} + Q_{n-2}$, $\{Q_n\}$ is the Fibonacci sequence.

EXERCISE 1.22. By Proposition 1.2, we have $n = qm + r$, $0 \le r < m$. Then $e = a^n = a^{qm+r} = (a^m)^q a^r = a^r$. By the minimality of m, $r = 0$.

EXERCISE 1.23. If $aa^{\#} = a^{\#}a = 1$, then $a^*aa^{\#} = a^{\#} = a^*$ which proves the uniqueness of the inverse of a when it exists. The rest is routine.

EXERCISE 1.24. Routine.

EXERCISE 1.25. Easy.

EXERCISE 1.26. Let $\alpha = [q_1, \ldots, q_{n-1}, \alpha_n]$ and $\alpha' = [q'_1, \ldots, q'_{m-1}, \alpha'_m]$. Note that $\alpha_n = \alpha'_m \Rightarrow \alpha_{n+k} = \alpha'_{m+k}$, for all $k \in \mathbb{N}$.

EXERCISE 1.27. Routine.

EXERCISE 1.28. Routine.

EXERCISE 1.29. Routine.

EXERCISE 1.30. By the assumption $a^{2^n} \equiv -1 \pmod{p}$, we have $a^{2^{n+1}} \equiv 1 \pmod{p}$. Here in the group $\mathbb{F}_p^\times = (\mathbb{Z}/p\mathbb{Z})^\times$, the order of $[a]$ must divide 2^{n+1}, i.e., $\mathrm{ord}[a] = 2^m$, $0 \le k \le n+1$. If $k \le n$, then $a^{2^n} \equiv 1 \pmod{p}$ and so $-1 \equiv 1 \pmod{p}$, a contradiction (because $p \ne 2$). Hence $k = n+1$. On the other hand, $a^{p-1} \equiv 1 \pmod{p}$ which implies that $2^{n+1} \mid p-1$.

EXERCISE 1.31. Let $a = 2$ and $n = 5$ in Exercise 1.30. If $p \mid F_5$, then $2^6 \mid p-1$, i.e., $p = 1 + k2^6$. Try k until 10:

k	1	2	3	4	5	6	7	8	9	10
$1 + k2^6$	65	129	193	257	321	385	449	513	577	641
prime?	no	no	yes	yes	no	no	yes	no	yes	yes

Here 641 is the first prime which divides $F_5 = 4294967297$. In fact, the prime decomposition of $F_5 = 641 \times 6700417$ (try!).

EXERCISE 1.32. $pa^2 = 2^{p-1} - 1 = (2^{(p-1)/2} + 1)(2^{(p-1)/2} - 1) \Rightarrow 2^{(p-1)/2} + 1 = b^2$ or $2^{(p-1)/2} - 1 = b^2$, $b \in \mathbb{N}$. Since b is odd, write $b = 2c + 1$, $c \ge 0$.

Case 1: $2^{(p-1)/2} = b^2 - 1 = 4c(c+1) \Rightarrow c = 1 \Rightarrow b = 3 \Rightarrow p = 7$.

Case 2: $2^{(p-1)/2} = b^2 + 1 = 2(2c^2 + 2c + 1) \Rightarrow c = 0 \Rightarrow b = 1 \Rightarrow p = 3$.

EXERCISE 1.33. $p^e = a^{p-1} - 1 = (a^{(p-1)/2} + 1)(a^{(p-1)/2} - 1) \Rightarrow a^{(p-1)/2} - 1 = 1$ (because the two factors of $a^{p-1} - 1$ above are relatively prime) $\Rightarrow$ a $= 2$, $p = 3$.

EXERCISE 1.34. Routine.

EXERCISE 1.35. Routine.

EXERCISE 1.36. Given $a_1, \ldots, a_l \in \mathbb{Z}$, we must show that the system

$$\begin{cases} x \equiv a_1 \quad (m_1) \\ \vdots \\ x \equiv a_l \quad (m_l) \end{cases}$$

has a solution. Let $m = m_1 \cdots m_l = m_i m_i^*$. Hence $(m_i, m_i^*) = 1$, $m_j \mid m_i^*$ $(i \neq j)$. Let x_i be a solution of $m_i^* x_i \equiv 1 \quad (m_i)$. Then $x = a_1 m_1^* x_1 + \cdots + a_l m_l^* x_l$ is a solution to the system.

EXERCISE 1.37. Easy.

EXERCISE 1.38. Induction on $n = \deg A$ (for $A \neq 0$). If $n < m$, put $Q = 0$ and $R = A$. If $n \geq m$, then let $A' = A - B(a_n/b_m)X^{n-m}$, where a_n (resp. b_m) is the coefficient of X^n (resp. X^m) of A (resp. B). Since $\deg A' < n$, we get $A' = Q'B + R$, $R = 0$ or $\deg R < m$ and so $A = QB + R$, with $Q = Q' + (a_n/b_m)X^{n-m}$.

EXERCISE 1.39. Similar to Theorem 1.1.

EXERCISE 1.40. We can take $r = 2$ as a primitive root mod 13.

x	1	2	3	4	5	6	7	8	9	10	11	12
2^x	2	4	8	3	6	12	11	9	5	10	7	1

a	1	2	3	4	5	6	7	8	9	10	11	12
$\operatorname{ind}_2 a$	0	1	4	2	9	5	11	3	8	10	7	6

$\{2, 6, 11, 7\}$ is the set of all primitive roots mod 13. Solutions of three equations are $x \equiv 10$, $x \equiv \pm 5$, and $x \equiv 8$, 11, 7 (mod 13), respectively.

EXERCISE 1.41. Easy.

EXERCISE 1.42. For k, $1 \leq k \leq p - 1$, put

$$N = \binom{p}{k} = \frac{p(p-1)\cdots(p-k+1)}{k!}.$$

Then $p \mid k!N$. As $p \nmid k!$, we have $p \mid N$.

EXERCISE 1.43. Routine.

EXERCISE 1.44. Obvious when $x = e$. For $x \neq e$, write $x = g^k$, where g is a generator of G. Hence $n \nmid k$. If we put $\chi_1(g) = \zeta$, then $\chi_1(x) = \zeta^k \neq$

1. Therefore

$$S=\sum_{\chi\in\hat{G}}\chi(x)=\sum_{\chi}(\chi_1\chi)(x)=\chi_1(x)S=0$$

because $\chi_1(x)\neq 0$.

EXERCISE 1.45. The pair $\{17,2\}$ is an example which satisfies the condition.

Chapter 2

EXERCISE 2.1. Easy.

EXERCISE 2.2.

1. $f_\alpha=gh$, g, $h\in K[X]$, $0<\deg g$, $\deg h<\deg f_\alpha\Rightarrow 0=f_\alpha(\alpha)=g(\alpha)h(\alpha)\Rightarrow g(\alpha)=0$ or $h(\alpha)=0\Rightarrow f_\alpha \mid g$ or $f_\alpha \mid h$, a contradiction.

2. Suppose $f_\alpha=(X-\gamma)^2g$, $\gamma\in\bar{\mathbb{Q}}$, $g\in\bar{\mathbb{Q}}[X]$. Then the derivative $f'_\alpha=2(X-\gamma)g+(X-\gamma)^2g'\in\bar{\mathbb{Q}}[X]$ and $d=(f_\alpha,f'_\alpha)$ is a nonconstant polynomial $\in\bar{\mathbb{Q}}[X]$. On the other hand, as d is obtained from f_α, f'_α by the Euclidean algorithm in $K[X]$, we have $d\in K[X]$ which contradicts that f_α is irreducible in $K[X]$.

3, 4. By Theorem 1.14, φ induces an isomorphism $\varphi'\colon K[X]/(f_\alpha)\approx K[\alpha]$. By the argument in the proof of Theorem 1.12 ($\Leftarrow$) (applied to $K[X]$ in place of $\mathbb{Z}$), we see that $K[X]/(f_\alpha)$ is a field; hence $K[\alpha]=K(\alpha)$. If $n=\deg f_\alpha$, then, by the minimality of f_α, $\{1,\alpha,\ldots,\alpha^{n-1}\}$ is independent over K, which implies $[K(\alpha)\colon K]=n$.

EXERCISE 2.3. By Note B, we know that $\mathbb{C}$ is algebraically closed; hence for $f\in\bar{\mathbb{Q}}[X]\subset\mathbb{C}[X]$ there exists $\alpha\in\mathbb{C}$ such that $f(\alpha)=0$. We claim that $\alpha\in\bar{\mathbb{Q}}$. As is seen easily, we can assume that

$$\alpha^n+\alpha_1\alpha^{n-1}+\cdots+\alpha_{n-1}\alpha+\alpha_n=0,\qquad \alpha_i\in\bar{\mathbb{Z}}.$$

Now, since each α_i satisfies a relation of the form

$$\alpha_i^{n_i}+c_{i1}\alpha_i^{n_i-1}+\cdots+c_{i,n-1}\alpha_i+c_{in}=0,\qquad c_{ij}\in\mathbb{Z}.$$

if we call M the submodule of $\mathbb{C}$ generated by finitely many complex numbers $\alpha^j\alpha_1^{k_1}\cdots\alpha_n^{k_n}$, $0\le j\le n-1$, $0\le k_i\le n_i-1$, $1\le i\le n$, then we have $\alpha M\subset M$ and hence $\alpha\in\bar{\mathbb{Z}}$ (Proposition 1.20).

EXERCISE 2.4. Easy.

EXERCISE 2.5. There are $l-1$ primitive lth roots of 1, i.e., ζ^i, $1\le i\le l-1$. For $\sigma\in G(\bar{\mathbb{Q}}/\mathbb{Q})$, we have $(\zeta^\sigma)^l=(\zeta^l)^\sigma=1$, hence $\zeta^\sigma=\zeta^{j(\sigma)}$, $j(\sigma)\in\mathbb{Z}$, is a primitive root of 1. Thus, $K^\sigma=\mathbb{Q}(\zeta^\sigma)=\mathbb{Q}(\zeta^{j(\sigma)})=\mathbb{Q}(\zeta)=K$ which proves (1). (2) This is a famous theorem with many proofs. The shortest proof uses Eisenstein's criterion: If for a polynomial $P(X)=a_nX^n+a_{n-1}X^{n-1}+\cdots+a_0\in\mathbb{Z}[X]$ there is a prime p such that

(a) $p \nmid a_n$, (b) $p \mid a_i$, $0 \leqq i \leqq n-1$, and (c) $p^2 \nmid a_0$, then $P(X)$ is irreducible in $\mathbb{Q}[X]$. (Proof of this statement is a good exercise in itself.) Now the question of the irreducibility of $f_\zeta(X)$ is equivalent to that of $P(X) = f_\zeta(X+1) = [(X+1)^l - 1]/X = X^{l-1} + \binom{l}{1}X^{l-2} + \cdots + \binom{l}{2}X + l$. By Exercise 1.42, we see that (a), (b), and (c) hold for $P(X)$ with $p = l$. (3) The mapping $\sigma \mapsto j(\sigma)$ induces a homomorphism $G(K/\mathbb{Q}) \to (\mathbb{Z}/l\mathbb{Z})^\times$ whose kernel is trivial because λ generates $K/\mathbb{Q}$. To see the surjectivity, let $a \in \mathbb{Z}$ be such that $(a, l) = 1$. Then ζ^a shares the same minimal polynomial with ζ [by (2)] and so there is a $\sigma \in G(K/\mathbb{Q})$ such that $a \equiv j(\sigma)$ (l) [Proposition 2.2(4)] which proves (3). (4) By definition 1.20, the Gauss sum takes value in K. Use Proposition 1.25. (5) Since $G(K/\mathbb{Q}) \approx \mathbb{F}_l^\times$ is cyclic of order $l-1$, an even number, it has only one subgroup of order 2 (Proposition 1.14). By Theorem 2.3, $[K: G^\flat] = 2$ and $G^\flat$ is the only one subfield of K with this property. Since $l \neq 2$ we have $\zeta \neq \zeta^{-1} = \bar{\zeta}$ and so σ: $\zeta \mapsto \bar{\zeta}$ is the unique element of order 2 in $G(K/\mathbb{Q})$ and we find that $k = \mathbb{Q}(\zeta + \zeta^{-1}) = G^\flat$.

Exercise 2.6. Routine.

Exercise 2.7. $f_\alpha(X) = \prod_{i=1}^n (X - \alpha^{(i)}) \Rightarrow f'_\alpha(\alpha^{(i)}) = \prod_{j \neq i} (\alpha^{(i)} - \alpha^{(j)})$.

Exercise 2.8. Easy.

Exercise 2.9. Use Exercise 2.6. To see that $N_{K/\mathbb{Q}}(\zeta - 1) = l$, notice that $f_{\zeta-1}(X) = f_\zeta(X+1)$.

Exercise 2.10. Easy.

Exercise 2.11. Routine.

Exercise 2.12. Tedious but easy.

Exercise 2.13. If θ is a generator of $K^\times$, then $\theta_0 = \theta^{(q^f-1)/(q-1)}$ is a generator of $k^\times$.

Exercise 2.14. $H^0(K^\times) = H^1(K^\times) = 1$.

Exercise 2.15. Consider the matrices in $\mathrm{GL_n}(\mathbb{Z})$ of the following three types:

$$P_i = \begin{array}{cc} \begin{matrix} & & & i & & & \end{matrix} & \\ \begin{bmatrix} 1 & & & \vdots & & & \\ & \ddots & & \vdots & & & \\ & & 1 & & & & \\ & \cdots & & -1 & & \cdots & \\ & & & & 1 & & \\ & & & \vdots & & \ddots & \\ & & & \vdots & & & 1 \end{bmatrix} & \begin{matrix} \\ \\ \\ i \\ \\ \\ \\ \end{matrix} \end{array}$$

$$
Q_{i,j} = \begin{bmatrix}
 & & i & & j & & \\
1 & & \vdots & & \vdots & & \\
 & \ddots & \vdots & & \vdots & & \\
 & 1 & & & & & \\
\cdots & & 0 & \cdots & 1 & \cdots & \\
 & & 1 & & & & \\
 & & \vdots & \ddots & \vdots & & \\
 & & & & 1 & & \\
\cdots & & 1 & \cdots & 0 & \cdots & \\
 & & & & & 1 & \\
 & & \vdots & & \vdots & \ddots & \\
 & & & & & & 1
\end{bmatrix}
\begin{matrix} \\ \\ \\ \\ i \\ \\ \\ \\ j \\ \\ \\ \\ \end{matrix}
$$

$$
R_{i,j}(\lambda) = \begin{bmatrix}
 & & j & & \\
1 & & \vdots & & \\
 & \ddots & \vdots & & \\
\cdots & \ddots & \lambda & \cdots & \\
 & & \vdots & \ddots & \\
 & & & & 1
\end{bmatrix} i, \qquad i \neq j, \qquad \lambda \in \mathbb{Z}.
$$

For $X \in M_n(\mathbb{Z})$

1. $X \mapsto P_i X$: change the sign of the ith row of X,
   $X \mapsto XP_i$: change the sign of the ith column of X.
2. $X \mapsto Q_{i,j}X$: exchange the ith row of X with the jth row.
   $X \mapsto XQ_{i,j}$: exchange the ith column of X with the jth column.
3. $X \mapsto R_{i,j}(\lambda)X$: add λ times the jth row to the ith row.
   $X \mapsto XR_{i,j}(\lambda)$: add λ times the ith column to the jth column.

By $X \to Y$, $X, Y \in M_n(\mathbb{Z})$, we mean that Y is obtained from X by applying finitely many operations of types (1), (2) and (3). Prove that for any $A \in M_n(\mathbb{Z})$ one can find D, a diagonal matrix, such that

$$
A \to D = \begin{bmatrix}
d_1 & & & & 0 \\
 & \ddots & & & \\
 & & d_r & & \\
 & & & 0 & \\
 & & & & \ddots \\
0 & & & & 0
\end{bmatrix}, \qquad r = \operatorname{rank} A, \quad d_i \in \mathbb{N}, \quad d_i \mid d_{i+1}.
$$

EXERCISE 2.16. Any $\alpha \in k$ satisfies $\alpha^2 - (t_{k/\mathbb{Q}}\alpha)\alpha + N_{k/\mathbb{Q}}\alpha = 0$.

EXERCISE 2.17. Use Exercise 2.16.

Exercise 2.18. We know that $D_{k/\mathbb{Q}}(\zeta)=(-1)^{(l-1)/2}l^{l-2}$ (Exercise 2.9) and $D_{k/\mathbb{Q}}(\zeta)=\Delta_k[\mathfrak{o}_k\colon \mathbb{Z}[\zeta]]^2$ (Proposition 2.19). Therefore, for this Exercise, it is enough to prove that $\mathfrak{o}_k=\mathbb{Z}[\zeta]$. So assume that $\mathfrak{o}_k\neq\mathbb{Z}[\zeta]$. As $[\mathfrak{o}_k:\mathbb{Z}[\zeta]]$ is a power of l, there is an $\omega\in\mathfrak{o}_k$, $\omega\notin\mathbb{Z}[\zeta]$ such that

$$l\omega=x_0+x_1\zeta+\cdots+x_{l-2}\zeta^{l-2}\in\mathbb{Z}[\zeta],\qquad x_i\in\mathbb{Z}.$$

Since $\mathbb{Z}[1-\zeta]=\mathbb{Z}[\zeta]$, we can also write $l\omega=y_0+y_1(1-\zeta)+\cdots+y_{l-2}(1-\zeta)^{l-2}$, $y_i\in\mathbb{Z}$. Then $l=(1-\zeta)^{l-1}\lambda$ implies $y_0=(1-\zeta)\mu_0$, $\mu_0\in\mathfrak{o}_k$ and so $l\mid y_0$ as $N_{k/\mathbb{Q}}(1-\zeta)=l$. Similarly, we get, in turn, $l\mid y_1,\ldots,l/y_{l-2}$; this means that $\omega\in\mathbb{Z}[1-\zeta]=\mathbb{Z}[\zeta]$, a contradiction.

Exercise 2.19. Easy.

Exercise 2.20. Routine.

Exercise 2.21. If $\mathfrak{b}=\mathfrak{o}_k$, then put $\mathfrak{c}=\mathfrak{a}^{h-1}$, h being the class number of k. If $\mathfrak{b}\neq\mathfrak{o}_k$, let $\mathfrak{p}_1,\ldots,\mathfrak{p}_l$ be all distinct prime factors of $\mathfrak{b}$. First find $\mu\in\mathfrak{a}$ such that $\mu\notin\mathfrak{a}\mathfrak{p}_i$, for all i, $1\le i\le l$. To do this, put $\mathfrak{m}=\mathfrak{a}\mathfrak{p}_1\cdots\mathfrak{p}_l$, $\mathfrak{m}_i=\mathfrak{m}\mathfrak{p}_i^{-1}$, find $\mu_i\in\mathfrak{m}_i$ such that $\mu_i\notin\mathfrak{m}$, and put $\mu=\mu_1+\cdots+\mu_l$. Next, let $\mathfrak{c}$ be an ideal $\subset\mathfrak{o}_k$ such that $(\mu)=\mathfrak{a}\mathfrak{c}$. Verify that $\mathfrak{c}$ has the required property.

Exercise 2.22. By Exercise 2.18, we have $\mathfrak{o}_k=\mathbb{Z}[\zeta]$. Use this ζ in place of θ in Theorem 2.17.

Exercise 2.23. Use Exercise 2.17 and Theorem 2.17.

Exercise 2.24. Routine.

Exercise 2.25. Routine.

Exercise 2.26. When $\mathfrak{o}_k=[\omega_1,\ldots,\omega_n]$, $\Delta_k=d_{k/\mathbb{Q}}(\omega_1,\ldots,\omega_n)=d_k(\omega_1,\ldots,\omega_n)=\det(t_k(\omega_i\omega_j))$ (Exercise 2.25). Now, $\{\bar\omega_1,\ldots,\bar\omega_n\}$ forms a basis of $A=\mathfrak{o}_k/(p)$ over $\mathbb{F}_p$. For $\alpha\in\mathfrak{o}_k$, $(\alpha\omega_1,\ldots,\alpha\omega_n)=(\omega_1,\ldots,\omega_n)M(\alpha)$, $M(\alpha)\in M_n(\mathbb{Z})$ and so $(\bar\alpha\bar\omega_1,\ldots,\bar\alpha\bar\omega_n)=(\bar\omega_1,\ldots,\bar\omega_n)\bar M(\bar\alpha)$, $\bar M(\bar\alpha)=\overline{M(\alpha)}\in M_n(\mathbb{F}_p)$. Since $t_A(\bar\alpha)=\operatorname{tr}(\bar M(\bar\alpha))=\overline{\operatorname{tr}(M(\alpha))}=\overline{t_k(\alpha)}$, we get $d_A(\bar\omega_1,\ldots,\bar\omega_n)=\det(t_A(\overline{\omega_i\omega_j}))=\overline{\det(t_k(\omega_i\omega_j))}=\bar\Delta_k$.

Exercise 2.27. Easy.

Exercise 2.28. Easy.

Exercise 2.29. Easy.

Exercise 2.30. Easy.

Exercise 2.31. Define a mapping $\varphi_n\colon I_{\mathbb{Q}}(l)\to\mathbb{F}_l^\times/(\mathbb{F}_l^\times)^n$ so that for a prime $p\neq l$ we have $\varphi_n(p)=[p]\bmod(\mathbb{F}_l^\times)^n$. Check that $\operatorname{Ker}\varphi_n=H_n(l)$.

Exercise 2.32.

1. $(x,N)=(y,N)=1$, $x\equiv y\ (N)\Rightarrow(\text{for } p\mid \mathbb{N})\quad (x,p)=(y,p)=1$, $x\equiv y\ (p)\Rightarrow\lambda_p(x)=\lambda_p(y)\Rightarrow\eta_N(x)=\eta_N(y)$.

2. Let $N=\prod_p p$ and $M=\prod_q q$ where p and q are prime. Then $\eta_M(N)\eta_N(M)=\prod_p \lambda_p(M)\prod_q \lambda_q(M)=\prod_{p,q}\lambda_p(q)\lambda_q(p)$ and use the reciprocity for Legendre characters.

3, 4. These can be proved similarly.

EXERCISE 2.33.

1. $\Delta_k \equiv 1 \quad (4)$. $x\in\mathbb{N}$, $(x,\Delta_k)=1$, $x=2^\alpha y$, y odd, $y=\prod p$. For brevity, write $\chi_k=\chi$ and $\Delta_k=\Delta$. $\chi(x)=\chi(2^\alpha)\chi(y)=\eta_{|\Delta|}(2^\alpha)\prod_p \chi(p)=\eta_{|\Delta|}(2^\alpha)\prod_p \lambda_p(\Delta)=\eta_{|\Delta|}(2^\alpha)\eta_y(\Delta)$. Now

$$\begin{aligned}\eta_y(\Delta)&=\eta_y((\operatorname{sgn}\Delta)\,|\Delta|)=\eta_y(\operatorname{sgn}\Delta)\eta_y(|\Delta|)\\&=\eta_y(\operatorname{sgn}\Delta)\eta_{|\Delta|}(y)(-1)^{[(y-1)/2]\cdot[(|\Delta|-1)/2]}\\&=\begin{cases}\eta_{|\Delta|}(y)(-1)^{[(y-1)/2]\cdot[(\Delta-1)/2]}=\eta_{|\Delta|}(y)\\ \eta_y(-1)\eta_{|\Delta|}(y)(-1)^{[(y-1)/2]\cdot[(-\Delta-1)/2]}\end{cases} && \text{if } \Delta>0,\\&=\eta_{|\Delta|}(y)(-1)^{[(y-1)/2]\cdot[1-(\Delta+1)/2]}=\eta_{|\Delta|}(y) && \text{if } \Delta<0.\end{aligned}$$

Hence $\chi(x)=\eta_{|\Delta|}(2^\alpha)\eta_{|\Delta|}(y)=\eta_{|\Delta|}(x)$.

2. Similar to (1).

3. Case 1, $\Delta\equiv 1 \quad (4)$. Write $|\Delta|=pg$, p a prime, $g\in\mathbb{N}$, $(p,g)=1$. Find $a\in\mathbb{Z}$ such that $\left(\frac{a}{p}\right)=-1$. Find $x\in\mathbb{N}$ such that $x\equiv a \quad (p)$, $x\equiv 1 \quad (g)$. Then, using (1), we get $\chi(x)=-1$. Use (2) to handle Case 2, $\Delta\equiv 0 \quad (4)$.

4. Case 1, $\Delta\equiv 1 \quad (4)$, is easy. Case 2, $\Delta\equiv 0 \quad (4)$, requires a little more effort.

Chapter 3

EXERCISE 3.1. Let $M=\{1/2^\nu,\ \nu=0,1,2,\ldots\}\subset\mathbb{R}$. M is discrete but not finite.

EXERCISE 3.2. Let $N\subset M=[u_1,\ldots,u_n]$. If $n=0$, the matter is trivial. If $n=1$, then $M\approx\mathbb{Z}$ and the statement is nothing but Theorem 1.1. Assume that the statement is true for a submodule of $M^*=[u_1,\ldots,u_{n-1}]$. If $N\subset M^*$ then the statement follows from the induction hypothesis. If $N\not\subset M^*$, there is a $v_n=b_1u_1+\cdots+b_{n-1}u_{n-1}+b_nu_n\in N$ with $b_n\in\mathbb{N}$; we assume v_n is such that the natural number b_n is minimal. Let $v=a_1u_1+\cdots+a_nu_n$ be any element of N and let $a_n=b_nq+r$, $0\le r<b_n$. Since $v-qv_n\in N$, we have $v-qv_n\in M^*$ by the minimality of b_n and so $v=u+qv_n$ with $u\in N\cap M^*\subset M^*$. By induction $N\cap M^*$ is a free module of rank $\le n-1$. As $v=u+qv_n=0\Rightarrow u=0,\ q=0$, N is a free module of rank $\le n$.

EXERCISE 3.3. Let $\varphi: \mathbb{R}^n \to \mathbb{R}^n$ be the linear mapping defined by $\varphi(e_i) = v_i$, $1 \leq i \leq n$, where $\{e_i\}$ is the standard basis of $\mathbb{R}^n$. Notice that $A = J_\varphi$ = the Jacobian matrix for φ.

EXERCISE 3.4. Easy.

EXERCISE 3.5. Passing to polar coordinates $(x_{s+j} = r_j e^{i\theta_j})$ we have $\mathrm{vol}(B) = 2^s(2\pi)^t I$ and $I = 4^{-t}\rho^n J$ where J is a Dirichlet integral with the value

$$J = \frac{\Gamma(1)^s \Gamma(2)^t \Gamma(1)}{\Gamma(s+2t+1)} = \frac{1}{n!}$$

[E. T. Whittaker and G. N. Watson, *A Course of Modern Analysis,* 4th ed., p. 258, Cambridge University Press, London (1958)].

EXERCISE 3.6. Easy.

EXERCISE 3.7. Since $n = 2$, $r_2 = 1$, we find $M_k = (2/\pi)\sqrt{-\Delta_k} < 0.63663\sqrt{-\Delta_k}$. We have $M_k < 2$ for $-m = 1, 2, 3, 7$, and $h_k = 1$ in these cases. There remain five more cases to check: $-m = 11, 19, 43, 67, 163$. Note that $m \equiv 1\,(4)$ and $\Delta_k = m$ for these cases. Let us explain the case $-m = 43$. Suppose $h_k \geq 2$ and take an ideal $\mathfrak{a} \subset \mathfrak{o}_k$ such that $N\mathfrak{a} \leq M_k < 0.63663\sqrt{43} < 4.174663$. Hence $N\mathfrak{a} = 2, 3, 4$.

1. $N\mathfrak{a} = 2$. Then $\mathfrak{a} = \mathfrak{p}$, $N\mathfrak{p} = 2$, $\chi(2) = 1$. On the other hand, $\chi(2) = (-1)^{(43^2-1)/2} = (-1)^{231} = -1$, so this case does not happen.

2. $N\mathfrak{a} = 3$. In this case, $\mathfrak{a} = \mathfrak{p}$, $N\mathfrak{p} = 3$, $\chi(3) = 1$. On the other hand,

$$\chi(3) = \left(\frac{\Delta}{3}\right) = \left(\frac{-43}{3}\right) = \left(\frac{-1}{3}\right) = -1,$$

impossible.

3. $N\mathfrak{a} = 4 = 2^2$. If $\mathfrak{a} = \mathfrak{p}$, then since 2 is unramified for $k/\mathbb{Q}$, $\mathfrak{p} = 2 \sim 1$ which contradicts $\mathfrak{a} \not\sim 1$. Hence $\mathfrak{a} = \mathfrak{p}^2$ or $\mathfrak{a} = \mathfrak{p}\mathfrak{q}$, $\mathfrak{p} \neq \mathfrak{q}$. If $\mathfrak{a} = \mathfrak{p}^2$, then $2^2 = N\mathfrak{a} = (N\mathfrak{p})^2$, i.e., $N\mathfrak{p} = 2$ which is impossible by (1). If $\mathfrak{a} = \mathfrak{p}\mathfrak{q}$, then $2^2 = N\mathfrak{a} = N\mathfrak{p}N\mathfrak{q}$ and so $N\mathfrak{p} = N\mathfrak{q} = 2$ which is impossible again.

EXERCISE 3.8. Since $n = 2$, $r_2 = 0$, we have

$$M_k = \frac{2!}{2^2}\sqrt{|\Delta_k|} = \frac{1}{2}\sqrt{\Delta_k}.$$

Hence $h_k = 1$ whenever $\Delta_k < 16$; this is the case for $m = 2, 3, 5, 13$.

EXERCISE 3.9. As $m = -ab \equiv 2, 3\ (4)$, we have $\Delta_k = 4m$ and $\mathfrak{o}_k = [1, \sqrt{m}]$. Verify that $\mathfrak{a} = (a, \sqrt{m})$ is not principal.

EXERCISE 3.10. Easy.

EXERCISE 3.11. Generalize Exercise 2.5(1), (2) for arbitrary cyclotomic field.

EXERCISE 3.12. Let m be the number of roots of unity in k. Then, on the one hand, $2l \mid m$ but, on the other hand, $\varphi(m) \leq l-1$.

EXERCISE 3.13. Easy.

EXERCISE 3.14. $\gamma \in W_k \Rightarrow \gamma^N = 1 \Rightarrow (\gamma^{(i)})^N = 1 \Rightarrow |\gamma^{(i)}| = 1$ for all i. Conversely, suppose that $|\gamma^{(i)}| = 1$ for all i. Since $x(\mathfrak{o}_k)$ is a lattice in $E^{r_1, r_2} = \mathbb{R}^n$ (Proof of Theorem 3.2), $x(\gamma)$ must be in a finite set of $\mathbb{R}^n$. As x is injective, it follows that there are only a finite number of γ's such that $|\gamma^{(i)}| = 1$. Since for any $j \in \mathbb{N}$, absolute values of all conjugates of γ^j are 1 like γ, we must have $\gamma^a = \gamma^b$ for $a > b$, $a, b \in \mathbb{N}$, and so $\gamma^{a-b} = 1$, i.e., $\gamma \in W_k$.

EXERCISE 3.15. $\mathfrak{o}_K^\times \supset \langle \zeta \rangle \mathfrak{o}_k^\times$ is obvious. One sees easily that $\langle \zeta \rangle \cap \mathfrak{o}_k^\times = 1$ because $l \neq 2$. It remains to show that $\mathfrak{o}_K^\times \subset \langle \zeta \rangle \mathfrak{o}_k^\times$. For any $\epsilon \in \mathfrak{o}_K^\times$, its complex conjugate $\bar{\epsilon}$ is also in $\mathfrak{o}_K^\times$. Since $G(K/\mathbb{Q})$ is abelian, $\overline{\epsilon^{(i)}} = (\bar{\epsilon})^{(i)}$. It follows that $|(\bar{\epsilon}/\epsilon)^{(i)}| = 1$; hence $\bar{\epsilon}/\epsilon \in W_K = \{\pm\zeta^\nu\}$ (Exercise 3.12, Exercise 3.14). As $\bar{\epsilon}/\epsilon = \pm\zeta^\nu = \pm\zeta^{\pm\nu+l}$, we may assume that $\nu = 2g$, $g \in \mathbb{Z}$. Put $\epsilon_0 = \zeta^g \epsilon \in \mathfrak{o}_K^\times$. Then $\bar{\epsilon}_0 = \zeta^{-g}\bar{\epsilon} = \zeta^{-g}(\pm 1)\zeta^{2g}\epsilon = \pm\zeta^g \epsilon = \pm\epsilon_0$. If $\bar{\epsilon}_0 = \epsilon_0$, then we are done. Suppose, on the contrary, that $\bar{\epsilon}_0 = -\epsilon_0$. Write $\epsilon_0 = c_0 + c_1\zeta + \cdots + c_{l-2}\zeta^{-(l-2)}$ (Exercise 2.18). Then $\bar{\epsilon}_0 = c_0 + c_1\zeta^{-1} + \cdots + c_{l-2}\zeta^{-(l-2)}$. Let $\mathfrak{l} = (1-\zeta)$, this being the prime ideal in $\mathfrak{o}_K$ such that $(l)_K = \mathfrak{l}^{l-1}$ (Exercise 2.22). As $\zeta \equiv 1\ (\mathfrak{l})$, $\zeta^{-1} \equiv l\ \ (\mathfrak{l})$, we have $\epsilon_0 \equiv \bar{\epsilon}_0 = c_0 + \cdots + c_{l-2}\ \ (\mathfrak{l})$; but $\bar{\epsilon}_0 = -\epsilon_0$ implies $2\epsilon_0 \equiv 0\ \ (\mathfrak{l})$, a contradiction (because $l \neq 2$).

EXERCISE 3.16. From the inequalities

$$\int_n^{n+1} \frac{dx}{x^s} < \frac{1}{n^s} < \int_{n-1}^n \frac{dx}{x^s},$$

we get

$$\int_1^{N+1} \frac{dx}{x^s} < \sum_{n=1}^N \frac{1}{n^s} < 1 + \int_1^N \frac{dx}{x^s}. \tag{*}$$

Since

$$\int_1^\infty \frac{dx}{x^s} = \left[\frac{x^{1-s}}{1-s}\right]_1^\infty = \frac{1}{s-1}$$

is finite $(s > 1)$, $\zeta_\mathbb{Q}(s) = \sum_{n=1}^\infty 1/n^s$ converges for $s > 1$. Next, letting $N \to \infty$ in (*), we have

$$\frac{1}{s-1} < \zeta_\mathbb{Q}(s) < 1 + \frac{1}{s-1},$$

hence $1 < (s-1)\zeta_\mathbb{Q}(s) < s$ and so $\lim_{s \to 1+0} (s-1)\zeta_\mathbb{Q}(s) = 1$.

Exercise 3.17. Easy.

Exercise 3.18. Let $\mathfrak{a}_0 \cap \mathbb{Z} = a\mathbb{Z}$, $a > 0$. Then one verifies that $x(a) \in C \cap \Gamma$, i.e., $c \cap \Gamma \neq \emptyset$.

Exercise 3.19. As $C_1 = C \cap V_1 = \{x \in C; |F(x)| \leq 1\}$, it is enough to prove that $S = \{x \in C_1; |F(x)| = 1\}$ is bounded.

Exercise 3.20. Let

$$u = \left(\frac{1}{r_1 + r_2}, \ldots, \frac{1}{r_1 + r_2}\right) \in \mathbb{R}^{r_1+r_2} \quad \text{and} \quad \lambda = (\lambda_1, \ldots, \lambda_{r_1+r_2}) \in H^r \subset \mathbb{R}^{r_1+r_2}.$$

We find $\langle u, \lambda \rangle = 0$, orthogonality. Let $\Gamma = [u, l^{\#}(\epsilon_1), \ldots, l^{\#}(\epsilon_r)]$, a lattice in $\mathbb{R}^{r_1+r_2}$. Verify that $V(\Gamma) = R_k$.

Exercise 3.21. Easy.

Exercise 3.22. Easy.

Exercise 3.23. Routine.

Exercise 3.24. Routine.

Exercise 3.25. l is the only prime ramified for $k/\mathbb{Q}$ and $l = \mathfrak{l}^{l-1}$ (completely ramified, Exercise 2.22). Hence $f_l = g_l = 1$ and $T(s) = 1$.

Exercise 3.26. $m = 4$ and there is only one $\mathfrak{p}$ such that $\mathfrak{p} \mid 4$, i.e., the $\mathfrak{p}$ with $2 = \mathfrak{p}^2$. For this $\mathfrak{p}$ we have $f_2 = g_2 = 1$ and so $T(s) = 1$.

Exercise 3.27. By Exercise 3.26, we have $\zeta_k(s) = \zeta_{\mathbb{Q}}(s) L(s, \chi)$. Compare coefficients of Dirichlet series on both sides.

Exercise 3.28. For any $\epsilon > 0$, take n_0 so that $1/n_0^{s_0} < \epsilon/2B$. If $n \geq n_0$, $s \geq s_0$, then $1/n^s \leq 1/n_0^{s_0} < \epsilon/2B$. For $M > N \geq n_0$, we have

$$\sum_{\nu=N}^{M} \frac{a_\nu}{\nu^s} = \sum_{\nu=N}^{M} \frac{A_\nu - A_{\nu-1}}{\nu^s} = \sum_{\nu=N}^{M} \frac{A_\nu}{\nu^s} - \sum_{\nu=N-1}^{M-1} \frac{A_\nu}{(\nu+1)^s}$$

$$= -\frac{A_{N-1}}{N^s} + \sum_{\nu=N}^{M-1} A_\nu \left(\frac{1}{\nu^s} - \frac{1}{(\nu+1)^s}\right) + \frac{A_M}{M^s}.$$

Hence

$$\left|\sum_{\nu=N}^{M} \frac{a_\nu}{\nu^s}\right| \leq B\left(\frac{1}{N^s} + \sum_{\nu=N}^{M-1} \left(\frac{1}{\nu^s} - \frac{1}{(\nu+1)^s}\right) + \frac{1}{M^s}\right) \leq \frac{2B}{N^s} < \epsilon.$$

Exercise 3.29.

$$|R(s, \chi)| < \sum_p \sum_{n=2}^{\infty} \frac{1}{p^{ns}} \leq \sum_p \sum_{n=2}^{\infty} \frac{1}{p^n} = \sum_p \frac{1}{p^2 - p} < \sum_p \frac{1}{(p-1)^2} < b.$$

Chapter 4

EXERCISE 4.1. Routine.

EXERCISE 4.2. Easy.

EXERCISE 4.3. Let f_x be the characteristic function of the one point set $\{x\}$, $x \in \mathbb{F}_p$, and let A be the matrix determined by the relation

$$(\ldots, Tf_x, \ldots) = (\ldots, f_y, \ldots)A, \qquad A = (a_{yx}).$$

Applying Tf_x on ξ, we find $A = (\zeta^{xy})$ by using $\psi_y(x) = \zeta^{xy}$. Then we find that

$$A^2 = \begin{bmatrix} l & 0 & \cdots & \cdots & 0 \\ 0 & 0 & \cdots & \cdots & l \\ 0 & 0 & \cdots & l & 0 \\ & & \cdot^{\cdot^{\cdot}} & & \\ 0 & l & & & 0 \end{bmatrix}$$

and so $\det A = \pm i^{l(l-1)/2} l^{l/2}$. It remains to show that $\det A = i^{l(l-1)/2} \times$ (a positive number). To do this, use the Vandermonde theorem to get

$$\det A = \prod_{0 \le x < y \le l-1} (\zeta^y - \zeta^x) = \prod_{x<y} e^{\pi i(x+y)/l} \prod_{x<y} (e^{\pi i(y-x)/l - \pi i(x-y)/l})$$

and verify that the first factor in the right-hand side equals 1 and the second factor equals $i^{l(l-1)/2} \times$ (a positive number).

EXERCISE 4.4. Easy.

EXERCISE 4.5. Easy.

EXERCISE 4.6. First derive that

$$\log(1 - \zeta^{-\xi}) = \log|1 - \zeta^{-\xi}| + i\pi\left(\frac{1}{2} - \frac{\xi}{l}\right) = \log\left(2\sin\left(\frac{\pi\xi}{l}\right)\right) + i\pi\left(\frac{1}{2} - \frac{\xi}{l}\right).$$

Next, compute

$$S_\chi = \sum_{\xi=1}^{l-1} \bar{\chi}(\xi) \log(1 - \zeta^{-\xi}).$$

Case 1, $\chi =$ an even character. Apply the change of variable $\xi \mapsto -\xi$. As $\chi(-1) = 1$, we have $S_\chi = \sum_{\xi=1}^{l-1} \bar{\chi}(\xi) \log(1 - \zeta^{\xi})$ and so

$$2S_\chi = 2\sum_{\xi=1}^{l-1} \bar{\chi}(\xi)\ \log\left(2\sin\left(\frac{\pi\xi}{\zeta}\right)\right)$$

because $\sum_{\bar{\chi}}(\xi) = 0$.

Case 2, $\chi =$ an odd character. By the change of variable $\xi \mapsto -\xi$ we have, this time, $S_\chi = -\sum \bar{\chi}(\xi)\log(1-\zeta^\xi)$ and then $2S_\chi = 2\sum \bar{\chi}(\xi)\pi i(\frac{1}{2} - \xi/l)$. Use $\sum \bar{\chi}(\xi) = 0$, again.

EXERCISE 4.7. Since $199 \equiv 7 \pmod 8$, $h_k = \sum_{0<\xi<99.5} \lambda_{199}(\xi) = 54 - 45 = 9$.

EXERCISE 4.8. It is enough to prove that $\mathfrak{m} \subset [a, b + c\omega]$. So take any $x + y\omega \in \mathfrak{m}$ and write $y = qc + r$, $0 \le r < c$. Then $x + y\omega - q(b + c\omega) = (x - qb) + r\omega \in \mathfrak{m}$, and, by the minimality of c, we have $r = 0$, i.e., $y = qc$. Now, $x + y\omega = x + qc\omega = x - qb + q(b + c\omega) \Rightarrow x - qb \in \mathfrak{m} \cap \mathbb{Z} = \mathbb{Z}a \Rightarrow x = sa + qb$, $s \in \mathbb{Z} \Rightarrow x + y\omega = sa + qb + qc\omega = sa + q(b + c\omega) \in [a, b + c\omega]$.

EXERCISE 4.9. Easy.

EXERCISE 4.10. If $\mathfrak{m}$ is an ideal, then $a\omega \in \mathfrak{m}$ and so $c \mid a$ by the minimality of c. As for $c \mid b$, start with $\omega(b + c\omega) \in \mathfrak{m}$ and consider the cases $\mathfrak{m} \equiv 2, 3 \pmod 4$ and $m \equiv 1 \pmod 4$, separately.

EXERCISE 4.11. ($\Rightarrow$) $N(b + \omega) = (b + \omega)(b + \omega') \in \mathfrak{m} \cap \mathbb{Z} = \mathbb{Z}a \Rightarrow a \mid N(b + \omega)$. ($\Leftarrow$) It is enough to prove that $a\omega$, $(b + \omega)\omega \in \mathfrak{m}$. Now, $a\omega = -ab + ab + a\omega = -ab + a(b + \omega) \in \mathfrak{m}$. Next, $(b + \omega)\omega = (b + \omega)(b + s) - aa' \in \mathfrak{m}$ where $s = \omega + \omega'$ and $N(b + \omega) = aa'$.

EXERCISE 4.12. Use Exercise 4.11.

EXERCISE 4.13. Easy.

EXERCISE 4.14. We may assume that $\mathfrak{a}$ is primitive: $\mathfrak{a} = [a, b + \omega] = (a, b + \omega)$. Hence $\mathfrak{a}' = [a, b + \omega'] = (a, b + \omega')$. Then $\mathfrak{a}\mathfrak{a}' = (a, b + \omega)(a, b + \omega') = (a^2, a(b + \omega), a(b + \omega'), N(b + \omega)) = a(a, b + \omega, b + \omega', d) = a\mathfrak{b}$, with $N(b + \omega) = ad$. Now $a^2 = (N\mathfrak{a})^2 = N\mathfrak{a}N\mathfrak{a}' = N(\mathfrak{a}\mathfrak{a}') = N(a\mathfrak{b}) = a^2 N\mathfrak{b} \Rightarrow N\mathfrak{b} = 1 \Rightarrow \mathfrak{b} = \mathfrak{o}_k$. Hence $\mathfrak{a}\mathfrak{a}' = (a) = (N\mathfrak{a})$.

EXERCISE 4.15. $a \mid N(b + \omega) \Rightarrow a_i \mid N(b + \omega) \Rightarrow \mathfrak{a}_i = [a_i, b + \omega]$ is a primitive ideal, $i = 1, 2$. Verify that $\mathfrak{a}_1\mathfrak{a}_2 \subset \mathfrak{a}$ and compare norms of both sides.

EXERCISE 4.16. Notice first that $M_k < l$. ($\Rightarrow$) Assume that $P(b)$, $0 \le b \le l - 2$, are all prime. For $p \in S_k$ (Proposition 4.3), we have $p = N\mathfrak{p}_p$, $\mathfrak{p}_p = [p, b + \omega]$, $0 \le b < p$. As $p \le M_k < l$, we have $0 \le b \le l - 2$. $1 \sim [P(b), b + \omega]$ and $p \mid N(b + \omega) = P(b) \Rightarrow P(b) = p$. Hence $1 \sim [p, b + \omega] = \mathfrak{p}_p$, $p \in S_k$, which implies that $h_k = 1$. ($\Leftarrow$) Notice that $P(b) = P(l - 2) < l^2$. Hence if $P(b_0)$ is not prime for a b_0, $0 \le b_0 \le l - 2$, and if p is the smallest prime dividing $P(b_0)$, then we have $p \le l - 1 < l$ by $P(b_0) < l^2$. Verify that $\mathfrak{p} = [p, b_0 + \omega]$ is a prime ideal which is not principal.

EXERCISE 4.17. Easy.

EXERCISE 4.18. Routine.

EXERCISE 4.19. For $\mathfrak{a} \in_N I_k$, put $\mathfrak{b} = \mathfrak{o}_k + \mathfrak{a} + \mathfrak{a}^{1+\sigma} + \cdots + \mathfrak{a}^{1+\sigma+\cdots+\sigma^{n-2}}$, $n = [k : \mathbb{Q}]$.

EXERCISE 4.20. $k = \mathbb{Q}(\sqrt{221})$.

EXERCISE 4.21. $h_k = h_k^+ = 2^{t-1}h_k^*$. For $m = 1 - 2^{225} \equiv 1\,(4)$, $N\omega = 2^{223}$ (223 is prime), $\Delta_k = m$, and $t = 14$. Let $\mathfrak{a} = [2, \omega]$. As $\mathfrak{a} \nsim 1$, we find $223 \mid h_k^*$ and hence $2^{13} \times 223 = 1826816 \mid h_k$.

EXERCISE 4.22. Easy.

EXERCISE 4.23. (1) $1+\sqrt{2}$, (2) $2+\sqrt{3}$, (3) $5+2\sqrt{6}$, (4) $8+3\sqrt{7}$.

EXERCISE 4.24. (1) $\frac{1}{2}(1+\sqrt{5})$, (2) $\frac{1}{2}(3+\sqrt{13})$.

EXERCISE 4.25. Just do!

EXERCISE 4.26.

$$x^2 + \frac{b}{a}x + \frac{c}{a} = (x-\alpha)(x-\alpha') = x^2 + \frac{b'}{a'}x + \frac{c'}{a'} \Rightarrow \frac{b}{a} = \frac{b'}{a'}, \qquad \frac{c}{a} = \frac{c'}{a'}.$$

EXERCISE 4.27. Let

$$\alpha_1 = \frac{p\alpha + q}{r\alpha + s}, \qquad \sigma = \begin{pmatrix} p & q \\ r & s \end{pmatrix} \in G.$$

Then

$$\sigma\begin{pmatrix} \alpha \\ 1 \end{pmatrix} = \begin{pmatrix} \alpha_1 \\ 1 \end{pmatrix}.$$

$$0 = a\alpha^2 + b\alpha + c = (\alpha, 1)A\begin{pmatrix} \alpha \\ 1 \end{pmatrix} = \lambda^2(\alpha_1, 1){}^t\sigma^{-1}A\sigma^{-1}\begin{pmatrix} \alpha_1 \\ 1 \end{pmatrix},$$

$$\lambda = r\alpha + s, \qquad A = \begin{pmatrix} a & \frac{b}{2} \\ \frac{b}{2} & c \end{pmatrix}$$

$$\begin{aligned} {}^t\sigma^{-1}A\sigma^{-1} = A_1 = \begin{pmatrix} a_1 & \frac{b_1}{2} \\ \frac{b_1}{2} & c_1 \end{pmatrix} &\Rightarrow (\alpha_1, 1)A_1\begin{pmatrix} \alpha_1 \\ 1 \end{pmatrix} \\ &= a_1\alpha_1^2 + b_1\alpha_1 + c_1 = 0 \Rightarrow \det A_1 \\ &= \det A = b^2 - 4ac = D(\alpha) = D(\alpha_1) \Rightarrow \alpha_1 = \sigma(\alpha) \in X_2(D). \end{aligned}$$

EXERCISE 4.28. Easy.

EXERCISE 4.29. One way is obvious. Conversely, assume that $\alpha \in R_2(\Delta)$. Then, by (4.59) and (4.60), $\alpha + \alpha' = -b/a > 0$ and $\alpha\alpha' = (b^2 - \Delta)/4a^2 = c/a < 0$.

EXERCISE 4.30. Similar to Exercise 4.26.

EXERCISE 4.31. Easy.

EXERCISE 4.32. Use (4.63).

Notes

A. Peano Axioms

In Section 1.1, we denoted by $\mathbb{N}$ the set $\{1, 2, 3, \ldots\}$ of all natural numbers. We did so under the assumption that everybody can feel the totality of natural numbers intuitively. As $\mathbb{N}$ is only a letter, what we did is merely a matter of convenience. To develop the theory of natural numbers less intuitively, one usually starts with the following axioms due to Peano (1858–1932):

I. $1 \in \mathbb{N}$,
II. there is a mapping $\varphi: \mathbb{N} \to \mathbb{N}$,
III. φ is injective,
IV. $1 \notin \varphi(\mathbb{N})$,
V. if a subset A of $\mathbb{N}$ satisfies the following two conditions:
(a) $1 \in A$, (b) $\varphi(A) \subset A$,
then $A = \mathbb{N}$.

Once axioms are stated, it is natural to ask the existence and the uniqueness of objects (the set $\mathbb{N}$ satisfying I–V). In our case, one can prove the uniqueness.† However, to prove the existence of the set $\mathbb{N}$ seems to be extremely difficult; the set $\mathbb{N}$ is *infinite* and studies such as cosmic physics, cerebral physiology, or metaphysical ontology seem to be helpless for such an attempt. If one wants to avoid the vicious circles of analyzing a product of human thoughts such as $\mathbb{N}$ by using human thoughts, the best way must be to believe in its existence.

From $\mathbb{N}$ we pass to $\mathbb{Q}$ (rationals) via $\mathbb{Z}$ (integers) purely algebraically. We obtain $\mathbb{R}$ (reals) by completing $\mathbb{Q}$ with respect to the distance in $\mathbb{Q}$ defined by the absolute value. Of course, the best way is to assume that

† E.g., p. 176 of S. Iyanaga, *System of Numbers,* Vol. 1, Iwanami, 1972 (in Japanese).

everybody can feel the totality $\mathbb{R}$ of real numbers as the straight line. The passage from $\mathbb{R}$ to $\mathbb{C}$ (complex numbers) is again purely algebraic.

Our standpoint in this book is quite naive; think of $\mathbb{Z}$ as a subset of $\mathbb{R}$ (the straight line) and extract its properties as a commutative ring but not a field.

B. Fundamental Theorem of Algebra

This theorem asserts that $\mathbb{C}$ is algebraically closed, i.e., for any polynomial $f(X) \in \mathbb{C}[X]$, $\deg f \geq 1$, there is an $\alpha \in \mathbb{C}$ such that $f(\alpha) = 0$. It was proved by C. F. Gauss (1777 4/30–1855 2/23) in the fall of 1797. While he was a student at University of Göttingen (1795–1798) Gauss made important discoveries in mathematics including the famous *Disquisitiones arithmeticae.* However, his first proof of the fundamental theorem of algebra was submitted to the University of Helmstedt as Dissertation and published in 1799 under the title *Demonstratio nova theorematis omnen functionem algebraicam rationalem integram unius variabilis in factores reales primi vel secundi gradus revolvi posse.* Actually, Gauss's first proof is the first proof of the theorem and so the word *nova* in the title is misleading. The polynomial (integral rational function) $f(X)$ in the Dissertation has real coefficients: $f(X) \in \mathbb{R}[X]$. However, it is easy to extend the result for any $f(X) \in \mathbb{C}[X]$; in fact, write $f(X) = g(X) + ih(X)$, $g, h \in \mathbb{R}[X]$, and consider

$$r(X) = f(X)\bar{f}(X) = g(X)^2 + h(X)^2 \in \mathbb{R}[X].$$

Then, for $\alpha \in \mathbb{C}$, $r(\alpha) = 0 \Rightarrow f(\alpha) = 0$ or $f(\bar{\alpha}) = 0$ which proves the theorem for any $f(X) \in \mathbb{C}[X]$. After the Dissertation, Gauss published his second and third proof in 1816 and his fourth proof in 1850.

One of the shortest proof we have today depends on a theorem in the theory of complex variables due to Liouville (1809–1882): a bounded integral function is a constant. In fact, if $f(X) \in \mathbb{C}[X]$, $\deg f \geq 1$, has no roots in $\mathbb{C}$, then $1/f$ is integral and bounded as $f(z) \to \infty$ when $z \to \infty$, hence $1/f$ or f is a constant, a contradiction.

Anyway, it is delightful to know that one reaches the algebraic closure of $\mathbb{R}$ by a quadratic extension $\mathbb{C} = \mathbb{R}(i)$ where the ground field $\mathbb{R}$ is far from being algebraically closed; e.g., ± 1 are the only roots of 1 in $\mathbb{R}$.

C. Zorn's Lemma

Let X be a set. X is an *ordered set* if there is defined a relation $<$ (order) among elements in X satisfying the following conditions:

1. $x < x$, $\forall x \in X$,
2. $x < y$, $y < x \Rightarrow x = y$,
3. $x < y$, $y < z \Rightarrow x < z$.

If, in addition to (1)–(3), we have

4. for any $x, y \in X$, either $x < y$ or $y < x$,

then X is a *totally ordered set.*

An ordered set X is called an *inductively ordered set* if every totally ordered subset Y of X has an upper bound, i.e., for any $y \in Y$, there is a $b \in X$ such that $y < b$.

ZORN'S LEMMA. *Every inductively ordered set $X \neq \emptyset$ has at least one maximal element m, i.e., an element m such that $m < x, x \in X \Rightarrow x = m$.*

REMARK. Although the statement is called "lemma" it is rather an "axiom" because we have no intention of proving it.

As an example, let us prove that any vector space V over a field K has a basis. First of all, a subset S in V is called *independent* if any set of finitely many vectors in S is independent. Let X be the family of all independent subsets of X. In X, define the relation $<$ by

$$S < S' \Leftrightarrow S \subset S', \qquad S, S' \in X.$$

Clearly, X is an ordered set. Let $Y = \{T_\lambda; \lambda \in \Lambda\}$ be any totally ordered subset of X. Then $T = \bigcup_\lambda T_\lambda$ belongs to X and, as $T_\lambda < T$ for all T_λ in Y, X is inductively ordered. By Zorn's lemma, there is a maximal element M in X. We claim that M is a basis for V, i.e., M is independent and any vector in V is a linear combination of finitely many vectors in M. In fact, if some $v \in V$ fails to be such a linear combination, then $S = \{M, v\} \in X$ with $M < S$, $M \neq S$, which contradicts the maximality of M.

D. Quadratic Fields and Quadratic Forms

Fixing once and for all a quadratic field $k = \mathbb{Q}(\sqrt{m})$, consider the set

$$Q(\Delta_k) = \{f = ax^2 + bxy + cy^2;\ a, b, c \in \mathbb{Z},\ b^2 - 4ac = \Delta_k,\ f > 0 \text{ if } \Delta_k < 0\},$$

where f is a binary quadratic form with integral coefficients a, b, and c, with discriminant b^2-4ac, and $f>0 \Leftrightarrow f(x, y)>0$, for all $(x, y) \neq (0, 0)$. As we shall see later, $Q(\Delta_k) \neq \varnothing$. For $f, g \in Q(\Delta_k)$, put

$$f \pm g \overset{\text{def}}{\Leftrightarrow} g(z) = f(\gamma z), \qquad \gamma \in \mathrm{SL}_2(\mathbb{Z}) = \{\alpha \in \mathrm{GL}_2(\mathbb{Z}), \det\alpha = 1\},$$

this being an equivalence relation in $Q(\Delta_k)$. Let $\tilde{Q}(\Delta_k)$ denote the quotient space $Q(\Delta_k)/\pm$. Now, let $\mathfrak{a} = [\alpha, \beta]$ be a fractional ideal of k. We agree to orient the basis so that

$$\begin{cases} \begin{vmatrix} \beta & \beta' \\ \alpha & \alpha' \end{vmatrix} > 0 & \text{if } \Delta_k > 0, \\ \dfrac{1}{i}\begin{vmatrix} \beta & \beta' \\ \alpha & \alpha' \end{vmatrix} > 0 & \text{if } \Delta_k < 0. \end{cases}$$

For example, the order of the canonical basis of a primitive ideal $\mathfrak{a} = [a, h+\omega]$ is the right one. In general, for a fractional ideal $\mathfrak{a} = [\alpha, \beta]$ put

$$f_{\mathfrak{a}} = \frac{1}{N\mathfrak{a}} N(x\alpha + y\beta).$$

Then $f_{\mathfrak{a}} \in Q(\Delta_k)$. It follows that

$$\mathfrak{a} \pm \mathfrak{b} \Leftrightarrow f_{\mathfrak{a}} \pm f_{\mathfrak{b}}$$

where $\pm$ for ideals is the one defined in Section 4.6. We then find that the mapping $\mathfrak{a} \mapsto f_{\mathfrak{a}}$ induces a bijection i_k:

$$i_k \colon H_k^+ \approx \tilde{Q}(\Delta_k).$$

Since the left-hand side is a group, the right-hand side becomes a group, too. This is the "group" of Gauss; in *Disquisitiones arithmeticae,* he defined in $\tilde{Q}(\Delta_k)$ a product called composition and proved that $\tilde{Q}(\Delta_k)$ forms a finite abelian group.

$$\begin{array}{lccccl} \text{(group)} & H_k^+ & \xrightarrow[\approx]{i_k} & \tilde{Q}(\Delta_k) & & \text{("group")} \\ \text{(character)} & \chi_i^+ & \searrow & \swarrow & \psi_i & \text{("character")} \\ & & \{\pm 1\} & & & \end{array}$$

In Theorem 4.9, we associated with each i, $1 \leq i \leq t$, t being the number of distinct prime factors of Δ_k, the character χ_i^+ of H_k^+, and proved that

$$\bigcap_{i=1}^{t} \operatorname{Ker} \chi_i^+ = (H_k^+)^2.$$

Each coset of $H_K^+/(H_k^+)^2$ is a genus, each genus consists of $h_k^* = [(H_k^+)^2]$ classes, and $h_k^+ = 2^{t-1} h_k^*$. All these results are in *Disquisitiones* in the language of quadratic forms and we have only translated them into the language of ideals via our dictionary i_k.

Although we did not touch it at all, we can define the notion of genera of quadratic forms by using p-adic numbers as a classification of quadratic forms weaker than the classification by $f \stackrel{+}{\sim} g$. This method can be generalized almost endlessly to classify polynomial mappings and leads us to the idea of the adelization of algebraic groups.†

† See T. Ono, *Algebraic Groups and Number Theory*, Sugaku Expositions, American Mathematical Society Vol. 1 (1988), pp. 95–112.

List of Mathematicians

This list includes names and dates of birth and death for mathematicians quoted in the text, except those still living.

Euclid (330?–275? B.C.)
Diophantus (246?–330?)
Fibonacci (1174?–1250?)
Mersenne (1588–1647)
Fermat (1601–1665)
Pell (1611–1685)
Goldbach (1690–1764)
Euler (1707–1783)
Lagrange (1736–1813)
Legendre (1752–1833)
Fourier (1768–1830)
Germain (1776–1831)
Gauss (1777–1855)
Abel (1802–1829)
Jacobi (1804–1851)
Dirichlet (1805–1859)
Liouville (1809–1882)
Kummer (1810–1893)
Galois (1811–1832)
Catalan (1814–1894)
Eisenstein (1823–1852)
Kronecker (1823–1891)
Riemann (1826–1866)
Dedekind (1831–1916)
Weber (1842–1913)
Frobenius (1849–1917)
Peano (1858–1932)
Hurwitz (1859–1919)
Hilbert (1862–1943)
Minkowski (1864–1909)
Takagi (1875–1960)
Artin (1898–1962)
Hasse (1898–1979)
Herbrand (1908–1932)
Chevalley (1909–1984)

Bibliography

Here are some books, mainly on number theory, chosen from my own library and, with the exception of the last two general references, arranged in alphabetical order by author's name.

1. E. Artin, in: *The Collected Papers of Emil Artin* (S. Lang and J. Tate, eds.), Addison-Wesley, Reading (1965).
2. A. Baker, *A Concise Introduction to the Theory of Numbers,* Cambridge University Press, Cambridge (1984).
3. Z. I. Borevich and I. R. Shafarevich, *Number Theory,* Academic Press, New York (1966).
4. J. Cassels and A. Fröhlich (eds.), *Algebraic Number Theory,* Thompson Book Co., Washington D.C. (1967). (Reprint by Academic Press, New York, 1986.)
5. R. Courant and H. Robbins, *What is Mathematics?,* Oxford University Press, Oxford (1941).
6. H. Davenport, *The Higher Arithmetic,* 5th ed., Cambridge University Press, Cambridge (1982).
7. L. E. Dickson, *Introduction to the Theory of Numbers,* Dover, New York (1957).
8. L. E. Dickson, *History of the Theory of Numbers,* 3 volumes, Chelsea, New York (1971).
9. P. G. L. Dirichlet, *Mathematische werke,* I and II, Chelsea, New York (1969).
10. P. G. L. Dirichlet, *Vorlesungen über Zahlentheorie,* Chelsea, New York (1968).
11. M. Eichler, *Introduction to the Theory of Algebraic Numbers and Functions,* Academic Press, New York (1966).
12. G. Fujisaki, *Introduction to Algebraic Number Theory,* 2 volumes, Shokabo, Tokyo (1957) (in Japanese).

13. C. F. Gauss, *Disquisitiones arithmeticae,* Yale University Press, New Haven (1966).
14. L. Gårding, *Encounter with Mathematics,* Springer, New York (1977).
15. G. H. Hardy and E. M. Wright, *An Introduction to the Theory of Numbers,* Oxford University Press, Oxford (1960).
16. G. H. Hardy, *Ramanujan,* Cambridge University Press, Cambridge (1940).
17. H. Hasse, *Bericht über neuere Untersuchungen und Probleme aus der Theorie der algebraischen Zahlkörper,* I, Ia, II, Jber. Deutsch. Math. Verein., Leipzig (1926, 1927, 1930).
18. H. Hasse, *Zahlentheorie,* Akademie-Verlag, Berlin (1949).
19. H. Hasse, *Vorlesungen über Zahlentheorie,* Springer, Berlin (1950).
20. H. Hasse, *Über die Klassenzahl Abelscher Zahlkörper,* Akademie-Verlag, Berlin (1952).
21. E. Hecke, *Vorlesungen über die Theorie der Algebraischen Zahlen* (English translation, Springer, Chelsea, New York 1970).
22. E. Hecke, *Mathematische Werke,* Vandenhoeck & Ruprecht, Göttingen (1959).
23. D. Hilbert, *Gesammelte Abhandlungen,* 3 volumes, Springer, Berlin (1970).
24. L. K. Hua, *Introduction to Number Theory,* Springer, (1982).
25. K. Ireland and M. Rosen, *A Classical Introduction to Modern Number Theory,* Springer, Berlin (1982).
26. M. Ishida, *Introduction to Algebra,* Jikkyo, Tokyo (1978) (in Japanese).
27. M. Ishida, *Algebraic Number Theory,* Morikita, Tokyo (1974) (in Japanese).
28. K. Iwasawa, *Theory of Algebraic Functions,* Iwanami, Tokyo (1952) (in Japanese).
29. K. Iwasawa, *Local Class Field Theory,* Iwanami, Tokyo (1980) (in Japanese).
30. S. Iyanaga (ed.), *Number Theory,* North-Holland, Amsterdam (1975).
31. S. Iyanaga, *System of Numbers,* 2 volumes, Iwanami, Tokyo (1972, 1978) (in Japanese).
32. G. J. Janusz, *Algebraic Number Fields,* Academic Press, New York (1973).
33. K. Katayama, *Introduction to Number Theory,* Jikkyo, Tokyo (1975) (in Japanese).
34. Y. Kawada, *Algebraic Number Theory,* Kyoritsu, Tokyo (1957) (in Japanese).
35. Y. Kawada, *Number Theory,* 3 volumes, Iwanami, Tokyo (1983) (in Japanese).

36. F. Klein, *Vorlesungen über die Entwicklung der Mathematik im 19. Jahrhundert*, Chelsea, New York (1956).
37. M. Kuga, *Galois's dream,* Nippon Hyoron sha, Tokyo (1968) (in Japanese).
38. L. Kronecker, *Mathematische Werke,* 5 volumes, Chelsea, New York (1968).
39. T. Kubota, *Introduction to Number Theory,* Asakura, Tokyo (1971) (in Japanese).
40. E. Landau, *Vorlesungen über Zahlentheorie,* 2 volumes, Chelsea, New York (1950, 1955).
41. S. Lang, *Algebra,* Addison-Wesley, Reading (1967).
42. S. Lang, *Introduction to Diophantine Approximations,* Addison-Wesley, Reading (1966).
43. S. Lang, *Algebraic Number Theory,* Addison-Wesley, Reading (1970).
44. R. L. Long, *Algebraic Number Theory,* Marcel Dekker, New York (1977).
45. D. A. Marcus, *Number Fields,* Springer, New York (1977).
46. G. B. Mathews, *Theory of Numbers,* Chelsea, New York (1982).
47. H. Minkowski, *Gesammelte Abhandlungen,* I and II, Chelsea, New York (1967).
48. H. Minkowski, *Geometrie der Zahlen,* Chelsea, New York (1953).
49. L. J. Mordell, *Diophantine Equations,* Academic Press, New York (1969).
50. T. Nagell, *Introduction to Number Theory,* Chelsea, New York (1964).
51. K. Nakamura, *Modern History of Mathematics,* Nippon Hyoron sha, Tokyo (1980) (in Japanese).
52. W. Narkiewicz, *Elementary and Analytic Theory of Numbers,* Polish Sci. Publ., Warszawa (1974).
53. C. D. Olds, *Continued Fractions,* Random House, New York (1963).
54. T. Ono, *Variations on a Theme of Euler,* Jikkyo, Tokyo (1980) (in Japanese).
55. O. Perron, *Die Lehre von den Kettenbrüchen,* Chelsea, New York (1950).
56. G. Polya, *How to Solve it,* Doubleday, Garden City (1957).
57. P. Ribenboim, *13 Lectures on Fermat's Last Theorem*. Springer, New York (1979).
58. B. Riemann, *Gesammelte Mathematische Werke,* Dover, New York (1953).
59. P. Samuel, *Théorie Algébrique des nombres,* Hermann, Paris (1967).
60. J.-P. Serre, *Corps locaux,* Hermann, Paris (1962).
61. J.-P. Serre, *A Course in Arithmetic,* Springer, New York (1970).

62. D. Shanks, *Solved and Unsolved Problems in Number Theory,* Chelsea, New York (1978).
63. K. Shoda, *A Manual of Number Theory,* Kyoritsu, Tokyo (1949) (in Japanese).
64. G. Shimura and Y. Taniyama, *Complex multiplication of abelian varieties and its application to number theory,* Publ. Math. Soc. Japan 6, Tokyo (1961).
65. C. L. Siegel, *Analytische Zahlentheorie,* I, II, Göttingen, Berlin (1963–1964).
66. C. L. Siegel, *Gesammelte Abhandlungen,* I, II, III, Springer, Berlin (1966).
67. H. J. S. Smith, *Report on the Theory of Numbers,* Chelsea, New York (1965).
68. D. E. Smith, *A Source Book in Mathematics,* 2 volumes Dover, New York (1959).
69. H. M. Stark, *An Introduction to Number Theory,* The MIT Press, Cambridge (1978).
70. D. J. Struik, *A Concise History of Mathematics,* Dover, New York (1948).
71. J. Suetsuna, *Analytic Number Theory,* Iwanami, Tokyo (1950) (in Japanese).
72. T. Takagi, *An Introduction to Analysis,* Iwanami, Tokyo (1961) (in Japanese).
73. T. Takagi, *Lectures on Algebra,* Kyoritsu, Tokyo (1983) (in Japanese).
74. T. Takagi, *Lectures on Elementary Number Theory,* Kyoritsu, Tokyo (1983) (in Japanese).
75. T. Takagi, *Algebraic Number Theory,* Iwanami, Tokyo (1971) (in Japanese).
76. T. Takagi, *A Historical Story of Modern Mathematics,* Kyoritsu, Tokyo (1984) (in Japanese).
77. B. L. van der Waerden, *Algebra,* 2 volumes, Frederick Ungar, New York (1970).
78. B. A. Venkov, *Elementary Number Theory,* Wolters–Noordhoff, Groningen (1970).
79. I. M. Vinogradov, *Elements of Number Theory,* Dover, New York (1954).
80. H. Wada, *World of Numbers,* Iwanami, Tokyo (1984) (in Japanese).
81. L. C. Washington, *Introduction to Cyclotomic Fields,* Springer, New York (1982).
82. H. Weber, *Lehrbuch der Algebra,* 3 volumes, Chelsea, New York (1891, 1894, 1896).
83. A. Weil, *Number Theory for Beginners,* Springer, New York (1985).
84. A. Weil, *Basic Number Theory,* Springer, New York (1967).

85. A. Weil, *Oeuvres Scientifiques,* Collected Papers, 3 volumes, Springer, New York (1979).
86. A. Weil, *Number Theory,* Birkhäuser, Basel–Boston (1984).
87. H. Weyl, *Algebraic Theory of Numbers,* Princeton University Press, Princeton (1940).
88. H. Weyl, *Gesammelte Abhandlungen,* Springer, Berlin (1968).
89. E. T. Whittaker and G. N. Watson, *A Course of Modern Analysis,* 4th ed., Cambridge University Press, Cambridge (1958).
90. D. B. Zagier, *Zetafunktionen und quadratische Körper,* Springer, Berlin (1981).
91. O. Zariski and P. Samuel, *Commutative Algebra,* 2 volumes, D. Van Nostrand, New York (1958, 1960).
92. *Encyclopedic Dictionary of Mathematics,* Mathematical Society of Japan, The MIT Press, Cambridge (1977).
93. *Mathematical Reviews* (Monthly), American Mathematical Society, Providence.

Comments on the Bibliography

1. As I said before, these books are chosen from my bookshelf; most of them I myself bought since the 1950s but some are gifts from authors. Of course, there are many more books on number theory and the above list is only a small and perhaps one-sided collection.

2. The following are some classic publications: 9, 10, 13, 22, 23, 38, 47, 48, 55, 58, 66, 67, 82, 85, 88, and 89.

3. The following are those books I always kept at the side of my desk for reference while writing this book: 3, 13, 15, 21, 25, 27, 30, 34, 41, 42, 43, 53, 69, 72, 74, 75, 76, 83, and 92.

4. The following have been adopted as textbooks in classes taught by myself and by my colleagues: 2, 6, 40, and 83 (Elementary Number Theory); 3, 4, 43, 45, and 84 (Algebraic Number Theory); 41, 77, and 90 (Algebra).

Index

www.ingramcontent.com/pod-product-compliance
Ingram Content Group UK Ltd.
Pitfield, Milton Keynes, MK11 3LW, UK
UKHW020125290926
15471UKWH00003B/298
* 9 7 8 1 4 6 1 2 7 8 7 2 6 *